DYNAMICS OF FOREST ECOSYSTEMS IN CENTRAL AFRICA
DURING THE HOLOCENE
Past – Present – Future

Palaeoecology of Africa

International Yearbook of Landscape Evolution
and Palaeoenvironments

Volume 28

Editor in Chief

J. Runge, Frankfurt, Germany

Editorial board

Dynamics of Forest Ecosystems in Central Africa during the Holocene

Past – Present – Future

Jürgen Runge

Centre for Interdisciplinary Research on Africa (CIRA/ZIAF),
Johann Wolfgang Goethe University, Frankfurt am Main, Germany

Taylor & Francis
Taylor & Francis Group
LONDON / LEIDEN / NEW YORK / PHILADELPHIA / SINGAPORE

Taylor & Francis is an imprint of the Taylor & Francis Group, an informa business

© 2008 Taylor & Francis Group, London, UK

Typeset by Vikatan Publishing Solutions (P) Ltd, Chennai, India
Printed and bound in Great Britain by Antony Rowe Ltd (a CPI-group company), Chippenham, Wiltshire.

Published by: Taylor & Francis/Balkema
                P.O. Box 447, 2300 AK Leiden, The Netherlands
                e-mail: Pub.NL@tandf.co.uk
                www.balkema.nl, www.taylorandfrancis.co.uk, www.crcpress.com

Library of Congress Cataloging-in-Publication Data
Dynamics of forest ecosystems in Central Africa during the Holocene: past - present - future / editor Jürgen Runge.
    p. cm.
  Includes bibliographical references and index.
  ISBN 978-0-415-42617-6 (hardcover: alk. Paper)
1. Palaeoecology – Africa, Central, 2. Palaeoecology – Holocene. 3. Palaeoecology – Pleistocene.
4. Palaeobotany – Africa, Central. 5. Palaeobotany – Holocene. 6. Palaeobotany – Pleistocene.
7. Forest ecology – Africa, Central. I. Runge, Jürgen.

QE720. 2. A352D96 2008
560'. 45340967 – dc22   2007013654

ISBN: 978-0-415-42617-6 (Hbk)
ISBN: 978-0-203-93042-7 (Ebook)

Contents

Contributors

Samuel Aime Abossolo

Department of Geography, University of Yaoundé I, BP 755, Yaoundé, Cameroon. Email: abossoamai@yahoo.fr.

Pierre Agbani

Faculty of Agronomic Sciences, Laboratory of Applied Ecology, University of Abomey-Calavi, BP 526, Cotonou, Benin. Email: pagbani@yahoo.fr.

Didier Agonyissa

Faculty of Agronomic Sciences, Laboratory of Applied Ecology, University of Abomey-Calavi, BP 526, Cotonou, Benin. Email: dagonyissa@yahoo.fr.

Joseph Armathée Amougou

Department of Geography, University of Yaoundé I, BP 755, Yaoundé, Cameroon. Email: joearmathe@yahoo.fr.

Eva Becker

Institute of Physical Geography, Johann Wolfgang Goethe University Frankfurt am Main, Altenhöfer Allee 1, D-60438 Frankfurt, Germany. Email: ev.becker@em.uni-frankfurt.de.

Brahim Damnati

University of Abdelmalek Essadi, Faculté des Sciences et Techniques de Tanger, Department of Earth Sciences, Natural Resources and Risks Observatory (ORRNA), BP 416, Tangier, Morocco. Email: b_damnati@yahoo.fr.

Frédéric Dumay

Laboratory of Zonal geography for Development, University of Reims Champagne-Ardenne, Reims, France. Email: frederic.dumay@univ-reims.fr.

Joachim Eisenberg

Institute of Physical Geography, Johann Wolfgang Goethe University Frankfurt am Main, Altenhöfer Allee 1, D-60438 Frankfurt, Germany. Email: j.eisenberg@em.uni-frankfurt.de.

Kah Elvis Fang

Department of Geography, University of Yaoundé I, BP 755, Yaoundé, Cameroon. Email: kah_elvis@yahoo.fr.

Pierre Giresse

Laboratoire d'Etudes des Géo-Environnements Marins, EA3678, Université de Perpignan, 52 Av. Paul Alduy, F-66860 Perpignan, France. Email: giresse@univ-perp.fr.

Dethardt Goetze

Department of Botany, University of Rostock, Wismarsche Str. 8, D-18051 Rostock, Germany. Email: dethardt.goetze@uni-rostock.de.

Karen Hahn–Hadjali

Institute for Ecology and Geobotany, Johann Wolfgang Goethe University Frankfurt am Main, Siesmeyerstraße 70, D-60323 Frankfurt am Main, Germany. Email: hahn-hadjali@bio.uni-frankfurt.de.

Klaus Josef Hennenberg

Department of Botany, University of Rostock, Wismarsche Str. 8, D-18051 Rostock, Germany. Email: klaus.hennenberg@uni-rostock.de.

Thorsten Herold

Institute of Physical Geography, Johann Wolfgang Goethe University Frankfurt am Main, Altenhöfer Allee 1, D-60438 Frankfurt, Germany. Email: thorsten.herold @arcor.de.

Alexa Höhn

Institute for Archaeological Sciences, Department of African Archaeology and Archaeobotany, Johann Wolfgang Goethe University Frankfurt am Main, Grüneburgplatz 1, D-60629 Frankfurt am Main, Germany. Email: a.hoehn@em. uni-frankfurt.de.

Michel Icole

CEREGE, Europôle du petit Arbois, BP 80, F-13545 Aix-en-Provence Cedex, France. Email: icole@cerege.fr.

Stefanie Kahlheber

Institute for Archaeological Sciences, Department of African Archaeology and Archaeobotany, Johann Wolfgang Goethe University Frankfurt am Main, Grüneburgplatz 1, D-60629 Frankfurt am Main, Germany. Email: kahlheber@em. uni-frankfurt.de.

Boniface Kankeu

Institut de Recherches Géologiques et Minières (IRGM), BP 4110, Yaoundé, Cameroon. Email: bonifacekankeu@yahoo.fr.

Kostantin König

Institute of Physical Geography, Johann Wolfgang Goethe University Frankfurt am Main, Altenhöfer Allee 1, D-60438 Frankfurt, Germany. Email: k.koenig@em. uni-frankfurt.de.

Marcel Koko

Laboratoire de Climatologie, de Cartographie et d'Etudes Géographiques (LACCEG), Département de Géographie, Faculté des Lettres et Sciences Humaines, Université de Bangui, BP 1037, Bangui, RCA. Email: marceleza_koko@yahoo.fr.

Arsène Igor Kondayen

Institut Supérieur de Développement Rural (ISDR), Université de Bangui, BP 1450, Bangui, RCA. Email: kondayen@yahoo.fr.

Annick Koulibaly

Laboratory of Botany, University of Cocody, BP 582, Abidjan, Ivory Coast. Email: koulannick@yahoo.fr.

Conny Meister

Institute of Pre- and Protohistory and Medieval Archaeology, Eberhard-Karls University Tübingen, Burgsteige 11, D-72070 Tübingen, Germany. Email: conny.meister@ uni-tuebingen.de.

Katharina Neumann
Institute for Archaeological Sciences, Department of African Archaeology and Archaeobotany, Johann Wolfgang Goethe University Frankfurt am Main, Grüneburgplatz 1, D-60629 Frankfurt am Main, Germany. Email: k.neumann@em.uni-frankfurt.de.

Marion Neumer
Institute of Physical Geography, Johann Wolfgang Goethe University Frankfurt am Main, Altenhöfer Allee 1, D-60438 Frankfurt, Germany. Email: neumer@em.uni-frankfurt.de.

Félix Ngana
Laboratoire de Climatologie, de Cartographie et d'Etudes Géographiques (LACCEG), Département de Géographie, Faculté des Lettres et Sciences Humaines, Université de Bangui, BP 1037, Bangui, (RCA). Email: nganaf@yahoo.fr.

Simon Ngos III
Department of Earth Sciences, University of Yaoundé I, BP 812, Yaoundé, Cameroon. Email: sngos@yahoo.com.

Cyriaque-Rufin Nguimalet
Laboratoire de Climatologie, de Cartographie et d'Etudes Géographiques (LACCEG), Département de Géographie, Faculté des Lettres et Sciences Humaines, Université de Bangui, BP 1037, Bangui, RCA; CNRS–Laboratoire de Géographie Physique « P. Birot », 1, Place Aristide Briand, 92 195 Meudon Cedex, France. Email: cnguimalet@yahoo.fr.

Bienvenu Dénis Nizesete
Department of History, University of Ngaoundéré, BP 553, Ngaoundéré, Cameroon. Email: nizesete1@yahoo.fr.

Jude Mphoweh Nzembayie
Department of Geography, University of Yaoundé I, BP 755, Yaoundé, Cameroon. Email: mphowehjude@yahoo.fr

Stefan Porembski
Department of Botany, University of Rostock, Wismarsche Str. 8, D-18051 Rostock, Germany. Email: stefan.porembski@uni-rostock.de.

Lehné Rouwen
Institute for Geoscience, Johannes Gutenberg University Mainz, Becherweg 21, D-55099 Mainz, Germany. Email: lehne@uni-mainz.de.

Jürgen Runge
Institute of Physical Geography, Johann Wolfgang Goethe University Frankfurt am Main, Altenhöfer Allee 1, D-60438 Frankfurt, Germany. Email: j.runge@em.uni-frankfurt.de.

Ayobami T. Salami
Space Applications and Environmental Science Laboratory (SPAEL), Institute of Ecology & Environmental Studies, Obafemi Awolowo University, Ile-Ife, Nigeria. Email: ayobasalami@yahoo.com.

Mark Sangen
Institute of Physical Geography, Johann Wolfgang Goethe University Frankfurt am Main, Altenhöfer Allee 1, D-60438 Frankfurt, Germany. Email: m.sangen@em.uni-frankfurt.de.

Marco Schmidt
Institute for Ecology and Geobotany, Johann Wolfgang Goethe University Frankfurt am Main, Kuhwaldstraße 55, D-60486 Frankfurt, Germany. Email: marco.schmidt@senckenberg.de.

Astrid Schweizer

Institute for Archaeological Sciences, Departement of African Archaeology and Archaeobotany, Johann Wolfgang Goethe University Frankfurt am Main, Grüneburgplatz 1, D-60629 Frankfurt am Main. Email: schweizer@em.uni-frankfurt.de.

Michel Servant

IRD, Centre de recherche Ile de France, 32, Av. Henri Varagnat, F-93143 Bondy Cedex, France.

Frank Sirocko

Institute for Geoscience, Johannes Gutenberg University Mainz, Becherweg 21, D-55099 Mainz, Germany. Email: sirocko@uni-mainz.de.

Maurice Taieb

CEREGE, Europôle du petit Arbois, BP 80, F-13545 Aix-en-Provence Cedex, France. Email: taieb@cerege.fr.

Mesmin Tchindjang

Department of Geography, University of Yaoundé I, BP 755, Yaoundé, Cameroon. Email: mtchind@yahoo.fr.

Anselme Wakponou

Department of Geography, University of Ngaoundéré, BP 454, Ngaoundéré, Cameroon. Email: wakponouanselme@yahoo.fr.

Annika Wieckhorst

Institute for Ethnology and Africa Studies, Johannes Gutenberg University Mainz, Forum 6, D-55088 Mainz, Germany. Email: wieckhor@uni-mainz.de.

David Williamson

CEREGE, Europôle du petit Arbois, BP 80, F-13545 Aix-en-Provence Cedex, France. Email: davwill@cerege.fr

Foreword

This volume marks a change in the editorship of 'Palaeoecology of Africa' as I relinquish my duties as editor because of retirement. It is very gratifying to be able to announce that Professor Jürgen Runge from the University of Frankfurt am Main, Germany, has willingly consented to take on the responsibilitiy of Palaeoecology of Africa as editor-in-chief from volume 28 onwards. Interdisciplinary research on Africa has a long tradition at the University of Frankfurt. Hiob Ludolf (1624–1704) established with his 'Historia Aethiopia' scientific research on Ethiopia and the natural scientist Eduard Rüppell (1794–1884) laid the foundation of the renowned collections of the Senckenberg Museum. In 2003 the Centre for Interdisciplinary Research on Africa (ZIAF/CIAR) of the university was founded and partnerships with African institutions were arranged. The University of Frankfurt am Main will be the right place to continue with Palaeoecology of Africa.

The series started in 1966 with the compilation of eight reports which have been published over the years 1950–1963 under the title of Palynology in Africa written by Eduard van Zinderen Bakker. It was soon realized that the results of the study of fossil pollen and spores is so strongly correlated with many other disciplines that it would be more useful also to include biogeography, archaeology, geology, geomorphology, pedology and related fields which have a bearing on the study of the past and present environment. In the course of the time the area covered by the reports has been extended from South Africa to the whole of the continent, its surrounding islands and Antarctica. Since then Palaeoecology of Africa has become an independent international medium for palaeoenvironmental studies in Africa. Until the 1990s the flow of manuscripts which reaches Palaeoecology of Africa indicates the need for a publication of this type. The goals of Palaeoecology of Africa to make the publication truly African in coverage, fostering contributions from an equitably distributed international authorship, while maintaining high academic standards, were achieved. Meanwhile, many other journals are devoted to interdisciplinary articles also dealing in part with the palaeoecology in Africa. In response to addressing the key knowledge gaps and scientific and policy-related recommendations specific to Africa, Palaeoecology of Africa in future will provide insights and broad overviews of landscape evolution and palaeoenvironments and assessments of the general ecological, environmental and landscape features of various ecosystems in Africa including changes during the Cainozoic. The growing interest in the sustainable use of natural resources in Africa—relief, soils, water, vegetation, atmosphere—and a general awareness of global change issues has led to increased interest in environmental research and a need for public information on the subjects involved. To serve these requirements, the 'new' *Palaeoecology of Africa* will be an *International Yearbook of Landscape Evolution and Palaeoenvironments*.

With these issues in mind it is my pleasure to transfer the leadership of Palaeoecology of Africa to Professor Runge and I wish that under his able guidance Palaeoecology of Africa will be recognized not only in Africa but worldwide and thereby providing an opportunity for a meeting of minds and active exchange of knowledge and ideas across scientific and cultural borders.

Regensburg, February 2007
Klaus Heine

Preface and Introduction

'Palaeoecology of Africa' is back! Since the last volume (27) of the series had been published in 2001 with the workshop contributions of the Durban INQUA conference 1999, a six years long 'creative' pause followed to think over and to restructure this traditional journal.

Founded in 1966 by Professor Eduard Meine van Zinderen Bakker (1907–2002) and assisted by Professor Joey Coetzee and other guest editors 18 volumes had been published up to 1986. As Professor Klaus Heine pointed out in his 'Handover' foreword to this volume, the original focus of 'Palaeoecology of Africa and the surrounding islands and Antarctica' was mainly palynological, however, during the years the interdisciplinary character of palaeoclimatic and palaeoenvironmental research became obvious, and therefore the scope of 'PoA' turned into a much broader perspective. Professor Heine took over responsibility for the series in 1988 and edited nine volumes of 'Palaeoecology of Africa' that covered many different aspects of the palaeoenvironmental evolution of the African continent during the Quaternary. Quite regularly the scientific contributions of the Biennial SASQUA (South African Society for Quaternary Research) conferences were published within the series.

In 2005 the former editor in chief of the series proposed to take over the editorship of 'Palaeoecology of Africa' because of his retirement from the chair of Physical Geography at Regensburg University (Germany). Assisted by the publishers group 'Taylor & Francis' (UK) and the 'Frankfurt Centre for Interdisciplinary Research on Africa' founded in 2003 (CIRA/ZIAF, www.ziaf.de), a new conception and a modern layout for the continuation of 'Palaeoecology of Africa' was designed in the meantime, and it is now ready to start.

The series main title was changed to 'PALAEOECOLOGY OF AFRICA—*International Yearbook of Landscape Evolution and Palaeoenvironments*'. It is planned to have one book a year. The alignment and scope of the 'new' PoA should focus a bit more on geographical and on landscape aspects and somewhat less on palynology and botany. However, as the journal deals with 'ecology' in a broad sense, also the 'classical' PoA papers are welcome. The editorial board of 'Palaeoecology' was equally restructured allowing to take profit from the scientific input and expertise of other dynamic colleagues interested in the topic.

The 19 papers gathered together in this volume are mainly the outcome of an interdisciplinary workshop organized by the German Research Foundation (DFG), Research Unit 510 on 'Environmental and Cultural Change in West- and Central Africa' which took place from March 5–7, 2006 in the Goethe-Institute at Cameroon's capital Yaoundé. The main theme of the workshop focused on the 'Dynamics of forest ecosystems in Central Africa during the Holocene: Past–Present–Future'.

Tropical rain forests in Africa and in other low latitude regions around the world are recognized by a broader public as evolutionary old, species-rich and stable ecosystems over time. Another perspective originates in the fact, that these biodiversity 'hot spots' are often endangered by humans exploring biological and geological resources located inside. Since long already, there is widespread recognition and evidence that the assumed stability of rain forests in earth history on the landscape scale can not be generally confirmed. Natural shifts in climate during the Late Pleistocene and also in the Holocene repeatedly caused strong modifications and area dynamics of rain forest and savanna-woodland ecosystems over the last 40.000 years that temporarily led to an almost total disappearance of the closed forest and to an extension of more open, savanna-like vegetation patterns. In contradiction to the postulated evolutionary stability of the rain forests—on the environmental and landscape scale—a striking tendency for rapid modifications due to environmental changes can be stated. Aside of the spatio-temporal oscillations of the forested areas in Central Africa, this volume also focuses on possible relationships between environmental changes and human as well as cultural innovations, like hypothesises on the 'migration of Bantu' inside Central and Southern Africa. The latter point is one of the core topic questions of the interdisciplinary DFG-Research Unit 510—composed of Archaeologists, Archaeobotanists and Geographers—introduced by Jürgen Runge is the first paper in this volume. The second contribution of the same author ('Of Deserts and Forests') gives a brief overview of what is known on the Late Quaternary and Holocene environmental landscape dynamic in Central Africa evidenced by different proxy-data sources and references (Figure 1).

The third paper by Alexa Höhn and members of the palaeobotanical subproject (S. Kahlheber, K. Neumann, A. Schweizer) within the Research Unit 510 introduces new findings of extended fieldwork in the rain forest in Southern Cameroon on how climatically degraded forests (aridification?) might have been settled by farming comunities during the first millenium BC. Conny Meister, an archaeologist within the DFG-Research Unit 510 gives in his contribution (4) an overview of the new archaeological discoveries of the team supervised by Manfred K.H. Eggert in Southern Cameroon. Mesmin Tchindjang and his collaborators from Yaoundé University try in their contribution (5) to correlate podzolic and alluvial soil features in the Batié region of Cameroon in relation to the Late Holocene changes evidenced from other proxy-data sources as lake sediments. In paper (6) Mark Sangen introduces results from fieldwork of the geographical subproject within the DFG-Research Unit 510 inside the Ntem River's delta in Southern Cameroon where he and his colleagues discovered fluvial to partly lacustrine sedimentary records going back up to almost 50.000 BP. The contribution (7) of Simon Ngos III and European science collaborators (F. Sirocko, R. Lehné, P. Giresse and M. Servant) carried out coring in two crater lakes on the Adamawa plateau and interpreted by numerous analyses on the lakes' sediments their significance to Holocene environmental changes in the surroundings of Nagoundéré (Cameroon). Within paper (8) the regional interest on Holocene palaeoenvironments is shiftet to the Ngotto Forest in the southwest of the Central African Republic (CAR, Figure 1) where Marion Neumer, Eva Becker and Jürgen Runge studied pedo-geomorphological forms under today's rain forest vegetation and developed new hypothesises on how these might have emerged under climatic conditions different from those of today. Paper nine, written by Anselme Wakponou, Dénis Bienvenu Nizesete and Frédéric Dumay, looks for evidence on Holocene vegetation dynamics by consulting many different references to highlight the environmental shift in Northern Cameroon especially under the recently strengthening influence of humans. In East Africa, Kenya, the contribution (10) of Brahim Damnati, Michele Icole, Marice Taieb and David Williamson makes an efford to apply and test the organic carbon content and the overall carbonate stratigraphie on sediment cores coming from Lake Magadi (Figure 1).

The subsequent chapters (11–15) are concerned with recent dynamics of the forest-savanna border, its ecological processes, and the application of innovative surveying methods (remote sensing) how to document such landscape changes. Often the spatio-temporal shifts inside these forest-savanna-mosaics are strongly determined by the influence of man by setting up bush fires, hunting and logging. Dethardt Goetze and his colleagues Klaus Josef Hennenberg, Stefan Porembski and Annick Koulibaly present an extended review (11) on the scientific literature on that topic, and they introduce their work in Ivory Coast as a showcase for forest-savanna dynamics. Chapter (12) which originated from the interdisciplinary German BIOTA research project (www.biota-africa.de) directed by Konstantin König, Jürgen Runge, Marco Schmidt, Pierre Agbani, Karen Hahn-Hadjali, Didier Agonyissa and Annika Wieckhorst, is concerned with changes in biodiversity in rural, semi-humid areas due to modified land use and popoulation growth in Northern Benin. The importance of monitoring

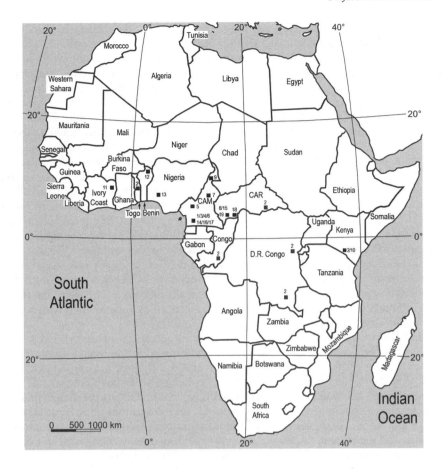

Location map of study sites

1 - Runge, J., DFG-Research Unit 510 on 'Ecological and Cultural Change in West and Central Africa'.
2 - Runge, J., Of Deserts and Forests: insights into Central African Palaeoenvironments since the Last Glacial Maximum.
3 - Höhn, A., et al., The environment of farming communities in southern Cameroon.
4 - Meister, C., Recent archaeological investigations in the tropical rain forest of SW Cameroon.
5 - Tchindjang, M. et al., The Batié palaeopodzol and its palaeoclimatic and environmental significance.
6 - Sangen, M., New evidence on palaeoenvironmental conditions in SW Cameroon since the Late Pleistocene.
7 - Ngos III, S. et al., The evolution of the Holocene palaeoenvironment of the Adamawa region of Cameroon.
8 - Neumer, M. et al., Palaeoenvironmental studies in the Ngotto Forest.
9 - Wakponou, A. et al., Extension of former tree cover in the today's sudano-sahelian milieu.
10 - Damnati, B. et al., Carbonate stratigraphy and Palaeoenvironments from Lake Magadi, Kenya.
11 - Goetze, D., et al., Forest-savanna dynamics in Ivory Coast.
12 - König, K., et al., The impact of land use on species distribution changes in North Benin.
13 - Salami, A., Potentials of NigeriaSat-1 for Sustainable Forest Monitoring in Africa.
14 - Herold, T. and Runge, J., Landscape and vegetation patterns studied by remotely sensed data anaysis near Ebolowa.
15 - Runge, J., Recent dynamics of forest-savanna boundaries at Ngotto Forest.
16 - Kankeu, B. and Runge, J., Late Neoproterozoic Palaeogeography of Central Africa.
17 - Eisenberg, J., The geomorphic evolution of the Ntem River in and below its interior delta, SW Cameroon.
18 - Nguimalet, C.- R., Forest clearings and urban floods in densely populated districts of Bangui.
19 - Nguimalet, C.- R. et al., Non Woody Forest Products (NWFPs) and food safety.

Figure 1. Location map of study sites and short title according to PoA-chapters.

sustainable forest development by applying remote sensing techniques is introduced by Ayobami Salami for Nigeria by using the new sensor of NigeriaSat-1 within chapter (13). A comparable case study from the surroundings of the town of Ebolowa (Cameroon) is introduced by Thorsten Herold and Jürgen Runge (14) by classifying landscape and vegetation patterns within the rain forest questioning the existance of palaeoenvironmentally significant open spaces like 'savane incluse' within the forest. Another remote sensing based study (15) on the spatio-temporal dynamics of the limits of Ngotto Forest and the neighbouring savanna ecosystemes from 1973 to 2002 using conventional LANDSAT-MSS/TM and ETM imageries, and the efforts of government controlled sustainable forest management policy, is demonstrated by Jürgen Runge in the Central African Republic (CAR, Figure 1). Chapters 16 and 17 highlight the importance of geological, tectonical and geomorphological knowledge that contributes to the identification of suitable sites within the landscape where good proxy-data availability is given and where sediment accumulations can be expected. Boniface Kankeu and Jürgen Runge establish in their contribution (16) a connection between the Precambrian palaeogeographical shape of Central Africa and its importance for the making of the 'younger' Pleistocene and Holocene landscapes. In the same context Joachim Eisenberg (17) examines—again in the framework of the physiogeographical subproject of the DFG-Research Unit 510—the importance of neotectonics for the geomorphological evolution of the Ntem interior delta and its suitability as a 'sediment trap' for acting as a proxy-data archive within the rain forest of southern Cameroon.

Finally, Cyriaque-Rufin Nguimalet discusses the effects of deforestation of rain forest since the beginning of the French colonization around the today's Central African Republic's capital of Bangui (18) on floodings and inundations in overpopulated 'slum' districts. Another contribution (19) from the colleagues of the University of Bangui (C.-R. Nguimalet, A.I. Kondayen, M. Koko and F. Ngana) makes an approach to understand the importance of 'non woody forest products' (NWFPs) and they show how these forest plant and animal products—aside of commercial timber exploitation—can contribute to the welfare and the income of people living in and around the Central African rain forest. This latter point illustrates that 'Palaeoecology of Africa' is now also concerned with aspects of sustainable use of rain forest resources and the future of these highly sensitive ecosystems.

Many thanks go to all colleagues for submitting their papers. Several reviewers supplied helpful comments and suggestions to improve the manuscripts. Formatting of the papers to the new PoA layout was supported by several student assistants, in particular by Thorsten Herold and Erik Hock to whom I am most greatful. Special thanks go to Ursula Olbrich for revising numerous figures and doing cartographic work. The 'Deutsche Forschungsgemeinshaft (DFG)', the 'Deutsche Gesellschaft für Technische Zusammenarbeit (GTZ)' and the Frankfurt Centre for Interdisciplinary Research on Africa (CIRA/ZIAF) most commendably gave financial support to realize this publication.

Finally, I take this opportunity to place on record my gratitude to the former editor Klaus Heine for proposing me as the new editor of PoA, and I do wish and hope that this re-edition of the series may become a success.

Jürgen Runge
Frankfurt am Main
March 2007

CHAPTER 1

DFG-Research Unit 510 on 'Ecological and Cultural Change in West and Central Africa', Yaoundé workshop report, and outlook for 2007–2009

Jürgen Runge

Centre for Interdisciplinary Research on Africa (CIRA/ZIAF), Johann Wolfgang Goethe University, Frankfurt am Main, Germany

ABSTRACT: The main interest of the 2003 established interdisciplinary Research Unit 510 on 'Ecological and Cultural Change in West and Central Africa' of the German Research Foundation (Deutsche Forschungsgemeinschaft, DFG), composed of Archaeologists, Archaeobotanists and Physical Geographers from the Universities of Tübingen and Frankfurt, is to elucidate correlations between Late Holocene climatic changes, development of land-use strategies and cultural innovations in a high temporal resolution. In close cooperation with African partners, data on settlement patterns, subsistence, climate and landscape history will be combined for selected areas of West and Central Africa. In the beginning, regional case studies concentrated on northeast Nigeria and southern Cameroon were carried out. The major working areas are situated in the ecological transitional zones between Sahara/Sahel (desert-savanna boundary) and Sudan/Guinea (savanna-rain forest boundary) which are both very sensitive to climatic fluctuations. Due to the large-scale perspective and the important role of the research area, the investigations might be relevant in a Panafrican context. This chapter introduces some core topics of the different subprojects located in the rain forest of southern Cameroon, and gives a brief report of the first interdisciplinary workshop of the research unit held in the Goethe Institute in Yaoundé (Cameroon) in March 2006; finally an outlook on the future research activities of this group planned for the period 2007–2009 is added.

1.1 INTRODUCTION

Inter- and transdisciplinary research—cooperation and exchange of ideas between scientists of different disciplines—is nowadays an indispensable precondition for the better unterstanding of the complex interrelationsships in spatio-temporal processes in landscape evolution. Therefore, the DFG (Deutsche Forschungsgemeinschaft) Research Unit 510—established in 2003—investigates the possible links between Late Holocene environmental changes and significant cultural transformations in West and Central Africa. Consequently, the main research topic concentrates on whether and in what manner prehistoric cultural and economic innovations were correlated with changes in vegetation, landscape and climate. One major temporal focus lies on the period roughly between around 1.000 BC and 500 AD. This time span is believed to cover what is frequently called the 'stone to metal transition', i.e. the transition from Later Stone Age or 'Neolithic' cultures to those of Early Iron Age type. In this transitional period large farming communities and hierarchical societies emerged in West Africa. At the same time, the Central African rain forest was settled by ceramic-producing, Bantu languages speaking populations. The main research centers of the Research Unit 510 are located in northern and central Nigeria as well as in southern and eastern Cameroon.

In 2003 DFG granted four subprojects focussing from different scientific viewpoints on the above outlined interdisciplinary problems: (1) Cultural changes in the first millennium BC and AD in Central and Northeast Nigeria (Peter Breunig); (2) Environmental and cultural change: savanna, rain forest and culture in southern and eastern Cameroon (Manfred K.H. Eggert); (3) Late–Holocene Vegetation history and the development of agriculture in West– and Central Africa (Katharina Neumann), and (4) Rain forest–Savanna–Contact (ReSaKo): Late Pleistocene, Holocene and recent landscape sensitivity of the rain forest–savanna boundary in equatorial Africa and its influence on human and cultural changes (Jürgen Runge). In 2006 the Research Unit was successfully evaluated and DFG granted a second period of research on the topic lasting from 2007–2009.

1.2 SELECTION OF RESEARCH AREAS AND METHODS

In the framework of the subproject ReSaKo or 'Rain forest–Savanna–Contact' (RU-555/14-1 /14-2) of the DFG-Research Unit alluvial sedimentary records of tropical fluvial systems were used as proxy-data archives for exploring the Late Holocene palaeoenvironmental conditions in southern Cameroon. It was aimed to investigate these sediments concerning their palaeoenvironmental information content across the catchments of the Nyong and Ntem Rivers by applying geomorphological and pedological methods. These catchments, draining an over 60.000 km² sized area into the Atlantic, have been chosen as study sites because they are recently located in tropical rain forest as well as on the rain forest-savanna margin (Figure 1). The formerly dynamics of

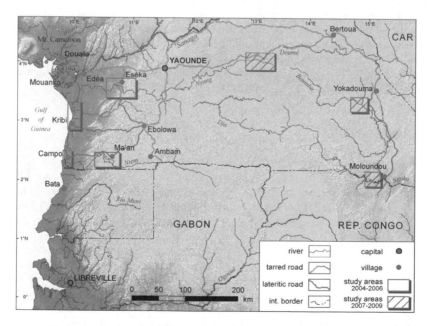

Figure 1. Location map of southern Cameroon with studied field sites 2004–2006 and future field sites 2007–2009 (cartography: J. Eisenberg).

this margin is supposed to be determined and evidenced in both regions, especially for the Holocene when around 3.000 yrs. BP strong ecological and cultural changes ('hypothesis of a migration of the Bantu people') might have occurred.

Recent concepts of landscape sensitivity have indicated that also in relatively short periods many ecosystems can experience large scale modifications in extension and structure because of radical environmental changes. With a focus on several river basins in southern Cameroon (Nyong, Ntem) the conspicuous coincidence between an aridification of climate which started in Central Africa 3.000 years ago and the settlement expansion of ceramic manufacturing Bantu tribes into the rain forest is being investigated. In co-operation with archaeologists (Prof. M.K.H. Eggert, University Tübingen), archaeobotanists (PD Dr. K. Neumann) and the Institute of Prehistory and Early History (Prof. P. Breunig) the study examines cause and effect of ecological changes and discusses its consequences on settlement expansion against the background of the contemporaneous initiating development of innovative land use systems. The "mobile" and land surface sparing economy of the hunters and gatherers as well as nomadic land-use systems are replaced by "sedentary" surface consuming land-use. On the basis of new data it will in the course of the project be checked if there are comprehensible connections between climate change and land-use and of what kind they are.

1.3 STATE OF THE ART

Late Pleistocene human adaptations and Holocene cultural innovations in the tropics of subsaharan Africa seems to be the least well known of any major area of the Old World (Brooks and Robertshaw, 1990). Still up today, there is limited evidence available that may allow a preliminary examination of such ecosystems and human/cultural adaptations closely connected to the Late Pleistocene and Holocene landscape as well as human/cultural history (Lanfranchi and Schwartz, 1990; Maloney, 1998). As it was evidenced during the first project phase 2004–2006 in the southern Cameroon rain forest—especially for the Ntem interior delta—floodplains and alluvial deposits are characterized by very complex sedimentary structures, often interrupted by formerly cut- and fill-features (palaeochannels), that can be interpreted in the context of former environmental changes (Runge *et al.*, 2006b; Sangen, in press). Because of the links between catchment and floodplain, it is obvious that alluvial sediments reflect overall drainage area conditions and therefore are an important source of information for the interpretation of the palaeoenvironment. At the coring site of Abong on the Ntem River in a depth of 340–360 cm fossil wood gave a surprisingly old uncalibrated radiocarbon age of 48.232 ± 6.411 BP (see M. Sangen in this volume).

The archaeologic findings within southern Cameroon are still very rare and less is known on how prehistoric people—of course already *Homo sapiens*—lived as hunter-gatherers in the transitional regions between dense rain forests and open savanna woodland ecosystems. The last 70.000 years or so, that are corresponding with the northern hemisphere Würm/ Weichsel glaciation (oxigene isotope stages 5e-2), were often characterized by repeated shifts in climate and therefore changes in vegetation cover. It is therefore necessary to draw a brief overview of what is known about the time from the Middle Stone Age (MSA) to the Late Stone Age (LSA) and finally to the Neolithic and Iron Age.

According to Lanfranchi and Schwartz (1990) in Congo–Brazzaville the Maluekien (70–40 ka BP) was a relatively dry and cold episode and MSA artefacts were frequently discovered in layers of coarse material (stone-lines) within multilayered soil-sediments. The Maluekien was followed by the Nijilien (40–30 ka BP) which was again humid. This supported the return of closed forests into the region. Up to now there is no evidence for palaeolithic industries for this time. On the Batéké plateau north of Kinshasa and

Brazzaville tropical podzols developed in the Kalahari Sands. These pedogenetic features prove striking modifications of the environment (Schwartz, 1988). Shortly before, during and after the maximum of the earth's last glaciation between 30–12 ka BP (Léopoldvillien) the climate reverted once again to very dry (up to 50% less precipitation) and cold (reduction of annual temperatures in the range of 4–6 °C, see Runge, 2001) conditions with an open, morphodynamically highly sensitive landscape. Fluvial systems like larger streams and also smaller rivers showed high riverbed mobility and because of a probable strong seasonality of climate, slope sediments (hillwash) and alluvia were deposited to a huge extend and they also often were redeposited inside the river channels (cut and fill structures). Within this open environment isolated forest relicts around and along rivers were a common feature (Runge, 2001) Late Stone Age (LSA) industries of the Lupembien (approx. 30.000–12.000 BP) have been discovered. From 12 ka BP onwards the climate shifted back again to humid and warm conditions (Kibangien A) which caused a rapid recolonisation of the low latitudes with dense forest. Stone cultures with microlithes are known for the Tshitolien (12.000–2.350 BP) within forested or at least woody biomass dominated landscapes. Open savanna environments showed comparable features of LSA industries summarized under the expression Wiltonien. The so-called 'First Millenium Crisis' around 3.000 yrs. BP (Kibangien B) with an extraordinary strong aridification and savannization process triggered the shrinkage of forests while savannas started again to expand (Maley, 2002). It is interesting to recognize that this environmental modification obviously marks the beginning of the Neolithic, which seems—as a very popular (not proven!) hypothesis—to coincide with the establishment of agropastoral societies and the migration of pottery making Bantu and/or other humans into Central and Southern Africa (Clist, 1987).

It has to be pointed out, that it is not at all sure, if these follow-up environmental and technological stages and transitions of cultures, that have temporarily brought in 'innovations' into former culturally 'empty' areas of Africa, is correct and well applicable to the larger regional frame of the study. Perhaps endogenic development is of greater importance than the implementation from outside. Therefore, the often sketched environmental and cultural models seem to be too deterministic in general, and perhaps also too 'eurocentristic' in their conception. Lanfranchi and Clist (1991) are describing archaeological findings within alluvial sediments that support the assumption of a possible 'parallel' development of technology and cultures instead of a simplistic follow up conception. This shows clearly the necessity of a broader temporal consideration of former ecosystem dynamics. The observation of Tshitolien stone microliths and Neolithic pottery in one sediment layer between two alluvial beds, for example underlines the importance of the conception to intensify the research on alluvial stratigraphies within tropical fluvial sediments, and to deliver new palaeoenvironmental, palaeobotanical and also archaeological data.

However, not all tropical streams within the study region contain alluvia—often rivers run-off directly on the Precambrian basement (Figure 2)—therefore it is of great importance to understand the transport and accumulation processes of sediments inside catchment areas and subsequently to identify sites along the river course where fluvial sediments suitable for palaeoenvironmental and archaeological interpretation have been deposited over time. The examination of the local neotectonic situation and the overall river network evolution in southern Cameroon (Lucazeau *et al.*, 2003; Ngako *et al.*, 2003; Eisenberg, in press) is a fundamental precondition for this study and directly contributes to the overall palaeoenvironmental objectives of the project. One PhD-work within the project studies former and recent tectonic processes and the Ntem River network evolution in southern Cameroon (see chapter 17 by J. Eisenberg in this volume); this helps to identify those key-sites for field work, where best stratigraphic archives of Late Pleistocene and Holocene landscape history can be expected.

Figure 2. Precambrian rock outcrops inside the alluvial 'sediment trap' of the Ntem River (southern Cameroon) (Photo: M. Sangen).

Floodplains with thick alluvia evolve over time and as it has been shown (Thomas and Thorp 2003), it is most likely that also tropical rivers store sediments for periods of at least 10^4 years (Thomas, 1998) or even more (Runge *et al.*, 2006a). This is sufficiently long enough to record the major climate changes before, around and after the Last Glacial Maximum (LGM) as well as Holocene environmental shifts in the catchments. Many parts of the humid tropics including the southern Cameroon study area were significantly drier for several millennia during this period, but became rapidly wetter during the climate warming at the end of the Pleistocene (Giresse *et al.*, 1994; Maley and Brenac, 1998; Runge, 2001; Gasse, 2005). It is still not known if a Younger Dryas (YD) event with probably again drier and cooler conditions was also recognizable in lower latitudes (Taylor *et al.*, 1993; Lezine and Cazet, 2004). For both of these severe environmental interruptions during the Pleistocene, which significantly modified the entire geoecosystem of an area, it is not at all clear, to what extent they have climatically influenced cultural complexity of the palaeolithic people. The picture becomes slightly clearer when one takes a look at the striking Late Holocene coincidence between a further aridification of climate that reduced the rain forest's extension (Maley, 2002), and which started in Central Africa around 3.000 years ago ('First Millenium Crisis'). Perhaps this could have triggered the settlement expansion of farming and ceramic manufacturing populations into the rain forest (Schwartz, 1992).

1.4 CORE QUESTIONS AND RESEARCH GOALS OF THE RESAKO PROJECT

Because of the new findings during the first project phase (2004–2006) in southern Cameroon (see chapter 6 by M. Sangen and chapter 17 by J. Eisenberg in this volume), the scientific approach of the ReSaKo II-subproject (2007–2009)—with local participation of the Archaeology and Archaeobotany subprojects—will focus on the following core topics:

What was the nature and extent of Late Pleistocene and Holocene environmental changes in southern Cameroon and Nigeria (Nok), with particular reference to the amount of forest cover, and how do environmental changes during the intervals around 20 ka (LGM), 11 ka (YD) and 3 ka ('First Millennium Crisis') correlate to those in neighbouring areas like Central African Republic and Gabon? In close contact and scientific exchange with the archaeological (Breunig, Eggert) and the archaeobotanical (Neumann) subprojects, special attention will be drawn to the questions if there is evidence for the development of specialized economies and/or cultures in this region with highly sensitive transitional ecosystems (rain forest–savanna boundary), and how can it be compared to other regions (Sahel/Sudan)?

Did hunter-gatherers occupy the tropical forest before the onset of agriculture with permanent settlements, and how was this evolutionary process linked to changes in climate and vegetation; is there especially any non-deterministic evidence for the phenomenon that is often described as the so-called 'migration of Bantu'? From the main perspective of the physiogeographical subproject, it is aimed to get a better temporal and spatial resolution of the former environmental changes at the rain forest–savanna fringe by comparing the alluvia and the palaeoenvironmental proxy signals gathered from the Atlantic oriented rivers (Nyong, Ntem (2004–2006, Figure 3) and Sanaga (2007)) with those of the fluvial systems in the eastern part of southern Cameroon draining into the Congo basin, e.g. Dja- and Ngoko Rivers (2008). Through comparison of results from the western Atlantic and the eastern hinterland it will be discussed if the former regression of the rain forest has been a homogeneously one that took place in the whole study area, or if there might have been a certain patchiness of remaining forest islands within a network of embedded savanna ecosystems (Runge, 2001, 2002; Maley, 2002). This former environmental pattern will subsequently be discussed and interpreted with the archaeological findings (neolithic pottery and different types of palaeolithic stone-tools as microlithes).

1.5 YAOUNDE WORKSHOP REPORT

From the establishment of the interdisciplinary DFG-Research Unit 510 in 2003, it always was a major aim of research policy to communicate with and to integrate different local researchers from the host countries (Nigeria, Cameroon) into all project activities. Already since 1996, when a first interuniversity cooperation contract with Geographers from the University of Bangui in the Central African Republic (CAR) was ratified, all aspects of field work and research in Central Africa had been discussed and planned in close consultation between the European and the African scientists. In 2006 this very fruitful cooperation with Bangui already celebrated its 10th anniversary (see chapters 18 and 19 in this volume).

The trend to more Africa oriented research activities at Frankfurt University became visible in 2003, when the Centre for Interdisciplinary Research on Africa (CIRA/ZIAF, www.ziaf.de) was founded by numerous Frankfurt scientists with a profound interest in regional African topics. The author of this article was the foundation director of this interdisciplinary centre. Actually (2007) he is the vice-director of this institution. In the framework of the DFG-Research Unit 510 the 'Département de Géographie' in Cameroon's capital Yaoundé (Université de Yaoundé I) was incorporated in May 2005 into a cooperation agreement similar to that with Bangui. Up to now, the cooperation brought about an intense and qualified exchange and communication, especially with the following Cameroonian colleagues: M. Tchindjang (see chapter 5 in this volume), M. Tsalefac, P. Tchawa, J. Youta Happi. In November 2005, the head of the Yaoundé Geography Department, Professor Tsalefac, was invited to Germany on behalf of the Research Unit ReSaKo-subproject and stayed for a one week working visit at Frankfurt, before continuing to a visit as a guest lecturer at the Université d'Orléans in France.

Aside of these contacts with the University at Yaoundé, further productive exchange and help was offered by the 'Institute des Recherches Géologiques et Minières' (IRGM, B. Kankeu, see chapter 16 in this volume) J.C. Ntonga, D. Sighomnou, and L. Sigha-Nkamdlou (Hydrology).

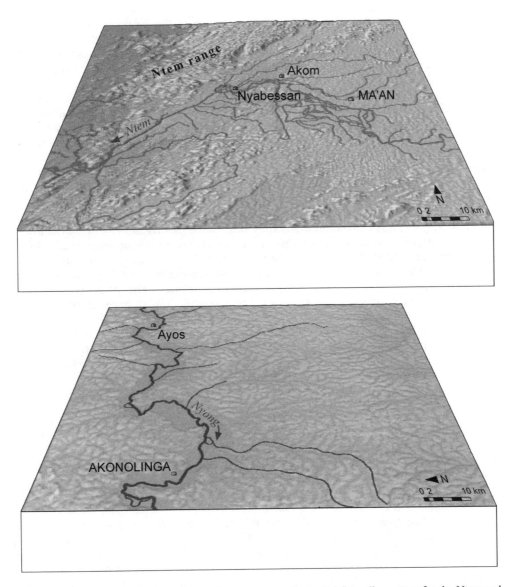

Figure 3. Digital terrain models illustrating the occurrence of extended flat valley sectors for the Ntem and Nyong River, probably containing alluvial sediments which can give evidence for palaeoenvironmental landscape conditions (cartography: J. Eisenberg).

A first promising success of the cooperation with the University of Yaoundé—and which is the basis for this publication—was the arrangement and realization of an international workshop in the Goethe Institute in Yaoundé on the topic 'Dynamics of forest ecosystems in Central Africa during the Holocene: Past–Present–Future' from 05.03.–07.03.2006, where next to members of the DFG-Research Unit 510 also colleagues from West and Central Africa, Asia and Europe participated. Essential topics and preliminary results of field work 2004–2006 were presented, discussed and placed into an international context.

Organizational support concerning the realization of this workshop was gained by the Goethe Institute (M. Friedrich) and the German Embassy in Yaoundé (S. Biedermann, R. Holzhauer). Additional DFG/BMZ funds for bilateral cooperation activities (445 KAM-121/1/06) supported the participation of other African and over-seas colleagues by granting travel subsidies.

Professor Yang Xiaoping from Beijing (China), an overseas delegate, reported in the newsletter of the International Association of Geomorphologists (IAG) about the DFG-Yaoundé workshop: 'Geomorphologists in Frankfurt (Germany) have been quite successful in deciphering knowledge about the ecological and cultural changes in West and Central Africa. A synthesis workshop was organized in the Goethe Institute Yaoundé, the capital city of Cameroon, in March 5–7, 2006 by Professor Jürgen Runge to have discussed the various evidences including geomorphological, sedimentological and pedological ones investigated by the multidisciplinary Research Unit coordinated by Professor Peter Breunig (Archaeology) under the auspices of the German Research Foundation (DFG). The group showed rapid, large-scale modifications of ecosystems in African tropics due to radical environmental changes. The workshop has served also as a much-appreciated platform for over 40 participants from Cameroon, Nigeria, Central Republic of Africa, Gabon, Germany, France, United Kingdom and China to exchange ideas and opinions on most recent studies about tropics. The presentations were focused on tropical geomorphology and landscape evolution, archaeological evidence of ecological changes, land use and environmental changes. The workshop's more than fruitful achievements in sciences will be published in the journal 'Palaeoecology of Africa' with the workshop organizer as its editor. The social recognition of the workshop is confirmed by the awards of generous travel grants from DFG and GTZ, by the invitation to a wonderful dinner party from the German Embassy in Yaoundé as well as by the report in 'Cameroonian TV'.

1.6 GENERAL SUMMARY, CONCLUSIONS AND OUTLOOK FOR 2007–2009

Since 2003 already, the DFG-Research Unit 510 investigates the relationship between significant cultural transformations and environmental changes in West and Central Africa. The research activities concentrate on whether, and in what manner, prehistoric cultural and economic innovations were correlated with changes in vegetation, landscape and climate. The major temporal focus lies on the Late Holocene to subrecent period roughly between around 1.000 BC and 500 AD. This time span is believed to cover what is now frequently called the 'stone to metal transition' or the transition from Later Stone Age or 'Neolithic' cultures to those of Early Iron Age type. In this transitional period large farming communities and hierarchical societies emerged in West Africa. Obviously at the same time, the Central African rain forest was settled by ceramic manufacturing, Bantu languages speaking populations.

The research areas are located in northern and central Nigeria and in southern and eastern Cameroon. Cultural change in northern Nigeria (Chad Basin) is evidenced by the appearance of large and often fortified proto-urban settlements, while in central Nigeria it is the famous art of the Nok culture which figures most prominently in this respect. Considering the results of research on settlement sites in Nigeria, a distinct population growth from 500 BC onwards is indicated which continued into the first millennium AD. It was accompanied and probably triggered by agricultural intensification. Both phenomena—population growth and agricultural intensification—are linked to a variety of other archaeologically traceable occurrences such as iron metallurgy, food storage, craft specialization, sophisticated art and expansion into new settlement areas. They characterize

Figure 4. Group portrait of the DFG-Research Unit 510 delegates in front of the Goethe Institute at Yaoundé, Cameroon, on March 7, 2006 (Photo: T. Herold). 1: Xiaoping Yang, 2: Alain Wabo, 3: Charles Domtchouang, 4: Dethardt Goetze, 5: Stefan Porembski , 6: Chiori Agwu, 7: Alfred Ngomanda, 8: Joachim Eisenberg, 9: Conny Meister, 10: Dominique Jolly, 11: Mark Sangen, 12: Jürgen Runge, 13: Francois Nguetsop, 14: Maurice Tsalefac, 15: Ayobami Salami, 16: Marcel Koko, 17: Silja Meier, 18: Stefanie Kahlheber, 19: Lindsay Norgrove, 20: Alexa Höhn, 21: Konstantin König, 22: Pascal Nlend Nlend, 23: Manfred K.H. Eggert, 24: Boniface Kankeu, 25: Denis Bienvenu Nizesete, 26: Cyriaque-Rufin Nguimalet, 27: Pierre Poukalé (†), 28: Katharina Neumann, 29: Félix Ngana, 30: Michele Delneuf (†), 31: Rosemarie Eggert.

a widespread complex which was discovered and exemplarily studied by the archaeological subproject of Peter Breunig in the first phase of the project.

On the basis of the present data it is assumed that the emergence of large settlements in the Chad Basin as well as its counterparts in the Nok area reflect a new strategy of organisation. It is hypothesised that this process was not stimulated only by environmental change but rather by internal, i.e. social developments as yet to be specified. This working hypothesis should be tested during the second phase 2007–2009 of the interdisciplinary project in concentrating even more strongly on environmental and cultural data. This will be done by systematically exploring the Nok culture territory in central Nigeria. At the same time the temporal frame will be enlarged from around 500 BC to approximately 500 BC to AD 500 in both northern and central Nigeria.

As far as southern Cameroon is concerned the results of the first two years of the project constitute an important step toward the goal of establishing a broader physiogeographical, archaeological and archaeobotanical database there. For the period between 2.400 BP and 1.750 BP several phenomena indicate a change in material culture and subsistence.

Significant in this respect is the discovery of early sites with iron or iron slag. This will have an important impact on the ongoing discussion about the presence of iron and the knowledge of its reduction from iron ore in the rain forest. As has been demonstrated by the Archaeologist's team around Manfred H.K. Eggert, the crucial time period for this seems to be the last half of the last millennium BC. One key site, Akonétye, provided the oldest graves with iron objects in Central Africa. It is apparent, however, that they date to the early Christian era (see chapter 4 by C. Meister in this volume).

The presence of pearl millet (*Pennisetum glaucum*) at two rain forest sites, Bwambé–Sommet and Abang Minko'o, dated to the last centuries BC, points to some mode of connection between the people of the Central African rain forest and those of the West African savanna. Furthermore, the basic cultural and material conditions which lay at the heart of rain forest economy at that time are yet not fully understood. Consequently, much additional palaeoenvironmental and archaeological work in both the rain forest proper and the contact zone of rain forest and savanna is required.

The main focus of the second phase of the Cameroon part of the project, however, will be the rain forest–savanna mosaic in eastern Cameroon. Specifically along the axis Nanga Eboko–Bertoua–Batouri additional fieldwork will be carried out. From that first and second hand knowledge of sites dating to the time interval of about 2.500 BP to 1.700 BP are already known. As during the project's first phase special attention will be paid to securing as much archaeobotanical and palaeoenvironmental evidence as possible from the sites in question. As far as the relationship between West and Central Africa during the first millennium BC is concerned it seems conceivable that the population growth in West Africa stimulated migrations between the savanna belt and the Central African rain forest. Several arguments can be put forward in favour of this hypothesis.

First, up to now there is no evidence that the settling of the Central African rain forest by sedentary farming and pottery-producing populations was the result of an autochthonous development. Secondly, the Sudano–Sahelian cultural changes seem to be contemporaneous with the settling of the Central African rain forest as verified by the discovery of sites dated to the respective period. Finally, a strong argument for the relationship between the two regions derives from economic similarities. The finds of *Pennisetum glaucum* at the two rain forest sites mentioned allow for the first time to establish a link between the drier West African savannas and the humid rain forest (Katharina Neumann). It seems that in the savanna region an intensified agricultural production with pearl millet and cowpea was developed which maybe subsequently spread into the rain forest from the middle of the first millennium BC onwards.

During the first project's phase 2003–2006 it was supposed that a generally drier climate existed during the first millennium BC. Considering the results achieved so far this view has to be modified. It seems that in the Nigerian savanna regions, at least for some centuries during the first millennium BC climatic conditions were more stable and somewhat more humid than today. This might explain the rapid growth of the settlements mentioned above. However, most of these settlements in the Chad Basin existed only for a short time. Thus, it cannot be totally excluded that climatic fluctuations caused some change in settlement activities. If there were such apparently small-scale modifications they still have to be detected in the palaeo record.

For the Central African rain forest, the general assumption that aridity was prevalent during the first millennium BC can be specified. Since pearl millet is a cereal from a distinctly drier vegetation zone and does normally not thrive under humid rain forest conditions, its presence points to a much more seasonal climate with longer dry seasons in the second half of the first millennium BC. This is in line with evidence for a contemporaneous disturbance of the characteristic rain forest vegetation and its replacement by pioneer formations. Based on the presence of pearl millet in early sites and its disappearance in the first half of the first millennium AD, it will be tested if a change in seasonality—the existence of longer or shorter dry seasons—was the driving force for the reduction of the rain forest in the first millennium BC and its subsequent reestablishment some centuries later.

Considerable human impact on landscape and vegetation can be assumed as a consequence of intensified land use. Erosion processes, the replacement of forest by secondary formations, savannization and the establishment of anthropogenic plant communities are the major human induced landscape changes to be expected from at least 500 BC onwards. It is a major objective in the project's second phase to detect and specify man-induced landscape changes in the palaeorecord.

As most of the archaeological sites are located in areas where no continuous sedimentological and palynological sequences can be obtained from lake sediments, the Research Unit exploits alternative proxy-data archives and interprets them in terms of vegetation and landscape development. There is evidence that tropical fluvial alluvia contain a great potential for landscape reconstruction. They have been extensively prospected and analysed with regard to their evolution and palaeoenvironmental proxy-data content by the geographical subproject (Jürgen Runge). This has been particularly achieved in the interior Ntem delta in southern Cameroon (Figures 1, 3). As far as the physiogeographical subproject is concerned its main goal is to establish a better temporal and spatial resolution of the former environmental changes at the rain forest–savanna fringe. This will be done by comparing the alluvia and the palaeoenvironmental signals gathered from rivers which are oriented toward the Atlantic (Nyong and Ntem in 2004 to 2006, Sanaga, scheduled for 2007) with those of the fluvial systems in the eastern part of southern Cameroon. The latter, the Dja and Ngoko Rivers, are draining into the Congo basin and should be investigated in 2008. By comparing the respective results from the western Atlantic and the eastern hinterland it will be discussed whether the regression of the rain forest in the first millennium BC was continuous and widespread or whether it was discontinuous in both time and space. If the latter was the case it is expected to find indications of forest 'islands' within savanna ecosystems.

Another important question concerns the nature and extent of Late Pleistocene and Holocene environmental changes in southern Cameroon and Nigeria (Nok area). Here, particular attention has to be paid to the amount of forest cover. It is also planned to investigate whether, and in what manner, environmental changes during the time slices around 20 ka (LGM), 11 ka (YD) and 3 ka ('First Millennium Crisis') in these regions correlate to the respective conditions in neighbouring areas like the Central African Republic and Gabon.

Since surprising and promising discoveries have been made during the analyses of the alluvia in the Cameroonian Ntem interior delta (Figure 3), it is necessary to widen the time frame in the second phase of the project. Fluvial to possibly palustrine and lacustrine sediments with ages of up to 50.000 years BP are important palaeoenvironmental archives (see chapter 6 by M. Sangen in this volume). They offer the unique possibility to study older climatically extreme periods, such as the LGM (18.000 BP) or the Younger Dryas (11.000 BP) and to compare the results with those of the first millennium BC. Fieldwork in the largely unexplored savanna–forest mosaic is of very high priority. Thus, the Research Unit 510 aims at a unique integrated archaeological–archaeobotanical and palaeoenvironmental transect reaching from the semi-arid Sahel (14°N) to the equatorial Guinea–Congo phytogeographical zone (2°N). Hopefully, this will enable the investigation of patterns evidencing environmental and cultural change from a supra-regional perspective.

ACKNOWLEDGEMENTS

Since 2003 the work of the Research Unit 510 has been generously supported by grants from the German Research Foundation (DFG) for which I am most grateful. I like to thank the subproject coordinators Katharina Neumann, Peter Breunig and Manfred H.K. Eggert and all their collaborators—who have partly contributed to this volume of 'Palaeoecology of Africa'—for their support and cooperation in Nigeria and in Cameroon.

REFERENCES

Breunig, P. and Neumann, K., 2002, Continuity or Discontinuity? The 1st millenium BC-crisis in West African prehistory. In: Tides of the Desert. Contributions to the Archaeology and Environmental History of Africa in Honour of Rudolph Kuper, edited by Lenssen-Erz, T. *et al. Africa Praehistorica*, **14** (Köln: Heinrich-Barth-Institut), pp. 491–505.

Brooks, A.S. and Robertshaw, P., 1990, The Glacial Maximum in Tropical Africa: 22.000–12.000 BP. In *The World at 18.000 BP*, **2**, edited by Gamble, C. and Soffer, O., Low latitudes, Unwin Hyman, pp.121–169.

Clist, B., 1987, Early bantu settlements in west central Africa: a review of recent research. *Current Anthropology*, **28**, 380–382.

Eggert, M.H.K., 2002, Southern Cameroon and the Settlement of the Equatorial Rainforest: Early Ceramics from Fieldwork in 1997 and 1998–99. *Africa Praehistorica*, **14**, pp. 507–522.

Eisenberg, J., in press, Neotektonische Prozesse und geomorphologische Entwicklung des Ntem-Binnendeltas, SW-Kamerun. *Zentralblatt für Geologie und Paläontologie, Teil 1*.

Elenga, H., Schwartz, D. and Vincens, A., 1994, Pollen evidence of late Quaternary vegetation and inferred climate changes in Congo. *Palaeogeography, Palaeoclimatology, Palaeoecology*, **109**, pp. 345–356.

Gasse, F., 2005, Continental palaeohydrology and palaeoclimate during the Holocene. *Comptes rendus Géoscience*, **337**, pp. 79–86.

Giresse, P., Maley, J. and Brenac, P., 1994, Late Quaternary palaeoenvironments in the Lake Barombi Mbo (West Cameroon) deduced from pollen and carbon isotopes of organic matter. *Palaeogeography, Palaeoclimatology, Palaeoecology*, **107**, pp. 65–78.

Lanfranchi, R. and Schwartz, D., 1990, *Paysages quaternaires de l'Afrique Centrale Atlantique* (Paris: ORSTOM éditions, Collections didactiques), pp. 1–535.

Lanfranchi, R. and Clist, B., 1991, *Aux origines de l'Afrique Centrale. Centre Culturels FranÁais d'Afrique Centrale, Centre International des Civilisation Bantu*. Libreville, pp. 1–268.

Lezine, A.-M. and Cazet, J.-P., 2004, High-resolution pollen record from core KW31, Gulf of Guinea, documents the history of the lowland forests of West Equatorial Africa since 40.000 yr ago. *Quaternary Research*, **64**, pp. 432–443.

Lucazeau, F., Brigaud, F. and Leturmy, P., 2003, Dynamic interactions between the Gulf of Guinea passive margin and the Congo River drainage basin: 2. Isostasy and uplift. *Journal of Geophysical Research*, **108** (B8), 2384, pp. 1–19.

Maley, J., 2002, A Catastrophic Destruction of African Forests about 2.500 Years Ago Still Exerts a Major Influence on Present Vegetation Formations. *IDS Bulletin*, **33**/I, pp. 13–30.

Maley, J. and P. Brenac, 1998, Vegetation dynamics, palaeoenvironments and climatic changes in the forests of western Cameroon during the last 28.000 years B.P. *Review of Palaeobotany and Palynology*, **99**, pp.157–187.

Maloney, B.K (edited), 1998: *Human activities and the tropical rain forest. Past, present and possible future* (Dordrecht: Kluwer Acad. Publ.), pp. 1–206.

Ngako, V., Affaton, P., Nnange, J.M. and Njanko, Th., 2003, Pan-African tectonic evolution in central and southern Cameroon: transpression and transtension during sinistral shear movements. *Journal of African Earth Sciences*, **36**, pp. 207–214.

Runge, J., 1995, New results on late Quaternary landscape and vegetation dynamics in eastern Zaire (Central Africa). *Zeitschrift für Geomorphologie*, N.F., Suppl.-Bd., **99**, pp. 65–74.

Runge, J., 1996, Palaeoenvironmental interpretation of geomorphological and pedological studies in the rain forest 'core-areas' of eastern Zaire (Central Africa). *South African Geographical Journal*, **78**, pp. 91–97.

Runge, J., 1997, Altersstellung und paläoklimatische Interpretation von Decksedimenten, Steinlagen (stone-lines) und Verwitterungsbildungen in Ostzaire (Zentralafrika). *Geoökodynamik*, **18**, pp. 91–108.

Runge, J., 1998, Rezente und holozäne Vegetations- und Klimadynamik an der Regenwald/ Savannengrenze in Nord–Kongo (Zaire) und der Zentralafrikanischen Republik. *Zentralblatt für Geologie und Paläontologie, Teil 1*, **1–2**, pp. 91–113.

Runge, J., 2000, Environmental and climatic history of the eastern Kivu area (DR Congo, ex Zaire) from 40 ka to the present. In *Southern Hemisphere Paleo- and Neoclimates* (IGCP 341), edited by Smolka, P. and Volkheimer, W, (Springer: Heidelberg), pp. 249–262.

Runge, J., 2001, Landschaftsgenese und Paläoklima in Zentralafrika. Physiogeographische Untersuchungen zur klimagesteuerten quartären Vegetations- und Geomorphodynamik in Kongo-Zaire (Kivu, Kasai, Oberkongo) und der Zentralafrikanischen Republik (Mbomou). *Relief, Boden, Paläoklima*, **17**, pp. 1–294.

Runge, J., 2002, Holocene landscape history and palaeohydrology evidenced by stable carbon isotope ($\delta^{13}C$) analysis of alluvial sediments in the Mbari valley (5°N/23°E), Central African Republic. *Catena*, **48**, pp. 67–87.

Runge, J., Eisenberg, J. and Sangen, M, 2005, Ökologischer Wandel und kulturelle Umbrüche in West– und Zentralafrika – Prospektionsreise nach Südwestkamerun vom 05.03.–03.04.2004 im Rahmen der DFG Forschergruppe 510: Teilprojekt "Regenwald–Savannen–Kontakt (ReSaKo)". *Geoökodynamik*, **26**, pp. 135–154.

Runge, J., Eisenberg, J. and Sangen, M., 2006a, Eiszeit im tropischen Regenwald: Der ewige Wald–eine Legende? Festgehalten über Jahrtausende: Umweltarchive in Süd–Kamerun. *Forschung Frankfurt*, **2**, pp. 34–37.

Runge, J., Eisenberg, J. and Sangen, M, 2006b, Geomorphic evolution of the Ntem alluvial basin and physiogeographic evidence for Holocene environmental changes in the rain forest of SW Cameroon (Central Africa)—preliminary results. *Zeitschrift für Geomorphologie, N.F, Suppl.* **145**, pp. 63–79.

Runge, J., 2007 (September), Congo River. In *Large Rivers—Geomorphology and Managment*, edited by Gupta, A., (Chicester, Singapore: Wiley and Sons).

Runge, J. and Neumer, M., 2000, Dynamique du paysage entre 1955 et 1990 à la limite forêt–savane dans le nord du Zaire, par l'étude de photographies aèriennes et de données LANDSAT-TM. In *Dynamique à long terme des écosystèmes forestières intertropicaux* (ECOFIT), edited by Sérvant, M. and Servant-Vildary, S (Paris:UNESCO, IRD), pp. 311–317.

Runge, J. and Nguimalet, C.-R., 2005, Physiogeographic features of the Oubangui catchment and environmental trends reflected in discharge and floods at Bangui 1911–1999, Central African Republic. *Geomorphology*, **70**, pp. 311–324.

Sangen, M., in press, Physiogeographische Untersuchungen zur holozänen Umweltgeschichte an Alluvionen des Ntem-Binnendeltas im tropischen Regenwald SW–Kameruns. *Zentralblatt für Geologie und Paläontologie, Teil 1.*

Schwartz, D., 1988, Histoire d'un paysage: Le Lousséké. Paléoenvironnements Quaternaires et podzolisation sur sables Batéké. *Editions de l'ORSTOM, Etudes et thèses*, pp. 1–285.

Schwartz, D., 1992, Assèchement climatique vers 3000 B.P. et expansion Bantu en Afrique centrale atlantique: quelques réflexions. *Bulletin de la société Géologique de France*, **163**, pp. 353–361.

Schwartz, D. and R. Lanfranchi, 1990, Les cadres paléoenvironnementaux de l'évolution humaine en Afrique Centrale atlantique. *L'Anthropologie*, **97**, pp. 17–50.

Taylor, K.C., Lamorey, G.W., Doyle, G.A., Alley, R.B., Grootes, P.M., Mayewski, P.A., White, J.W.C. and Barlow, L.K., 1993, The 'flickering switch' of Late Pleistocene climate change. *Nature*, **361**, pp. 432–436.

Thomas, M.F., 1998, Landscape sensitivity in the humid tropics—a geomorphological appraisal. In *Human Activities and the tropical rainforest*, edited by Maloney, B.K (Dordrecht: Kluwer Academic Publ.), pp. 17–47.

Thomas, M.F., 2004, Landscape sensitivity to rapid environmental change—a Quaternary perspective with examples from tropical areas. *Catena*, **55**, pp. 107–124.

Thomas, M.F. and Thorp, M.B., 2003, Palaeohydrological reconstructions for tropical Africa since the Last Glacial Maximum—evidence and problems. In *Palaeohydrology: Understanding Global Change*, edited by Gregory K.J. and Benito, G. (John Wiley and Sons), pp. 167–192.

CHAPTER 2

Of Deserts and Forests: insights into Central African Palaeoenvironments since the Last Glacial Maximum

Jürgen Runge

Centre for Interdisciplinary Research on Africa (CIRA/ZIAF), Johann Wolfgang Goethe University, Frankfurt am Main, Germany

ABSTRACT: Tropical rain forests in Africa are recognized by a broader public as evolutionary old, species-rich and stable ecosystems over time, often transfigured as the last 'Eden'. Another perspective on the tropical forests originates in the fact, that these biodiversity hotspots are often endangered by humans exploring biological and geological resources located inside. However, since long already, there is widespread recognition and evidence that the assumed earth history stability of rain forests on the landscape scale can not be generally confirmed, and that changes in climate and vegetation had been quite a common phenomena. Strong natural shifts in rainfall and to a lesser extent in temperature during the Late Pleistocene and also in the Holocene repeatetly caused modifications and area dynamics of rain forest ecosystems over the last 40.000 years that temporarily led to an almost total disappearance of the humid rain forests and to an extension of more open, semi-humid to even semi-arid savanna-like vegetation patterns. In contradiction to the earlier postulated evolutionary stability of rain forests, on the landscape scale, a striking tendency for rapid modifications due to climate change can be stated. This paper gives a brief overview of some of the major 'milestones' in supplying proxy-data information on former environmental changes in Central Africa.

2.1 INTRODUCTION

Due to its humid hot climate, its bad infrastructural setting and political and economical instability the Central African region is considered to be one of the world's most difficult working areas for scientific research. Therefore, compared with the rest of Africa this region is only insufficiently explored. From the palaeoenvironmental point of view the Central African rain forests and the adjoining semi-humid savannas and savanna woodland areas are almost a 'terra incognita', too. Still until the 1960s it was believed that the equatorial rain forest of Africa would be one of the stablest and unchanging ecosystems of the world. Informative climate witnesses in form of palaeosoils and sediments were not expected therefore, and also due to the intensive chemical weathering processes no traces of datable organic carbon within soils and sediments had for long not been considered to be a source of proxy-data under this tropical-humid climate.

This study presents some conclusions about the history of climate and landscape of Central Africa in the Quaternary with special reference to the Last Glacial Maximum (LGM). Results of field work which are relevant to palaeoclimate and landscape history and basic research aim to the description and understanding of a differentiated physio-geographical landscape genesis. A particular focus lies on the reconstruction of the palaeoecological conditions in Central Africa during and after the last glacial period. The different 'core area'-theories and conceptions, which try to reconstruct and explain the location of potential refugial areas of the rain forest ecosystem during environmental changes are also included and discussed.

2.2 ENVIRONMENTAL CHANGES IN CENTRAL AFRICA ACCORDING TO PROXY-DATA

2.2.1 Glacio-morphological traces

Glacio-morphological traces particularly moraines, glacio-fluvial deposits and palaeosoils, which are climate historically interpretable can be proven at the eastern border of the Congo Basin along the continental transition to the East African highland. There are several investigations (Hastenrath, 1984; Mahaney, 1990; Rosquist, 1990) about the glaciation history of Mount Kenya (5.194 m asl) and Mount Ruwenzori (5.109 m asl). One evidence for ancient glacio-morphodynamic processes in the surroundings of Mount Kenya are the glaciofluvial gravels of the Nanyuki formation, which spread at the mountain footslope. Deeply carved, gorge-like to U-shaped valleys interfere between the piedmont sediments and the glacier region. The 'Naro Moru Till' was radiometrically dated by tephra to a minimum age of 320 ± 20 ka (ka = 'kilo annum', 1.000 years) (Charsley, 1989). Palaeomagnetic measurements of the loess-like components of the Naro Moru glacial till gave an age of deposition of maximum 700 ka (Brunhes-epoch, Mahaney, 1990).

Morphologically diagnosable moraines at Mount Kenya in the Teleki Valley reach down to 2.850 m asl. This material is weakly weathered and covered by two palaeosoils. Thermoluminescence (TL) datings account for an age around 100 ka (Mahaney, 1990). This is to be rated as reference for a first strong glacier advance after the Teleki-Liki (Eemian) interglacial period. Other moraines that mark glacier advances at Mount Kenya exist between 3.100–3.400 m asl. The Liki glaciation with its major phases I, II and III is assigned to the LGM up to the Late Glacial (Mahaney, 1990). Organic deposits among the moraines of Liki I and Liki II resulted in a maximum ^{14}C-age of 12 ka for the end of the glacier advance in the LGM. The radiocarbon datings of Liki III, which is characterized by the morphologically best conserved moraines gave a radiocarbon age of 10 to 12,6 ka (Alexandre *et al.,* 1994). In addition even younger, apparently Holocene glacier fluctuations dated from 4,4 to 6,7 ka document at least sporadic younger glacial dynamics for the Liki III advance (Johansson and Holmgren, 1985).

At the in comparison with Mount Kenya less well researched Mount Ruwenzori the moraines in the Mobuku Valley reach down to 2.100 m asl. Livingstone (1962) analyzes a sediment core from Lake Mahoma (0°21'N/29°58'E) that is situated at 3.000 m asl and determines by ^{14}C-dating of the organic lake sediments an age of 14,7 ka, which he sets for the end of the LGM in the high levels of the western Rift Valley. Other morainic tracts like the ones from Orumubaho are yet not dated. Alexandre *et al.* (1994) presume due to the significant geomorphological similarity with the Liki III moraines at Mount Kenya that these glacial forms also give evidence for a minimum age of 10 ka.

A climatic interpretation of the glacio-morphological observations for the LGM (21–18 ka) and the Holocene optimum (Altithermal, 8–6 ka) in Central and Southern Africa was done by Partridge *et al.* (1999). Using the study of Rosquist (1990), Partridge *et al.* (1999) assume for the mountain areas of the western Rift a prominent glacier advance into lower sites by up to 2.110 m asl. This was caused by a lowering of temperature of 10 °C during the LGM for the mountain regions. The depression in precipitation in the central and eastern Congo Basin seems to be also quite high and is indicated with 50–60% against todays figures with reference to Livingstone (1971) and Runge and Runge (1995). For the East African Highlands (Mount Kenya and Mount Kilimandjaro) the depression of snow lines and glaciers turns out less clearly with 1.320 resp. 1.450 m asl. It is assumed that here the lowering of temperature is somewhat weaker and amounts to 7,4 °C to 8,5 °C during the LGM (Partridge *et al.*, 1999). During the Holocene between 8 and 6 ka a new glacier retreat of approx. 520 m asl occurred at Mount Ruwenzori and in the East African Upland (Mount Kenya) with a temperature of 1–2 °C higher compared to today. Contemporary the occurrence of precipitation in the eastern Congo Basin and in central East Africa increases by 10–20% in relation to today's values (Partridge *et al.*, 1999), which leads to a re-expansion of the forest.

2.2.2 Lake-level oscillations in the Central African Rift

Except for the barely explored inland lakes 'Tumba' and 'Mai-Ndombe' in the central Congo Basin the palaeoclimatic relevant lake systems are located in the eastern periphery of the survey area in the western Rift Valley. Lake-level oscillations in rift lakes may have climatic and/or tectonic causes. For the time period since the LGM Alexandre *et al.* (1994) underline that the climate is primarily responsible for the changing lake-levels, whereas tectonic movements as a cause (e.g. uplift or subsidence of the bottom of trench) can be declined.

Data about Late Quaternary and Holocene variations of precipitation and lake-levels exist for Lake Victoria by Kendall (1969) as well as for Lake Albert (ex Lake Mobutu) by Harvey (1976), Sowunmi (1991), Roberts and Baker (1993) and Ssemanda and Vincens (1993). For Lake Kivu indicatory studies were presented by Degens *et al.* (1973), Degens and Hecky (1974), Hecky (1978) and Haberyan and Hecky (1987). Lake Tanganyika was subject of a multidisciplinary research under the aspects of geoecology and landscape history in the context of the TANGAYDRO GROUP research programme (Coulter, 1991; TANGAYDRO GROUP, 1992a, 1992b; Tiercelin *et al.*, 1988, 1989, 1992; Vincens, 1993).

During the evaluation it is noticeable that the different large and deep lakes occurring in the western Rift Valley react partially quite inconsistent to the Late Glacial and Holocene environmental changes. In the quite large and deep Lake Tanganyika basin (max depth 1.470 m), lake-levels are clearly lowered between 23 and 25 ka under cold and arid conditions, which can be concluded from afro-montan pollen findings embedded in the lake's sediment. During the LGM between 25–17 ka after Vincens (1993) even until 15 ka the water level of Lake Tanganyika was still relatively high but obviously had no discharge by an outflow (Lukuga) to the western Congo Basin. Not until 17–15 ka and till around 12 ka continuing the lake-level falls clearly under semi-arid to arid and cold(?) environmental conditions. The maximum of aridity is reached at 14 ka with a maximal lowering of the lake-level of about 86 meters compared to today. At the beginning of the Holocene the lake-level fluctuations are without uniformity. Stronger volcanic influences in the wider area can be proved by tephra layers in the sediment delivered by the Rungwe volcanos at about 11 ka (Tiercelin *et al.*, 1988). In the Holocene from 10 to 5 ka the lake-level is again high and indicates more humid conditions in general. The spill-over of Lake Kivu into the Ruzizi River around 9,5 ka causes, temporally retarded, a particularly clear lake-level maximum at Lake Tanganyika between 6,7 and 6,8 ka. After an intermediate period between 5–3 ka the lake indicates a tendency to a lowering of the lake-level up to the present day.

Lake Albert, only 58 m deep, in the northeast of the western Rift Valley shows a relatively high lake-level between 27–25 ka, which indicates still moister conditions in the surrounding area. The more shallow Lake Albert probably dried up to a larger extent during the LGM (here about 25–19 ka) under arid and cold climatic conditions. Already between 19 and 14,5 ka it seems to become rapidly more humid; the lake-level rises strongly. The maximum of aridity around 14 ka determined at Lake Tanganyika appears also at Lake Albert until approx. 13 ka. Afterwards it gets continuously moister and warmer. The lake begins to fill again until the middle Holocene around 5 ka. After 5 ka continuing up to the present day the lake-level of the Albert system shows an intermediate pattern of inconsistent levels of the lake surface.

The lake-level of the volcanically originated 'Kivu reservoir' located between Lake Tanganyika and Lake Albert decreases around 12 ka dramatically by 310 meters! Afterwards with the beginning of the Holocene it becomes clearly warmer and more humid. The lake surface now rised far over its former level. The so far closed Lake Kivu system began to spill over around 9,5 ka (Hecky, 1978). In the south of Lake Kivu near Bukavu a new drainage via the Ruzizi River emerged. It is adjusted to the 700 m subjacent Lake Tanganyika lake-level. A decrease of the lake-level around 30 m about 4 ka makes the Lake Kivu to a closed system again. Until 1,2 ka the water level rises again strongly and the Ruzizi River is reactivated (Hecky, 1978).

2.2.3 Alluvia and slope surface sediments

For the interpretation of fluvial deposits on slopes and in valleys of the humid tropics under the aspects of climate and landscape change there is no universal conception existent. The alluvial deposits in the tropics are apparently more interlocked with colluvia of the interfluves and slopes than it is the case in the temperate regions. The more intensive rearrangement of coarse and fine material in the glacial periods on the slopes and in the valleys has led to a less prominent morphological development of fluvial terraces. Therefore, this complicates the spatio-temporal correlation of different terrace levels and alluvial deposits.

Alexandre *et al.* (1994) differentiate fluvial river deposits according to humid and arid climatic periods in the Shaba province (Biano Plateau and Lupembashi Valley, Alexandre– Pyre, 1971; Mbenza, 1983; Mbenza *et al.,* 1984). The observed alluvial accumulations in the spacious, flat valley systems are often underlain by solid bedrock, a compact gravel layer partially consolidated by iron solutions ('gravier sous berge'). They are covered by mainly sandy to clayish alternate beddings, which may be interrupted by single bands of coarser material (stone-lines). Alexandre *et al.* (1994) interpreted these sediments as residual debris that was exposed and redeposited in conjunction with linear erosion of the river under more humid conditions. Van Zinderen Bakker and Clark (1962) assign similar detrital accumulations at the Luembe River in northeastern Angola primarily to a notable arid climatic period. The more recent fluvial erosion under humid conditions has not yet led to a dissection of these detrital layers. Radiocarbon datings of the hanging layer following the 'gravier sous berge' strata resulted in Holocene ages of 6,3–6,8 ka (Van Zinderen Bakker and Clark, 1962; de Ploey, 1966–68, 1968; Alexandre *et al.,* 1994). Runge and Tchamié (2000) describe related phenomenons from the Niantin River valley in northern Togo (West Africa) and assume an Ogolian age around 18 ka for the accumulation of the vast pebbles accumulations and of other coarse material over solid bedrock. Here, the [14]C (AMS) datings gave however sub-recent ages of only several hundred years. This gives also evidence for a very effective morphodynamic activity of fluvial tropical systems by lateral erosion and redeposition of the complete alluvial body in the course of singular excessive events of discharge, Runge and Tchamié, 2000). In the Mbari Valley in the Central African Republic undisturbed alluvial sediments with a thickness of several meters can be dated to 8 ka (Runge, 2002). This corroborates the thesis of a continuous accumulation of sands and clays in the early Holocene under more humid climatic conditions. In Nigeria, at the northwestern limit of the volcanic Jos Plateau, Zeese (1991) defines two humid cool climatic periods with a strong valley erosion between 18–20 ka and around 11 ka (Younger Dryas?). Zeese (1991, 1996) postulates the filling up of valleys by alluvia with sandy accumulation up to 15 meters thick, with poorly graded fluvial and/or colluvial sediments during Late Glacial times. Preuss (1986a, 1986b) introduced a longer sedimentary sequence in the region of the Ruki River in the central Congo Basin (former Zaire, now DR Congo), which obviously took place under arid climatic conditions, e.g. with a LGM to post-LGM age of 19–17,7 ka. Already quite early De Ploey (1964a, 1965, 1968) created on the western limit of the Congo Basin in the Malebo-(Stanley) Pool near Kinshasa/Brazzaville a four-level sediment stratigraphy for the Late Quaternary: from 42 ka onwards strong erosion ('Maluekien') with subsequent spacious accumulation of sand until around 37 ka; during the 'Ndjilien' podzolization of the deposited sands under alternating humid warm to cool climatic conditions; the high glacial ('Léopoldvillien') is again cooler and dryer and shows by an alternation between humid to semi arid (?) conditions, a very strong fluvial activity with incision of interflow surfaces. Before the onset of the Holocene a strong mobility of the sediments with erosion and deposition of surface layers starts again around 16 ka and ends just before 11 ka in the Congo Basin. De Ploey augurs a temporary, climatic controlled thinning out of the vegetation cover with an increased morphodynamic activity. From 12–3 ka, during the so-called 'Kibangien', a predominant tendency to a renewed propagation of the rain forest persists in Central Africa.

Close to the surface, layers with predominant coarse material ('stone-lines') occur widely spread as stratigraphical markers of discontinuities in the development of landscapes in the Central Congo Basin and also in the surrounding swell regions (Waegemans, 1953; Stoops, 1967; Alexandre and Soyer, 1987; Runge, 1992; Runge, 1997). For the formation of surface layers and 'stone-lines' termites, particularly the so-called giant termites *Macrotermes subhyalinus* RAMBUR (ex *Bellicositermes bellicosus rex*) whose mounds reach up to 3–5 m in height and 10–60 m in diameter, play a decisive role. Because of the partially fossilized, large termite mounds within the present rain forest it can be assumed that these lifeforms originally existed under savanna-like and sparsely wooded environmental conditions and have been overgrown in the Holocene by the expanding forest respectively displaced largely by various ecological parameters. The rate of sedimentation in the submarine Congo estuary decreases between 11,2–10,3 ka and 10,3–8,4 from 160 cm to 35 cm per 1.000 years. This noticeable attenuation of the rate of sedimentation could be an indication for a climatic controlled, abruptly starting recession of the termite activity and the associated lower capacity of bioturbation in Central Africa (Runge and Lammers, 2001).

2.2.4 Traces of former arid conditions: a desert instead of a forest?

In the southern Congo Basin between 5–12 °S, especially in the Shaba province, with a current mean annual precipitation of 1.600–1.000 mm numerous traces of former processes of aeolian morphodynamics can be ovserved. Alexandre–Pyre (1971) reports that the expanded surface levels of the Upper Congo (Bandundu) to Shaba (Biano–Plateau and Lubumbashi) often carry thick sand caps ('épais manteau sableux'). The sand layers deposited obviously by the effect of wind often lie over a layer of rock of the 'grès polymorphe', which is classified as a tertiary, palaeogene formation by Cahen (1954) and Alexandre–Pyre (1971). The isolated occurrence of windkanter situated above the sand covers suggests a strongly arid, serir-like landscape type. In the valleys the sand caps were eliminated to a large extent by fluvial processes, so that the sandy soils can be particularly found on the interfluves and on the plateaus in 1.500–1.600 m asl.

The following palaeoclimatic expressive phenomena can be differentiated morphologically in southern Congo (Shaba) by: a) parallel, flat dune courses, which run longitudinal to the predominant wind direction (De Dapper, 1981, 1985; Thomas and Shaw, 1991a); b) smaller Nebkha dunes, which develop on the valley flanks of broad discharge channels and which interrupt temporarily the predominant longitudinal dune systems (Alexandre–Pyre, 1971); c) slightly arched, overlapping marginally transverse dunes, which run diagonally to the predominant wind direction (De Dapper, 1981); and finally, by d) pan-like, oval to round depressions close to the valley head.

According to Alexandre *et al.* (1994) the shape of sand dunes in the Shaba province (10°–11°S) belong, due to their homogenous striking from ESE to WNW, evolutionary to the northern Kalahari system (Thomas and Shaw, 1991a, 1991b). There are many differences of the dune courses north and south the Okavango delta in Botswana (Thomas, 1989) regarding the size and frequency of the longitudinal dunes in the southern Congo. A several kilometres expanded longitudinal dune system shows a distance between two dune ridges of about two kilometers in the Congo; in the south the dune ridges are located significantly more closely with distances around 350 m. Regarding the absolute height, which amounts only 1 m (!), the dunes in the southern Congo Basin are rather undistinctly developed resp. are not in a good condition anymore and are degraded strongly. The Kalahari dunes in the surrounding of the Victoria–Falls at the border to Zambia and Zimbabwe reach heights of 20 m in contrast (data of D.S.G. Thomas, quoting Alexandre *et al.,* 1994). The northernmost dune field, which is already situated close to the equator (4°–5°S) near Lusambo, strikes from WSW to ENE and probably relates to another evolutionary context than the southern Kalahari dune system.

Fossil blow out, deflation forms on the Kundelungu–Plateau in the DR Congo are documented by Soyer (1983). Alexandre *et al.* (1994) considers that the climatic conditions for the activity of such an aeolian process for granted starting from 300 mm of annual precipitation. Hövermann (1988) determines even lower 'critical precipitation values' for the climate-morphological development of significant arid land forms and soils in the Kalahari and Sahara. From the palaeoclimatic point of view the aeolian form complexes (e.g. pans and dunes) in Shaba with recent annual precipitations of 1.000–1.400 mm imply a dramatic formerly 'desertic' decrease of humidity and vegetation cover. For active arid morphodynamic processes with deflation forms 300 mm, for dune genesis limiting values around 150 mm of annual precipitation are assumed. The 50–60% decrease of precipitation in central and southeast Africa during the LGM suggested by Partridge *et al.* (1999) on the basis of the current precipitations would therefore not be sufficient in order to activate arid morphodynamic processes. Since no TL-datings from this area are existent up to now, Alexandre *et al.* (1994) state the question if the aeolian activity in Shaba has possibly rather a Tertiary than a Quaternary climate historic background.

2.2.5 Atlantic sea level variations

Similar to the Rift lakes at the eastern border of the Congo Basin it is difficult to distinguish clearly between tectonic epirogenetic changes and glacio-eustatic oscillations in sea level to be the release for changes of sea level at the Central African coast of the Atlantic Ocean. The narrow Central African coastal lowlands demonstrate counterrotating tectonic movements: the Angolian area shows slight tendencies of rising, whereas the coast of the Republic of Congo and Gabon is epirogenetically lowered (Giresse, 1980).

The glacial eustatic oscillations of sea level proven between Lobito (Angola) and Pointe Noire (Congo-Brazzaville) in the Late Quaternary display the following chronology. During Oxygene Isotope Stage 3, which was warmer and poor in global ice cover, the sea level rises from 42–37,7 ka above the present level ('Inchirienne' transgression). In the coastal hinterland swampy depressions and lagoon-like areas developed in which compact, black clays abundant in foraminifers were deposited (Malounguila-Nganga *et al.,* 1990). From pollen findings in the partially peaty deposits a natural milieu of a mangrove dominated ombrophil forest could be reconstructed (Caratini and Giresse, 1979). From 24,6 ka onwards, with the global increase of ice cover (Isotop Stage 2), the Ogolian regression was also recognizable at the Central African Atlantic coast (Alexandre *et al.,* 1994). Until 18 ka sea level depleted by 110 to 120 m under today's level (Caratini and Giresse, 1979; Giresse, 1980). Linear, yellow sandy deposits near Pointe Noire are interpreted as coastal dune formations blown out from the shelf during the LGM (Alexandre *et al.,* 1994). Gypsum deposits in lagoons and the complete absence of *Rhizophora* pollen in the sediments indicate a non-wooded, semi-arid to occasionally full arid (!) ecosystem at the Central African coast during the LGM (Caratini and Giresse, 1979; Lanfranchi and Schwartz, 1990). Jansen *et al.* (1984) determine by oceanic sediment cores near the Congo mouth, a strong terrigenous sedimentation within the coastal range until 14,5 ka. This is rated as a sign for a missing resp. strongly thinned out vegetation coverage in the hinterland of the wider Congo Basin. With the slow rising of the sea level starting from 13 ka, strengthened with the beginning of the Holocene around 10 ka, the base erosion level changes noticeably, which weakens the efficiency of fluvial morphodynamic processes near the coast. The contemporaneous development of a close vegetation cover in the Congo Basin in the Holocene (re-expansion of forest) reduces the terrestrial sediment input into the ocean increasingly. Until 5 ka the Atlantic sea level has reached its current level (Giresse *et al.,* 1979). The oscillations during the last 5.000 years are less distinct. Between 4–3 ka a lowering of the sea level around 1 meter, between 2–1,5 ka an increase of 0,5 meter is proven (Alexandre *et al.,* 1994).

2.3 DISCUSSION: ENVIRONMENTAL CHANGES IN CENTRAL AFRICA FROM THE LAST GLACIAL MAXIMUM (LGM) TO THE HOLOCENE

The 'Quaternary Environments Network (QEN)' by J. Adams und H. Faure gives an overview of some of the most important geo- and bio-scientific conclusions about the development of climate and vegetation of Central Africa (and other regions) for the Quaternary. For the Central African area the following scenario resulted on the basis of the preceded proxy-data discussion: The majority of the newer geomorphologic, zoological and palaeobotanical indications underline the fact that today's Central African rain forest spread in the Congo Basin was clearly reduced in extension during the LGM due to profound reductions in temperature and precipitation (Runge, 1992, 2001; Thomas, 1994; Kadomura, 1995, 1998). In many regions of Central Africa tree and grass savanna ecosystems adapted much better to aridity and replaced the rain forest within a short period of time. Today partially fossilized colonies of termite mounds north and south the equator still give evidence of the former dominance of savanna ecosystems in the Congo Basin (Alexandre *et al.,* 1994; Boulvert, 1996). By bioturbation and denudation under seasonal climates (e.g. change of rainy and dry season) the terrigenous sediment input to rivers that reached the ocean like the Congo strongly increased during the LGM (Giresse, 1980; Jansen *et al.,* 1984). Nevertheless spatially isolated areas must have existed in the Congo Basin and some swell regions during the dry high glacial period, which served the rain forest as retreat area or refugia ('core areas'). Spatial concepts of different refugia changed strongly during the last 35 years (Meggers *et al.,* 1973; Hamilton, 1982; Littmann, 1988; Colyn *et al.,* 1991). A fluvial rain forest 'refugium' within the catchment of the Congo and Oubangui River (Colyn *et al.,* 1992; Maley, 1995) and a larger number of isolated rain forest islands and expanded gallery forest systems ensured the survival of the rain forest species during arid periods. This also explains the rapid re-conolization of the Congo Basin with rain forest within some thousand years only. The rain forest is therefore a highly sensitive and spatio-dynamic ecosystem if it is studied on the landscape scale.

The extent of temperature and precipitation lowering is still discussed controversially. Older conceptions proceed on the assumption of a comparatively strong cooling down at the equator of 6–8 °C (Hamilton, 1982; Maley, 1987; Bonnefille et al., 1990). This may have to be attributed to the fact that often pollen profiles from mountain regions were consulted as a reference and the temperature gradients in the mountainous zones are more distinctive than in the low lands. Maley (1987) proposed a decrease in sea surface temperature in the Gulf of Guinea by 2–5 °C during the LGM (northward shifting of the cold Benguela current), which implicated cooling and reduced convective precipitation by upwelling of cold waters in the western Congo coast area.

Climate modelling and synoptic environmental reconstructions of the conditions during the LGM (Lautenschlager, 1991) and the Holocene climatic optimum (Partridge et al., 1999) in Central Africa come to the perception that climate cooled down only slightly approximately about 1–2 °C in annual average in the low lands on the equator during high glacial times. The depression of precipitation during the LGM however, seems to be much higher according to the models and to new analyses by ground surveys in Central Africa (Runge, 1997, 2001). A depletion of precipitation in the Congo Basin by 50–60% in relation to the current values are possible according to Partridge et al. (1999). The distinct, today fossilized, arid morphodynamical wealth of forms in the southern Congo Basin (Kasai and Shaba) cannot be explained conclusively even with such an increased aridity during the LGM.

During the Holocene climatic optimum around 8–7 ka the rain forest expands over its current borders to the north and the south of the equator. The effectivity of morphodynamic processes like denudation and lateral erosion of the rivers decreases. Therefore, the sediment input into the ocean declines significantly. The amount of precipitation is up to 10–20% higher than today and temperature may have risen between 1–2 °C during the Holocene climatic optimum (Partridge et al., 1999).

The Central African lakes react due to their different natural environments (catchment size, water depth) with temporal delay to the increasing aridity during the LGM. Only between 14–12 ka, already towards the end of the LGM, due to aridity low lake-levels can be recognized at Lake Albert, Lake Kivu and Lake Tanganyika (Harvey, 1976; Degens and Hecky, 1974; Vincens, 1993; Vincens et al., 1993).

In the middle Holocene around 5 ka the climatic optimum is already exceeded. The area of rain forest decreases again under the influence of declining precipitation. The rain forest is probably also repressed by the increasing human influence at that time in the course of the so-called 'Bantu migration' from approximately 3 ka on (Schwartz 1992; Roche, 1996; Wotzka, 1995). Bush fires, logging and enhanced activities of clearing caused particularly at the rain forest-savanna contact the development of forest-savanna mosaics. Maley (1992) indentifies, supported by palynologic findings, an indication that there exists a connection between the advance of early settlers into the rain forest and the changing climate. A natural climatic pejoration ('péjoration climatique') in Central Africa evidenced by arising grasses and herbal plants between 2,5–2 ka, locally replaced and opened the rain forest which could have been opened up more simply and favoured by the advancing population of the Bantu speaking population to the south (Batéké–Plateau, RP Congo; Schwartz, 1992) and to the highlands of East Africa (Uganda, Ruanda; Roche, 1996).

REFERENCES

Alexandre, J. and Soyer, J., 1987, Les Stone-lines, conclusions de la journée d'étude. *Revue internationale d'ecologie et de géographie tropicales,* **11**, pp. 229–239.

Alexandre, J., Aloni, K. and de Dapper, M., 1994, Géomorphologie et variations climatiques au Quaternaire en Afrique Centrale. *Revue internationale d'ecologie et de géographie tropicales,* **16**, pp. 167–205.

Alexandre-Pyre, S., 1971, Le plateau des Biano (Katanga). Géologie et Géomorphologie. *Académie Royale des Sciences d'Outre-Mer, Classe des Sciences Naturelles et Médicales, N.S.* **18**, 3, pp. 1–151.

Bonnefille, R., Roeland, J.C. and Guiot, J., 1990, Temperature and rainfall estimates for the past 40.000 years in equatorial Africa. *Nature,* **346**, pp. 347–349.

Boulvert, Y., 1996, Etude géomorphologique de la République Centralafricaine. Carte 1:1.000.000 en deux feuilles ouest et est. *ORSTOM, Notice explicative,* **110**, pp. 1–258.

Boyer, P., 1975, Les différents aspects de l'action des *Bellicositermes* sur les sols tropicaux. *Annales des Sciences Naturelles, Zoologie,* **12**, pp. 447–503.

Cahen, L., 1954, *Géologie du Congo Belge* (Liege), pp. 1–450.

Caratini, C. and Giresse, P., 1979, Contribution palynologique à la connaissance des environnements continentaux et marins du Congo à la fin du Quaternaire. *Comptes Rendus de l'Académie des Sciences,* **288**, pp. 379–382.

Charsley, T.J., 1989, Composition and age of older outwash deposits along the northwestern flank of Mount Kenya. In *Quaternary and environmental research on the East African mountains*, edited by Mahaney, W.C., (Rotterdam: Balkema), pp. 165–174.

Colyn, M., Gautier–Hion, A. and Verheyen, W., 1991, A re-appraisal of palaeoenvironmental history in Central Africa: evidence for a major fluvial refuge in the Zaire Basin. *Journal of Biogeography,* **18**, pp. 403–407.

Coulter, G.W. (ed.), 1991, Lake Tanganyika and its life. With contributions from J.-J. Tiercelin, A. Mondegeur, R.E. Hecky and R.H. Spigel. *Natural History Museums Publications*, (London, Oxford, New York), pp. 1–354.

de Dapper, M., 1981, The microrelief of the sandcovered plateaux near Kolwezi (Shaba, Zaire) II. The microrelief of the crest dilungu. *Revue internationale d'ecologie et de géographie tropicales,* **5**, pp. 1–12.

de Dapper, M., 1985, Quaternary aridity in the tropics as evidenced from geomorphological research using conventional panchromatic aerial photographs. Examples from peninsular Malaysia and Zaire. *Bulletin de la Société Belge de Géologie,* **3**, pp. 199–207.

de Ploey, J., 1964a, Cartographie geomorphologique et morphogenese aux environs du Stanley-Pool (Congo). *Acta Geographica Lovaniensia,* **3**, pp. 431–441.

de Ploey, J., 1964b, Nappes de gravats et couvertures argilo-sableuses au Bas-Congo: leur genèse et l'action des termites. In *Etudes sur les termites Africains,* edited by Bouillon, A. (Kinshasa: Colloque International, Université Lovanium, Léopoldville, 11.–16. Mai 1964), pp. 399–414.

de Ploey, J., 1965, Position géomorphologique, génèse et chronologie de certains dépôtssuperficiels au Congo occidental. *Quarternaria,* **7**, pp. 131–154.

de Ploey, J., 1966–1968, Report on the Quaternary of the Western Congo. *Palaeoecology of Africa,* **4**, 65–70.

de Ploey, J., 1968, Quaternary phenomena in the Western Congo. Proceedings of VII INQUA Congress. In *Means of correlation of Quaternary Successions*, edited by Morrison, B. *et al.* (Salt Lake City, University of Utah Press), pp. 501–515.

Degens, E.T. and Hecky, R.E., 1974, Paleoclimatic Reconstruction of Late Pleistocene and Holocene Based on Biogenic Sediments from the Black Sea and a Tropical African

24 Jürgen Runge

Lake. *Colloques Internationaux de C.N.R.S.,* **219** (les méthodes quantitatives d'étude des variations du climat au cours du Pléistocene), pp. 13–24.

Degens, E.T., von Herzen, R.P. and Wong, H.-K., 1971, Lake Tanganyika: Water Chemistry, Sediments, Geological Structure. *Naturwissenschaften,* **58**, 5, pp. 229–241.

Giresse, P., 1980, Carte sédimentologique du plateau continental du Congo. *ORSTOM, Notice explicative,* **85**, pp. 1–24.

Giresse, P., Kouyoumontzakis, G. and Moguedet, G., 1979, Le quaternaire supérieur du plateau continental Congolais. Exemple d'évolution paléoocéanographique d'une plateforme depuis environ 50.000 ans. *Palaeoecology of Africa and the surrounding islands,* **11**, pp. 193–218.

Haberyan, K.A. and Hecky, R.E., 1987, The late Pleistocene and Holocene stratigraphy and paleolimnology of lakes Kivu and Tanganyika. *Palaeogeography, Palaeoclimatology, Palaeoecology,* **61**, pp. 169–197.

Hamilton, A.C., 1982, Environmental History of East Africa. A study of the Quaternary, (London: Academic Press), pp. 1–328.

Harvey, T.J., 1976, *The palaeolimnology of Lake Mobutu Sese Seko, Uganda–Zaire: The last 28.000 years* (Ph.D. thesis, Duke University North Carolina).

Hastenrath, S., 1984, *The Glaciers of Equatorial East Africa*. Solid Earth Sciences Library, (Boston, Lancaster: Dordrecht), pp. 1–353.

Hecky, R.E., 1978, The Kivu–Tanganyika basin: the last 14.000 years. *Polske Archiwum Hydrobiologii* **25**, 1/2, pp. 159–165.

Hövermann, J., 1988, *The Sahara, Kalahari and Namib deserts: A geomorphological comparison*. Proceedings of the Symposium on the geomorphology of Southern Africa (Transkei), pp. 71–83.

Jansen, J.H.F., van Weering, T.C.E, Gieles, R. and van Iperen, J., 1984, Late Quaternary oceanography and climatology of the Zaire–Congo fan and the adjacent eastern Angola basin. *Netherlands Journal of Sea Research,* **17**, pp. 201–249.

Johansson, L. and Holmgren, K., 1985, Dating a moraine on Mount Kenya. *Geografiska Annaler,* **67**, pp. 123–128.

Kadomura, H., 1995, Palaeoecological and Palaeohydrological Changes in the Humid Tropics during the last 20.000 years, with reference to Equatorial Africa. In *Global Continental Palaeohydrology*, edited by Gregory, K.J., Starkel, L. and Baker V.R. pp. 177–202.

Kadomura, H., 1998, Environmental changes in the humid tropics during the last 20.000 years: synopses of data sets and perspectives. *Bullettin of the Graduate School of Letters, Rissho Universit,* **14** (Kumagaya, Japan), pp. 1–15.

Kendall, R.L., 1969, An ecological history of the Lake Victoria basin. *Ecological Monographs,* **39**, pp. 121–176.

Lanfranchi, R. and Schwartz, D., 1990, Paysages quaternaires de l'Afrique centrale atlantique. *Editions ORSTOM, Collection Didactiques*, pp. 1–535.

Lautenschlager, M., 1991, Simulation of the ice age atmosphere. January and July means. *Geologische Rundschau,* **80**, pp. 513–534.

Littmann, T., 1988, Jungquartäre Ökosystemveränderungen und Klimaschwankungen in den Trockengebieten Amerikas und Afrikas. *Bochumer Geographische Arbeiten,* **49**, pp. 1–210.

Livingstone, D.A., 1962, Age of Deglaciation in the Ruwenzori Range, Uganda. *Nature,* **194**, pp. 859–860.

Livingstone, D.A., 1971, A 22.000 years pollen record from the plateau of Zambia. *Limnology and Oceanography,* **16**, pp. 349–356.

Mahaney, W.C., 1990, *Ice on the equator. Quaternary geology of Mount Kenya, East Africa* (Wisconsin: Sister Bay).

Maley, J., 1987, Fragmentation de la forêt dense humide africaine et extension des biotopes montagnards au quaternaire recent: nouvelles données. *Palaeoecology of Africa,* **18,** pp. 307–336.

Maley, J., 1992, Mise en évidence d'une péjoration climatique entre ca. 2500 et 2000 ans B.P. en Afrique tropicale humide. Commentaires sur la note de D. Schwartz. *Bulletin de la Société Géologique de France,* **163,** pp. 363–365.

Maley, J., 1995, Les Fluctuations Majeures de la Forêt Dense Humide Africaine au Cours des Vingt Derniers Millenaires. In *L'alimentation en forêt tropicale: interactions bioculturelles et applications au développement,* edited by Hladik, C.M. *et al.,* pp. 1–12.

Malounguila-Nganga, D., Nguie, J. and Giresse, P., 1990, Les paléoenvironnements quaternaires du colmatage de l'estuaire du Kouilou (Congo). In *Paysages quaternaires de l'Afrique Centrale atlantique,* edited by Lanfranchi, R. and D. Schwartz, (Paris: ORSTOM Collections didactiques) pp. 89–97.

Mbenza, M., 1983, *Evolution de l'environnement géomorphologique de fonds de vallée au cours du quaternaire dans une région tropicale humide* (University of Liège: Thése Doctorand), pp. 1–278.

Mbenza, M., Roche, E. and Doutrelepont, H., 1984, Note sur les apports de la palynologie et de l'etude des bois fossiles aux recherches géomorphologiques sur la vallée de la Lupembashi (Shaba–Zaire). *Revue de Paléobiologie, Volume Spécial.,* pp. 149–154.

Meggers, B.J., Ayensu, E.S. and Duckworth, W.D. (eds.), 1973, *Tropical forest ecosystems in Africa and South America: a comparative Review* (Washington: Smithsonian Institution Press), pp. 1–350.

Partridge, T.C., Scott, L. and Hamilton, J.E., 1999, Synthetic reconstruction of southern African environments during the Last Glacial Maximum (21–18 kyr) and the Holocene Altithermal (8–6 kyr). *Quaternary International,* **57/58,** pp. 207–214.

Preuss, J., 1986a, Jungpleistozäne Klimaänderungen im Kongo–Zaire–Becken. *Geowissenschaften in unserer Zeit,* **4,** 6, pp. 177–187.

Preuss, J., 1986b, Die Klimaentwicklung in den äquatorialen Breiten Afrikas im Jungpleistozän. Versuch eines Überblicks im Zusammenhang mit Geländearbeiten in Zaire. *Marburger Geographische Schriften,* **100,** pp. 132–148.

Roberts, N. and Barker, P., 1993, Landscape stability and biogeomorphic response to past and future climatic shifts in intertropical Africa. In *Landscape Sensitivity,* edited by Thomas, D.S.G. and Allison, R.J. (Chichester, New York: Wiley and Sons), pp. 65–82.

Roche, E., 1996, L'Influence anthropopique sur l'environnement à l'age du fer dans le Rwanda ancien. *Revue internationale d'ecologie et de géographie tropicales,* **20,** pp. 73–89.

Rosquist, G., 1990, Quaternary glaciations in Africa. *Quaternary Science Reviews,* **9,** pp. 281–297.

Runge, J., 1992, Geomorphological observations concerning palaeoenvironmental conditions in eastern Zaire. *Zeitschrift für Geomorphologie, N.F., Suppl.-Bd.,* **91,** pp. 109–122.

Runge, J., 1997, Alterstellung und paläoklimatische Interpretation von Decksedimenten, Steinlagen (stone lines) und Verwitterungsbildungen in Ostzaire (Zentralafrika). *Geoökodynamik,* **18,** pp. 91–108.

Runge, J., 2001, Landschaftsgenese und Paläoklima in Zentralafrika. *Relief, Boden, Paläoklima,* **17,** pp. 1–294.

Runge, J., 2002, Holocene landscape history and palaeohydrology evidenced by stable carbon isotope (δ^{13}C) analysis of alluvial sediments in the Mbari valley (5°N/23°E), Central African Republic. *Catena,* **48,** pp. 67–87.

Runge, J. and Lammers, K., 2001, Bioturbation by termites and Late Quaternary Landscape evolution on the Mbomou plateau of the Central African Republic (CAR). *Palaeoecology of Africa,* **27,** pp. 153–169.

Runge, J. and Tchamié, T., 2000, Inselberge, Rumpfflächen und Sedimente kleiner Einzugsgebiete in Nord–Togo: Altersstellung und morphodynamische Landschaftsgeschichte. *Zentralblatt für Geologie und Paläontologie, Teil 1*, **5/6**, pp. 497–508.

Runge, J. and Runge, F., 1995, Late Quaternary palaeoenvironmental conditions in eastern Zaire (Kivu) deduced from remote sensing, morpho-pedological and sedimentological studies (Phytoliths, Pollen, C-14 data). *Publication Occasionelle CIFEG*, **31**, pp. 109–122.

Schwartz, D., 1992, Assèchement climatique vers 3.000 B.P. et expansion Bantu en Afrique centrale atlantique: quelques réflexions. *Bulletin de la Société Géologique de France*, **163**, pp. 353–361.

Sowunmi, M.A., 1991, Late Quaternary environments in equatorial Africa: Palynological evidence. *Palaeoecology of Africa and the surrounding islands*, **22**, pp. 213–238.

Soyer, J., 1983, Microrelief de buttes basses sur sols inondes saisonnierement au Sud–Shaba (Zaire). *Catena*, **10**, pp. 253–265.

Ssemanda, I. and Vincens, A., 1993, Végétation et climat dans le Bassin du lac Albert (Ouganda, Zaire) depuis 13.000 ans B.P.: Apport de la palynologie. *Comptes Rendus de l'Académie des Sciences*, **316**, pp. 561–567.

Stoops, G., 1967, Le profil d'altération au Bas–Congo (Kinshasa). *Pédologie*, **17**, 1, pp. 60–105.

TANGANYDRO GROUP, 1992a, Les sites hydrothermaux sous-lacustres à sulfures massifs du fossé Nord–Tanganyika, Rift Est–Africain: Expédition Tanganydro 1991. *Comptes Rendus de l'Académie des Sciences*, **315**, pp. 733–740.

TANGANYDRO GROUP, 1992b, Sublacustrine hydrothermal seeps in northern Lake Tanganyika, East African Rift: 1991 Tanganydro Expedition. *elf aquitaine production*, pp. 55–81.

Thomas, D.S.G., 1989, Reconstructing ancient arid environments. In *Arid Zone Geomorphology*, edited by Thomas, D.S.G. (London), pp. 311–334.

Thomas, D.S.G. and Shaw, P.A., 1991a, *The Kalahari environment* (Cambridge, New York: Cambridge Univ. Press), pp. 1–284.

Thomas, D.S.G. and Shaw, P.A., 1991b, 'Relict' desert dune systems: interpretations and problems. *Journal of Arid Environments*, **20**, pp. 1–14.

Thomas, M.F., 1994, *Geomorphology in the Tropics. A study of weathering and denudation in low latitudes*, (New York, Brisbane, Toronto: Wiley and Sons), pp. 1–460.

Tiercelin, J.-J., Mondeguer, A., Gasse, F., Hillaire–Marcel, C. *et al.*, 1988, 25.000 ans d'histoire hydrologique et sédimentaire du lac Tanganyika, Rift Est–africain. *Comptes Rendus de l'Académie des Sciences*, **307**, pp. 1375–1382.

Tiercelin, J.-J., Thouin, C., Kalala, T. and Mondeguer, A., 1989, Discovery of sublacustrine hydrothermal activity and associated massive sulfides and hydrocarbons in the north Tanganyika through, East African Rift. *Geology*, **17**, pp. 1053–1056.

Tiercelin, J.-J., Soreghan, M., Cohen, A.S., Lezzar, K.-E. *et al.*, 1992, Sedimentation in large rift lakes: example from the middle Pleistocene Modern deposits of the Tangayika trough East African Rift System. *elf aquitaine production*, pp. 84–111.

Vincens, A., 1993, Nouvelle séquence pollinique du Lac Tanganyika: 30.000 ans d'histoire botanique et climatique du Bassin Nord. *Review of Palaeobotany and Palynology*, **78**, pp. 381–394.

Vincens, A., Chalié, F., Bonnefille, R., Guiot, J. and Tierce, J.-J., 1993, Pollen-derived rainfall and temperature estimates from Lake Tanganyika and their implication for Late Pleistocene water levels. *Quaternary Research*, **40**, pp. 343–350.

van Zinderen Bakker, E.M. and Clark, J.D., 1962, Pleistocene climates and cultures in North–Eastern Angola. *Nature*, **196**, pp. 639–642.

Waegemans, G., 1953, Signification pédologique de la 'stone-line' *Bulletin agricole du Congo Belge et du Ruanda–Urundi,* **3**, 44, pp. 521–532.

Wotzka, H.-P., 1995, Studien zur Archäologie des zentralafrikanischen Regenwaldes. Die Keramik des inneren Zaire-Beckens und ihre Stellung im Kontext der Bantu-Expansion. *Africa Praehistorica,* **6**, pp. 1–582.

Zeese, R., 1991, Äolische Ablagerungen des Jungquartär in Zentral- und Nordnigeria. *Sonderveröffentlichungen des Geologischen Instituts der Universität zu Köln,* pp. 343–351.

Zeese, R., 1996, *Oberflächenformen und Substrate in Zentral- und Nordostnigeria. Ein Beitrag zur Landschaftsgeschichte. Berichte aus der Geowissenschaft* (University of Köln: Habil-Schrift), pp. 1–195.

CHAPTER 3

Settling the rain forest: the environment of farming communities in southern Cameroon during the first millennium BC

Alexa Höhn, Stefanie Kahlheber, Katharina Neumann and Astrid Schweizer

Institute of Archaeological Sciences, Johann Wolfgang Goethe University, Frankfurt am Main, Germany

ABSTRACT: In West and Central Africa important environmental and prehistoric changes took place in the first millennium BC. These are subjects of a research project involving geographers, archaeologists and archaeobotanists from the universities of Tübingen and Frankfurt. One of the main objectives is to investigate the environmental conditions under which farming and pottery-producing people settled in southern Cameroon in the first millennium BC. Archaeological excavations have been conducted at Bwambé-Sommet in the vicinity of Kribi at the Atlantic coast and at Akonétye, Minyin and Abang Minko'o close to Ambam, which is located 165 km to the southeast of Kribi (Figure 1). From the structures excavated a large body of archaeobotanical material was retrieved. In addition several pollen cores were taken, the one of Nyabessan in the Ntem Delta near Ma'an (Figure 1) yielding material that included the time slice in question. The data analysed so far indicate that the settlements were situated within rain forest vegetation which, however, was disturbed and partly substituted by pioneer plant formations in the first millennium BC. Open formations existed in the vicinity of the settlements.

3.1 INTRODUCTION

Archaeological evidence points to the settlement of the rain forest regions in Central Africa by farmers and pottery-producing people in the first millennium BC with an emphasis on

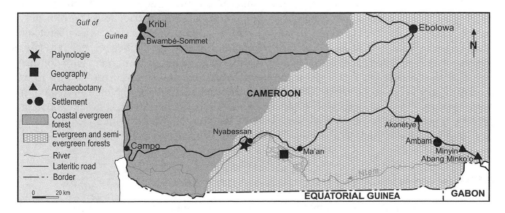

Figure 1. Research area.

the second half (summarised in Eggert *et al.*, 2006). In some publications the expansion of early ceramic traditions is linked to the evidence of historical linguistics and the expansion of the Bantu language family over much of the African continent south of the Sahara (e.g. Oslisly, 1994/95; Assoko Ndong, 2000/01, 2002). Both phenomena, however, are not necessarily dependent on each other and so links between them are difficult to prove (Eggert, 2005; Bostoen, 2006).

For roughly the same time period (3.000 to 2.500 BP) numerous hydrological and palynological data attest the partial breakdown of the rain forest in Central Africa (summarised by Maley, 2004; Vincens *et al.*, 1999). This climatically induced decline of the rain forest has been connected to "Bantu migration" by Schwartz (1992). Schwartz argues that the opening of the rain forest and its large-scale replacement by savannas favoured the immigration of Bantu speaking agriculturalists into this hitherto closed and seemingly hostile environment. The existence of savanna corridors had already been hypothesized by Letouzey (1968). Based on biogeographic and palaeoenvironmental data Maley (2004) has sketched a map with the possible distribution of residual rain forest areas, and open vegetation, mainly savannas.

Whatever language they were speaking, farming communities might have migrated within savannas into those areas covered by rain forest today. However, the question is if settlements were only established in savanna patches or if areas covered with forest were settled as well. To answer this question satisfactorily a combination of archaeological, archaeobotanical and palaeoenvironmental data is needed. Our sites are situated within the residual rain forest assumed for southern Cameroon, supposed to have been covered by a forest-savanna mosaic with forest dominating (Maley, 2004). Palaeovegetation proxy data, most of them with a coarse time resolution, were only available for the northern and southern edge of the Central African rain forest, but not for our region. So when our work started it was not sure if we were to expect savanna or forest environment. Correspondingly, the environmental aspect of our research focused on the question, if savanna did exist between Kribi and Ambam during the first millennium BC.

3.2 RESEARCH AREA

The research area lies in the South Province of Cameroon. It is roughly limited by the road between Kribi and Ebolowa in the north, the Transafricana from Ebolowa to the border with Gabon in the east, the Ntem River which is close to the border to Equatorial Guinea and Gabon in the south and the Atlantic Ocean to the west.

The climate is characterized by high humidity, a mean annual temperature around 25 °C and annual precipitation ranging from 1.500 mm in the interior to 3.000 mm in the coastal region. Rainfall distribution is bimodal with a short and a main rainy season stretching roughly from April to mid-June and from September to mid-November respectively (Atlas Cameroun, 1979). The vegetation is a Guineo-Congolian evergreen rain forest with dominant Caesalpiniaceae near the coast and a mosaic of evergreen and semi-evergreen rain forests around Ambam (Letouzey, 1968, 1985; Tchouto Mbatchou, 2004). In the interior delta of the Ntem, swamp forests are widely distributed (Tchouto Mbatchou, 2004). Although population density in the area is comparatively low, most of the forests have been modified by shifting cultivation and moderate logging activities (Zapfack *et al.*, 2002; Van Gemerden *et al.*, 2003).

3.3 STUDY SITES

Major archaeological fieldwork in 2004/05 was carried out on Bwambé-Sommet near Kribi on the Atlantic coast and at Akonétye, about 20 kilometres to the northeast of Ambam

(Figure 1). Parallel coring for palynological studies took place south of Ma'an in the interior delta of the Ntem River. In 2006 archaeological excavations were realized at Minyin and Abang Minko'o to the southeast of Ambam. The sites are characterized by a relative abundance of archaeological features, i.e. mainly pit structures and find concentrations. At all sites a couple of pit structures were dug and yielded a large amount of ceramics. Maximum dimensions of the pits were three meters in diameter and four meters in depth. Two graves were excavated in Akonétye. The archaeobotanical material was well preserved and could be retrieved in sufficient amounts.

The radiocarbon data of Bwambé-Sommet and Abang Minko'o lie in the second half of the first millennium BC. The ceramics of the two sites show considerable resemblance, albeit the sites are about 180 kilometres apart. Both sites did not provide even a fragment of a ground stone axe or adze. Likewise, no trace of iron implements or iron slag was found.

The sites Akonétye and Minyin are younger, dating to the first half of the first millennium AD. Their pottery is rather homogeneous and in both sites iron artefacts were found. For detailed descriptions of the archaeological results refer to Eggert *et al.* (2006) and Meister (this volume).

The palynological fieldwork concentrated on the Ntem interior delta, which is situated between the coast and the Ambam area. The delta is about 35 km long and 10 km wide at its broadest point. Within the delta the Ntem is an anastomosing river that divides into three main river channels (Runge *et al.*, 2006). Several cores were recovered from the numerous old river channels. The core from Nyabessan, at the downstream end of the delta (Figure 1), contained sediments dating to the first millennium BC and was focused on in the palynological analyses.

3.4 METHODS

To retrieve archaeobotanical macro-remains (charcoal, charred fruits and seeds), sediment samples of ten liters each were taken from all spits and cultural layers of the pits. Processing included a combination of flotation and wet sieving using 2,5, 1,0 and 0,5 millimetre meshes. In addition, large fruits, seeds and charcoal fragments were handpicked during the excavation. They were identified using the reference collections of the Frankfurt Archaeobotanical Laboratory, wood anatomical atlases (Normand, 1950–1960), and the Frankfurt DELTA anatomical database of African woods (Neumann *et al.*, 2001). The ecological interpretation of the woody species follows Tchouto Mbatchou (2004).

For palynological investigations overlapping sediment cores were taken to ensure complete sampling. Due to the relatively high groundwater level the botanical material was generally well preserved. Pollen samples were extracted from the sediment using a small sampler of a specific volume (0,28 cm³). They were prepared for analysis following Fægri and Iversen (1989). A total terrestrial pollen sum was employed for percentage calculation. Pollen and spores of the local aquatic and swamp vegetation (e.g. *Raphia* and Cyperaceae) were excluded from the pollen sum. Pollen percentage diagrams were constructed using Tilia and Tilia-graph computer programs (Grimm, 1992). For classification and identification of pollen types, the keys of Bonnefille and Riollet (1980), Caratini and Guinet (1974), Merville (1965), Sowunmi (1973, 1995), Ybert (1979) as well as the modern reference material of the Universities of Frankfurt and Montpellier were used.

For phytolith studies, sediment samples of five grams were prepared using a standard procedure following Runge (1999) and Pearsall (2000). Phytolith morphotypes were identified according to the descriptions and classification systems of Runge (1999, 2000), Piperno (1988), Kealhofer and Piperno (1998) and Barboni *et al.* (1999). From each sample, one slide was completely checked and the identifiable phytolith morphotypes semi-quantitatively recorded.

3.5 RESULTS

The results mainly refer to the sites Akonétye and Bwambé-Sommet, having been the first sites to be excavated and the analyses thus have been more exhaustive. Nonetheless first results from Abang Minko'o and Minyin were integrated whenever available.

3.5.1 Charcoal analysis

In the charcoal samples from Akonétye, Bwambé-Sommet and Abang Minko'o, 14 arboreal taxa have been identified (Table 1).

The interpretation is preliminary since exact identification of the ecologically significant species level was only rarely achieved due to the lack of reference material. The composition of the samples from Akonétye and Bwambé-Sommet reflects the ecological conditions at the sites: Taxa from the evergreen forest are more common at the coast, whereas taxa from the semi-evergreen forest dominate further inland.

At both sites the charcoal assemblages point to the presence of disturbed as well as undisturbed rain forest vegetation. Disturbances are inferred from charcoal of pioneer taxa such as *Alchornea* type (Figure 2) and *Manilkara/Malacantha* type and by two non-pioneer light demanding taxa, *Parinari* type and *Uapaca* type. Non-pioneer light demanding taxa are abundant in secondary forests because they require gaps in the canopy to develop to full maturity (Tchoutou Mbatchou, 2004). Mature rain forest vegetation is indicated by shade-bearers such as *Maesobotrya/Protomegabaria*, *Dichostemma* type and cf. *Coula*. These taxa grow, flower and fruit in the understorey of closed canopy of undisturbed forest (compare Tchoutou Mbatchou, 2004).

In Abang Minko'o so far only one type was identified, *Tetraberlinia* type which seems to be quite common in this sample. One fragment is derived from an Euphorbiaceae, the other

Table 1. Presence/absence of charcoal types: ++ – frequent, + – present.

Charcoal type	Bwambé-Sommet	Akonétye	Abang Minko'o
Euphorbiaceae, *Maesobotrya/Protomegabaria*	+	—	—
Euphorbiacea, *Dichostemma* type	++	—	—
Sapotaceae, *Malacantha/Manilkara* type	++	—	—
Caesalpiniaceae, *Berlinia* type	+	—	—
Caesalpiniaceae, *Tetraberlinia* type	—	—	++
Combretaceae, *Pteleopsis* type	+	—	—
Anacardiaceae/Burseraceae, *Canarium/Aucoumea* type	—	+	—
Euphorbiaceae, *Alchornea* type	—	+	—
Flacourtiaceae, cf. *Caloncoba*	—	+	—
Olacaceae, cf. *Coula*	—	+	—
Cola sp.	—	+	—
Chrysobalanaceae, *Parinari* type	+	++	—
Euphorbiaceae, *Uapaca* type	++	++	—
cf. *Lophira*	+	+	—

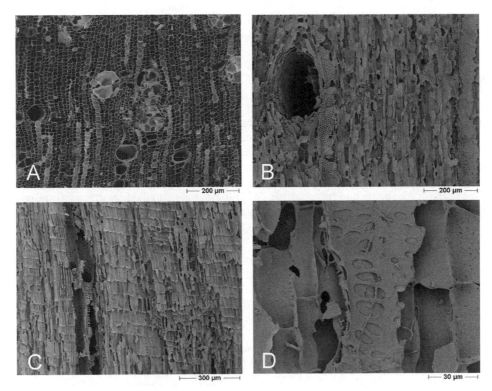

Figure 2. Photomicrographs of charcoal from *Alchornea* type (Euphorbiaceae).
A: transverse section, B: tangential longitudinal section,
C: radial longitudinal section, D: detail: enlarged ray-intervessel pitting.

fragments cannot be identified so far. *Tetraberlinia* species belong to forest vegetation. Some of them tend to become (co)dominant over large areas and such *Tetraberlinia* forests may extend over several square kilometres (Wieringa, 1999). The accumulation of *Tetraberlinia* type in this sample may be biased by anthropogenic selection of firewood. Further analyses have to be awaited before conclusions concerning the composition of the surrounding forest can be drawn.

3.5.2 Pollen analysis

Preliminary percentage curves for four ecological groups are depicted in figure 3. Included are taxa of pioneer, secondary and mature forest as well as taxa from wet habitats (swamp forests, mainly comprising ferns and Raphia). Separated by a solid bar, are the results of the pollen analysis of a surface sample at the coring site.

   The diagram shows a clear decrease of the local swamp taxa in the upper part of the core, above 250 centimetres. This points to a distinct change in the local hydrology and a turn from wetter to drier conditions roughly between 2.800 and 2.300 BP. Taxa of the mature evergreen rain forest also have lower percentage values in the upper part. Pioneer taxa are present throughout the diagram starting with values of 40 percent which indicates permanent disturbances of the river landscape. However, they show higher values in the upper part of the profile. This in turn points to at least partial destruction or substitution of mature as well as swamp forests by secondary plant communities. Grass pollen is insignificant in Nyabessan. There is no evidence that included savannas existed in the forest, let alone that they replaced the rain forest on a larger scale.

3.5.3 Fruits and seeds

So far, the analysis of macro-remains focused on the identification of major components of human nutrition including tree products and domesticated crops. Altogether four tree taxa, two crops species and one probable weed species are represented with fruits and/or seeds (Table 2). *Elaeïs guineensis* occurs in the assemblages of all four sites. *Canarium schweinfurthii* is much more rare, but still present in three of the site assemblages. *Coula edulis* and *Raphia* sp. were only recorded in Minyin and Akonétye, which proved to be richest in taxa. Especially in Akonétye, endocarp remains of oil palm make up the majority of the remains while the other taxa are present in small quantities or as isolated finds. The fruits and

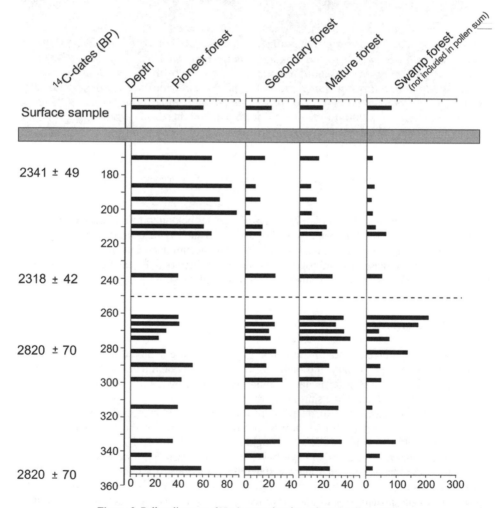

Figure 3. Pollen diagram of Nyabessan showing selected pollen groups.

Table 2. Presence of tree and crop taxa among the fruit and seed remains.

	Elaeis guineensis	*Canarium schweinfurthii*	*Coula edulis*	*Raphia* sp.	*Pennisetum glaucum*	*Vigna subterranea*
Bwambé-Sommet	●	●	—	—	●	—
Akonétye	●	●	●	●	—	●
Minyin	●	●	●	●	—	—
Abang Minko'o	●	—	—	—	●	—

seeds of these tree species are particularly rich in oil and fat, and are commonly collected by modern hunter-gatherers. *Elaeïs guineensis* and *Canarium schweinfurthii* are naturally growing in pioneer and secondary forests. The presence of their remains thus indicates disturbed habitats. *Coula edulis*, however, is a species of mature undisturbed forest, and *Raphia* palms grow in swamps and along rivers. It seems that especially the people of Akonétye and Minyin used a wide range of habitats for the collection of oil plants. As yet, there is no hint that any of these useful tree species were cultivated.

Regarding domesticated plants, the only crop at Akonétye is *Vigna subterranea*, the Bambara groundnut. Though represented by two seeds only, the presence of this pulse species is remarkable, as it is adapted to drier conditions of the savanna zone and currently does not belong to the typical crop inventory of rain forest agriculture. At Bwambé-Sommet and Abang Minko'o pearl millet (*Pennisetum glaucum*) was found, which proves to be most interesting since pearl millet is not cultivated in the rain forest in modern times. Rather, it constitutes the foremost staple crop of Sahelian agricultural systems.

3.5.4 Phytoliths

Phytoliths were studied from pit sediments of Bwambé-Sommet and Akonétye. They mainly contained morphotypes from grasses, Marantaceae, palms and other woody plants as well as charred particles. One objective was the search for banana phytoliths to corroborate the introduction and cultivation of this crop in the first millennium BC, as it has been stated by Mbida Mindzie *et al.* (2001) for the site Nkang near Yaoundé. However, up to now, banana phytoliths have not been found in the analysed samples.

Most common are those morphotypes that are typical for palms (Piperno, 1988; Runge, 1999) and several morphotypes originating from grasses. Charred particles are equally abundant. The spherical rugose morphotype generally attributed to woody plants, is consistently present but less common. In addition, there are a number of unknown types which could not be identified due to the lack of reference material.

The composition of the samples taken from the pits clearly differs from that of the surrounding natural sediment, where charred particles are missing. The phytolith composition of the pit samples is also distinctly different from modern reference samples of rain forest soils where the spherical rugose morphotype is usually present with percentages of 30–50% while grasses are only weakly represented (Alexandre *et al.*, 1997; Mercader *et al.*, 2000). This indicates that the pits were filled by anthropogenic activities and that the fill came from open spaces, probably from the settlements themselves.

A phytolith produced in the leaves of the Marantaceae plant family (Piperno, 1988: 94, plate 3, 4) was found in large quantities in the pit samples from Akonétye. Marantaceae (Figure 4) are very important in rain forest ecology and economy. Several species play a significant role in the regeneration cycles of forest trees (Maley, 1990) and Marantaceae leaves are used for many purposes (Dhetchuvi and Lejoly, 1996; Burkill, 1997). Thus, the presence of these phytoliths indicates the extensive use of Marantaceae leaves.

3.6 DISCUSSION

The results of the sites examined are unequivocal: the people we are dealing with lived in a rain forest habitat. Anthracological and palynological evidence does not indicate savannas. Quite to the contrary, the available data suggest that we have to expect rain forest vegetation between the Atlantic coast and the territory around Ambam during the first millennium BC and the beginning of the first millennium AD. This is in accordance with Maley (2004) who reconstructs a residual forest area between the Atlantic coast and Ebolowa/Ambam during the first millennium BC. However, a mosaic of forest and savanna as postulated for these areas by Maley (2004) cannot be proven for the Ma'an-Ambam region. The absence of savanna species in our palaeoecological records points to the existence of forest cover without savannas or savanna islands in the catchments of our archives during the first millennium BC.

This rain forest, however, had been altered by environmental changes before the people settled at the sites. Though forest persisted according to the Nyabessan core, the pollen record shows massive disturbance of the mature rain forest and expansion of pioneer formations, starting shortly after 2.800 BP. This disturbance is roughly contemporaneous with the forest deterioration indicated in the pollen profiles from western Cameroon

Figure 4. Marantaceae growing in the understorey of secondary forest, inspected by botanist Nolé Tsabang.

(Reynaud-Farrera *et al.*, 1996; Maley and Brenac, 1998) and Congo (Elenga *et al.*, 1996; Vincens *et al.*, 1999). With regard to hydrological sensitivity, disappearance of forest formations and the establishment of savannas can be observed at the driest sites to the north and south of the residual rain forest area, while at the most humid ones the forest seems to have been opened only locally in the last two millennia BC (Vincens *et al.*, 1999). The Ntem delta accordingly belongs to the humid sites, as does Lake Ossa situated about 150 kilometres to the north. There, as well, the forest had been disturbed but not replaced by vegetation dominated by grass after 800 BC (Reynaud-Farrera *et al.*, 1996). The structural changes following the climatic shift lead to an extension of secondary forests (Figure 5).

This is also reflected in the charcoal assemblages. In contrast to pollen the charcoal assemblages only mirror local vegetation. People of the respective villages collected firewood in different habitats—pioneer formations, secondary forest and closed rain forest. People will not go far, if they find sufficient amounts of firewood close to their settlement. Thus different stages of forest regeneration must have been present. From charcoal alone it cannot be concluded whether this diversity of habitats was a consequence of human impact or whether the settlers only exploited habitats that had already been established. The palynological data from the Ntem document vegetation change dating well before the sites excavated. In the wider region as well changes in the palynological data date a couple of hundred years before most of the sites discovered so far. It seems reasonable to assume that the vegetation change had been induced climatically rather than anthropogenically. Thus the charcoal assemblages might mirror the different habitats that had already existed when the settlements were founded, not excluding that under the human influence pioneer and secondary formations most probably gained more ground. We hope that the analyses of the charcoal assemblages from Abang Mink'o and Minyin, representing different settlement periods but situated not far from each other, may help us to gain some insight into the questions if and how the settlements changed the forest cover in the region.

It is intelligible that the establishment of settlements should have altered the vegetation in their vicinity. People kept paths and the surroundings of their dwellings open. Saplings of useful trees might have been protected but other trees and bushes most probably were kept low or were destroyed. Where trees lack, grasses can grow, so it is not surprising that evidence

Figure 5. Secondary forest, bordering banana plantation.

for grass vegetation is present in the phytoliths as well as the carpological assemblages from the pits of Bwambé Sommet and Akonétye. The grasses reflect the anthropogenic nature of the settlement vegetation, which mainly furnished the material for the filling of the pits.

The establishment of settlements in forest sites is in evident contrast to the hypothesis of Schwartz (1992) cited above. However, secondary forests are the habitat of a number of useful species like *Elaeïs guineensis* and *Canarium schweinfurthii*. It seems probable that the settlers used the opportunities offered by the structural forest changes. Remains of oil palm (*Elaeïs guineensis*) are very numerous not only in our sites but in many sites of the first millennium BC where they are often associated with *Canarium schweinfurthii* (e. g. Eggert, 1984; Mbida *et al.*, 2000; Lavachery, 2001; Clist, 2004/2005; Lavachery *et al.*, 2005; D'Andrea *et al.*, 2006). People exploited these plants for their oil rich fruits which would supplement a basic carbohydrate diet of cereals and possibly tubers.

The changes leading to the deterioration of the rain forest might be connected to stronger seasonality. The low values of swamp forest taxa in the Nyabessan diagram after 2.800 BP point to a marked change of hydrological conditions, e.g. lower groundwater tables. This might be due to decreasing precipitation or a change of seasonality. Maley (2004), after the evaluation of different paleoenvironmental data for Atlantic Central Africa, already suggested that presumably a longer dry season, probably accompanied by a modest decrease in annual precipitation, was the reason for the vegetation changes visible in the pollen records.

A stronger seasonal climate would also be a conclusive explanation for the successful cultivation of *Pennisetum glaucum* and *Vigna subterranea* in a rain forest environment during the first millennium BC. It is puzzling to find evidence in Bwambé-Sommet, Abang Minko'o and Akonétye for a subsistence strategy which is clearly not indigenous to the equatorial forest. Pearl millet and the Bambara groundnut are well adapted to arid conditions and their cultivation was developed in the semi-arid West African savannas. Under modern climatic conditions it appears most difficult or even impossible to cultivate them in the rain forest. Especially pearl millet is mainly cultivated in drier savanna regions, its wild ancestor being distributed in a belt at the southern fringe of the Sahara. The Sahelian origin of this cereal is still visible in its physiology with a C4 photosynthetic pathway and a short growing cycle, both adaptations to semi-arid conditions. High rainfalls, especially in the last growing period when ears are ripening, result in perishing of the grains. This implies that pearl millet cultivation cannot be productive in a rain forest environment. The findings of pearl-millet in the archaeobotanical record but without any savanna indicators in charcoal and pollen assemblages may be explained by hypothesizing that the dry season was long enough to allow pearl millet cultivation but at the same time short enough to sustain forest growth.

3.7 CONCLUSIONS

We conclude that the vegetation during the first millennium BC in southern Cameroon, from the coast to nearly 200 km inland, was dominated by forests. This is the first direct evidence for a residual forest area in this region, as hypothesised by Maley (2004) and Tchouto Mbatchou (2004). No unambiguous signs for the existence of savannas were found, and a mosaic of forest islands and savannas postulated for the residul forest areas cannot be confirmed so far. Our results show that the immigration of sedentary and farming populations did not stop at the borders of the rain forest. Even most humid and predominantly forested regions were settled during the first millennium BC. The (perturbed) forest was exploited for non-timber products as well as for firewood. It is quite puzzling, however, that people cultivated pearl-millet in a predominantly forested habitat. Seasonality must have been sufficiently pronounced in order to ensure the successful cultivation of the savanna species, but at the same dry seasons were short enough to sustain forest growth.

ACKNOWLEDGEMENTS

We would like to thank Manfred K. H. Eggert, Conny Meister and Pascal Nlend Nlend, as well as Joachim Eisenberg and Mark Sangen for fruitful fieldwork and discussions. Thanks are due to Barbara Eichhorn without her work identification of rain forest charcoal would have been much more difficult. The project is financed by the DFG.

REFERENCES

Alexandre, A., Meunier, J.-D., Lézine, A.-M., Vincens, A. and Schwartz, D., 1997, Phytoliths: indicators of grassland dynamics during the late Holocene in intertropical Africa. *Palaeogeography, Palaeoclimatology, Palaeoecology,* **136**, pp. 213–229.

Assoko Ndong, A., 2000/01, *Archéologie du peuplement holocène de la reserve de faune de la Lopé, Gabon,* **2**, (Université Libre de Bruxelles: Unpublished Ph.D. dissertation, Academic Year 2000–2001).

Assoko Ndong, A., 2002, Synthèse des données archéologiques récentes sur le peuplement à la Holocène de la réserve de faune de la Lopé, Gabon. *L'Anthropologie,* **106**, 2002, pp. 135–158.

Atlas Cameroun, 1979, Laclavère, G. (dir.), *Atlas de la République Unie du Cameroun. Les Atlas de Jeune Afrique.* (Paris: Editions Jeune Afrique).

Barboni, D., Bonnefille, R., Alexandre, A. and Meunier, J.D., 1999, Phytoliths as palaeoenvironmental indicators, West Side Middle Awash Valley, Ethiopia. *Palaeogeography, Palaeoclimatology, Palaeoecology,* **152**, pp. 87–100.

Bonnefille, R. and Riollet, G., 1980, *Pollens des Savanes d'Afrique Orientale* (Paris: CNRS).

Bostoen, K., 2006, What comparative Bantu pottery vocabulary may tell us about early human settlement in the Inner Congo Basin. *Afrique & Histoire,* **5**, pp. 221–263.

Burkill, H.M., 1997, *The Useful Plants of West Tropical Africa,* ed. 2, Vol. 4 (Families M–R) (Kew: Royal Botanic Gardens).

Caratini, C. and Guinet, P., 1974, Pollen et Spores d'Afrique tropicale. *Traveaux et Documents de Géographie Tropicale,* **16**, (Talence: Centre d'Etudes de Geographie Tropicale).

Clist, B.-O., 2004/2005, *Des premièrs villages aux premiers Européens autour de l'Estuaire du Gabon: Quatre millénaries d'interactions entre l'homme et son milieu,* **2**, (Université Libre de Bruxelles: Unpublished Ph.D. dissertation, Academic Year 2004–2005).

D'Andrea, A.C., Logan, A.L. and Watson, D.J., 2006, Oil palm and prehistoric subsistence in tropical West Africa. *Journal of African Archaeology,* **4**, pp. 195–222.

Dhetchuvi, M.M. and Lejoly, J., 1996, Les plantes alimentaires de la forêt dense du Zaïre, au nord-est du Parc National de la Salonga. In *L'alimentation en forêt tropicale. Interactions bioculturelles et perspectives de développement,* edited by Hladik, C.M., Hladik, A., Linares, O.F., Koppert, G.J.A. and Froment, A. (Paris: Editions Unesco), pp. 301–314.

Eggert, M.K.H., 1984, Imbonga und Lingonda: Zur frühesten Besiedlung des äquatorialen Regenwaldes. *Beiträge zur Allgemeinen und Vergleichenden Archäologie,* **6**, pp. 247–288.

Eggert, M.K.H., 2005, The Bantu problem and African archaeology. In *African Archaeology: A Critical Introduction.* Blackwell Studies in Global Archaeology, **3**, edited by Stahl, A.B. (Malden, MA, Oxford, Carlton: Blackwell), pp. 301–326.

Eggert, M.K.H., Höhn, A., Kahlheber, S., Meister, C., Neumann, K. and Schweizer, A., 2006, Pits, graves and grains: archaeological and archaeobotanical research in southern Cameroon. *Journal of African Archaeology,* **4**, pp. 273–298.

Elenga, H., Schwartz, D., Vincens, A., Bertaux, J., de Namur, C., Martin, L., Wirrmann, D. and Servant, M., 1996, Diagramme polinique Holocène du lac Kitina (Congo): mise en évidence de changements paléobotaniques et paléoclimatiques dans le massif forestier du Mayombe. *Comptes Rendus de l'Académie des Sciences* **323**, série 2a, pp. 403–410.

Fægri, K. and Iversen, J., 1989, *Textbook of Pollen Analysis*. 4th ed. (revised by K. Fægri, Kaland, P.E. and Krzywinski, K.), (Chichester: Wiley and Sons).

Grimm, E.C., 1992, TILIA and TILIA-graph: pollen spreadsheet and graphics programs. 8th International Palynological Congress, Aix-en-Provence 1992, Abstracts, p. 56.

Kahlheber, S., Höhn, A. and Neumann, K. (in press), Plant use in southern Cameroon between 400 BC and 400 AD. In *African Flora, Past Cultures and Archaeobotany,* edited by Fuller, D. and Murray, M.A., (Walnut Creek, CA: Left Coast Press).

Kealhofer, L. and Piperno, D., 1998, *Opal Phytoliths in Southeast Asian Flora.* Smithsonian Contributions to Botany, **88** (Washington: Smithonian Institution Press).

Lavachery, P., 2001, The Holocene archaeological sequence of Shum Laka Rock Shelter (Grass fields, Western Cameroon). *African Archaeological Review* **18**, pp. 213–247.

Lavachery, P., MacEachern, S., Bouimon, T., Gouem, B.G., Kinyock, P., Mbairo, J. and Nkokonda, O., 2005, Komé to Ebomé: Archaeological research for the Chad Export Project, 1999–2003. *Journal of African Archaeology* **3**, pp. 175–193.

Letouzey, R., 1968, *Étude Phytogéographique du Cameroun.* Encyclopédie biologique, **69** (Paris: Lechevalier).

Letouzey, R., 1985, *Notice de la Carte Phytogéographique du Cameroun au 1:500000.* (Yaoundé: Institut de la Carte Internationale de la Végétation, Toulouse & Institut de la Recherche Agronomique).

Maley, J., 1990, L'histoire récente de la forêt dense humide africaine: Essai sur le dynamisme de quelques formations forestières. In *Paysages quaternaires de l'Afrique centrale atlantique,* edited by Lanfranchi, R. and Schwartz, D., Editions de l'ORSTOM, Paris, pp. 367–382.

Maley, J., 2004, Les variations de la végétation et des paléoenvironnements du domaine forestier africain au cours du Quaternaire récent. In *L'Evolution de la végétation depuis deux millions d'années, edited by* Sémah, A.-M. and Renault-Miskovsky, J. Editions Artcom/Errance, Paris, pp. 143–178.

Maley, J. and Brenac, P., 1998, Vegetation dynamics, palaeoenvironents and climatic change in the forests of western Cameroon during the last 28000 years B.P. *Review of Palaeobotany and Palynology* **99**, pp. 157–187.

Mbida, C.M., Van Neer, W., Doutrelepont, H. and Vrydaghs, L., 2000, Evidence for banana cultivation and animal husbandry during the first millennium BC in the forest of southern Cameroon. *Journal of Archaeological Science* **27**, pp. 151–162.

Mbida Mindzie, C., Doutrelepont, H., Vrydaghs, L., Swennen, R.L., Beeckman, H., de Langhe, E. and de Maret, P., 2001, First archaeological evidence of banana cultivation in central Africa during the third millennium before present. *Vegetation History and Archaeobotany* **10**, pp. 1–6.

Meister, C. Recent archaeological investigations in the tropical rain forest of south-west Cameroon. (this volume).

Mercader, J., Runge, F., Vrydaghs, L., Doutrelepont, H., Ewango, C.E.N. and Juan-Tresseras, J., 2000. Phytoliths from archaeological sites in the tropical forest of Ituri, Democratic Republic of Congo. *Quaternary Research* **54**, pp. 102–112.

Merville, M., 1965, Le Pollen des Sapindacées d'Afrique Occidentale. *Pollen et Spores* **61**, pp. 465–489.

Neumann, K., Schoch, W., Détienne, P. and Schweingruber, F.H., 2001, *Woods of the Sahara and the Sahel. An Anatomical Atlas* (Bern: Paul Haupt).

Normand, D., 1950–1960. *Atlas des Bois de la Côte d'Ivoire,* vols. 1–3. Centre Technique Forestier Tropical, Nogent-sur-Marne.

Oslisly, R., 1994/95, The Middle Ogooué Valley: Cultural changes and Palaeoclimatic implications of the last four millennia. In *The Growth of Farming Communities in Africa from the Equator Southwards* (Azania 29/30, 1994/95), edited by Sutton, J.E.G. (Nairobi: British Institute in Eastern Africa), pp. 324–331.

Pearsall, D., 2000, *Palaeoethnobotany. A Handbook of Procedures.* 2nd ed. (San Diego: Academic Press).

Piperno, D.R., 1988, *Phytolith Analysis. An Archaeological and Geological Perspective*, (San Diego: Academic Press).

Reynaud-Farrera, I., Maley, J. and Wirrmann, D., 1996, Végétation et climat dans les forêts du Sud-Ouest Cameroun depuis 4770 ans BP: Analyse pollinique des sédiments du Lac Ossa. *Comptes Rendus de l'Académie des Sciences* **322**, 2a, pp. 749–755.

Runge, F., 1999, The opal phytolith inventory of soils in central Africa—Quantities, shapes, classicifation, and spectra. *Review of Palaeobotany and Palynology* **107**, pp. 23–53.

Runge, F., 2000, *Opal-Phytolithe in den Tropen Afrikas und ihre Verwendung bei der Rekonstruktion paläoökologischer Umweltverhältnisse,* (Paderborn).

Runge, J., Eisenberg, J. and Sangen, M., 2006, Geomorphic evolution of the Ntem alluvial basin and physiogeographic evidence for Holocene environmental changes in then rain forest of SW Cameroon (Central Africa)—preliminary results. *Zeitschrift für Geomorphologie, N.F. Supplement Volume,* **145**, pp. 63–79.

Schwartz, D., 1992, Assèchement climatique vers 3000 B.P. et expansion Bantu en Afrique centrale atlantique: quelques réflexions. *Bulletin de la Société Géologique de France* **163**, pp. 353–361.

Sowunmi, M.A., 1973, Pollen grains of Nigerian plants I. Woody species. *Grana* **13**, pp. 145–86.

Sowunmi, M.A., 1995, Pollen grains of Nigerian plants II. Woody species. *Grana* **34,** pp. 120–41.

Tchouto Mbatchou, G.P., 2004, *Plant Diversity in a Central African Rain Forest. Implications for Biodiversity Conservation in Cameroon,* (Ph.D. thesis, Wageningen University).

Van Gemerden, B.S., Shu, G.N. and Olff, H., 2003, Recovery of conservation values in Central African rain forest after logging and shifting cultivation. *Biodiversity and Conservation* **12**, pp. 1553–1570.

Vincens, A., Schwartz, D., Reynaud-Farrera, I., Alexandre, A., Bertaux, J., Mariotti, A., Martin, L., Meunier, J.-D., Nguetsop, F., Servant, M., Servant-Vildary, S. and Wirrmann, D., 1999, Forest response to climate changes in Atlantic Equatorial Africa during the last 4000 years BP and inheritance on the modern landscapes. *Journal of Biogeography* **26**, pp. 879–885.

Wieringa, J.J., 1999, *Monopetalanthus exit. A systematic study of Aphanocalyx, Bikinia, Icuria, Michelsonia and Tetraberlinia (Leguminosae, Caesalpinioideae)* (Wageningen: Agricultural University Papers), 1–320.

Ybert, J.P., 1979, Atlas des Pollens de Cote d'Ivoire. *Initiations Documentations Techniques,* **40**, (Paris: ORSTOM).

Zapfack, L., Engwald, S., Sonke, B., Achoundong, G. and Madong, B.A., 2002, The impact of land conversion on plant biodiversity in the forest zone of Cameroon. *Biodiversity and Conservation* **11**, pp. 2047–2061.

CHAPTER 4

Recent archaeological investigations in the tropical rain forest of South-West Cameroon

Conny Meister

Institute of Pre- and Protohistory and Medieval Archaeology, Eberhard-Karls University, Tübingen, Germany

ABSTRACT: Archaeological research in the Central African rain forest has increased slowly but steadily over the last three decades. This fortunate change is due to the activities of a few individuals, who proved that archaeological fieldwork in this region is not only possible but also very productive indeed. This paper summarizes recent archaeological fieldwork, be it survey or excavations, effected in southern Cameroon. The results presented are mainly based on fieldwork between 1997 and 2006 which was directed by M. K. H. Eggert. An important part of it was realized within the Research Unit 510 of the DFG (Deutsche Forschungsgemeinschaft). This Research Unit which was created in 2003 is a joint venture of the Universities of Frankfurt (Archaeology, Archaeobotany and Physical Geography) and Tübingen (Archaeology). It stands for an integrated approach exploring the changing interrelationship of environment and culture in the forest-savanna regions of West and Central Africa.

As far as Central Africa is concerned the archaeological research focussed mainly on the coastal and interior parts of the south province of the Republic of Cameroon. The features and finds presented in the following relate to the site of Bwambé-Sommet near Kribi on the Atlantic coast as well as to those of Abang Minko'o, Akonétye and Minyin in the vicinity of Ambam near the borders to Equatorial Guinea and Gabon. In addition, some brief comments are made on finds from Ndangayé and Boulou in the Lobéké National Park, situated in the far south-eastern part of Cameroon. Further investigations around Bertoua in the transitional area of forest and savanna are scheduled for 2007.

4.1 INTRODUCTION

The Research Unit 510 of the Universities of Frankfurt and Tübingen sponsored by the German Research Foundation explores the changing interrelationship of environment and culture in the forest-savanna regions of West and Central Africa. The cooperation of archaeologists, archaeobotanists and geographers stands for an integrated approach leading from archaeological classification and the establishment of regional chronologies to the analysis of significant environmental and climatic changes as well as alterations of subsistence strategies. This paper presents a summary of recent archaeological fieldwork in South-West Cameroon.

4.2 RESEARCH ACTIVITIES

In the late 1970s throughout 1987 grand scale archaeological work had been carried out in the inner Congo basin by Eggert (1984, 1987, 1992, 1993, 1994/95) and Wotzka (1995) in the realm of what was known as the 'River Reconnaissance Project'. Due to external circumstances the project was abandoned, only to be continued in Cameroon in 1997.

Having been able to demonstrate that the earliest farming, pottery-producing cultures in the inner Congo basin had to be connected with the so-called 'Imbonga Horizon' (Eggert, 1984, 1987), Eggert intended to continue his work in the northwestern part of the equatorial forest and did so by exploring parts of the Sanaga and Ngoko rivers in a survey in 1997 (Eggert, 2002). In addition a road survey was effected along the Nanga Eboko–Bertoua axis in the central part of Cameroon (Figure 1). This field research resulted in the discovery of a number of sites on the banks of the lower Sanaga and on the roads between Bertoua, Nanga Eboko and Edéa. Features in Ekak and Mfomakap (Figure 2) yielded flat based, incised ceramics, but only those from the latter site provided relatively early data i. e. from the first millennium BC (Table 1). The survey was successful since two early sites, Mouanko-Lobethal and Yatou (Figure 2), were discovered near the mouth of the Sanaga. The pottery found there displays significant characteristics as far as shape and decoration are concerned. In 1998/99, some excavation was conducted there. It provided a considerable amount of ceramics, the earliest of which was dated to the beginning of the first millennium AD (Table 1).

Excavations commenced in late 1998 and unveiled several connected pit structures and a grave. While in Lobethal 15 features where excavated, only five were dug at Yatou. The ceramics mainly consist of open bowls, usually incised, comb-stamped zigzag and wavy-line decoration, as well as short-necked globular pots with everted rims (see Eggert, 2002, Figure 8; 9,3.5.6; 10,1.4; 11,3.5; 2.3). Eggert (2002, 513) noted that this kind of pottery, although being more or less homogeneous in itself, was different from all other

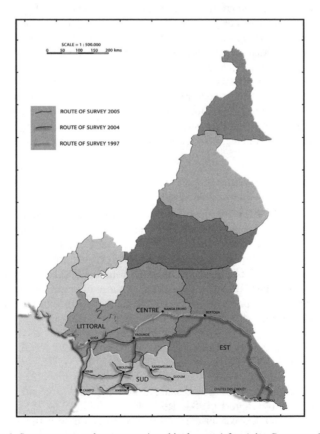

Figure 1. Survey routes and towns mentioned in the text (after Atlas Cameroun 1979).

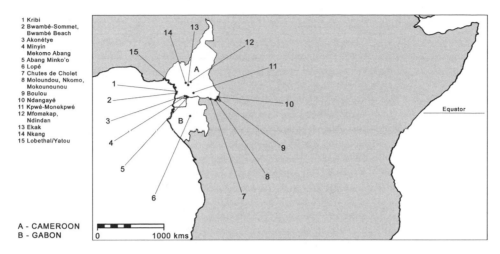

1 Kribi
2 Bwambé-Sommet,
  Bwambé Beach
3 Akonétye
4 Minyin
  Mekomo Abang
5 Abang Minko'o
6 Lopé
7 Chutes de Cholet
8 Moloundou, Nkomo,
  Mokounounou
9 Boulou
10 Ndangayé
11 Kpwé-Monekpwé
12 Mfomakap,
   Ndindan
13 Ekak
14 Nkang
15 Lobethal/Yatou

A - CAMEROON
B - GABON

Figure 2. Localities and sites mentioned in the text.

ceramics found in southern Cameroon. Part of the ceramics of the pit complex of Lobethal had been analysed in detail by Williams-Schmid in 2001, who distinguished an older from a younger ceramic tradition (Williams-Schmid 2001, 50). However, she was not able to demonstrate any connections to other sites. The association of Lobethal and Yatou ceramics with the pottery of the already excavated complexes of Nkang, Ndindan (Mbida Mindzie, 1995/96; 2002), Lopé (Assoko Ndong, 2000/01; 2002) and other sites in Gabon (Oslisly and Peyrot, 1992; Oslisly, 1994/95; Clist, 2004/05) is, as far as shape and decoration are concerned, all but strong. Oslisly (2006, 314), however, argued in favour of a similarity of open bowls from the site of Lobethal (pit 4: Eggert, 2002, Figure 10.3) to those from the site Campo 1 (Oslisly, 2006, Figure 10; 11). Judging from a photograph in Oslisly's article (2006, Figure 11), other vessels might also show similarities in shape and decoration (see Eggert, 2002, Figure 9,1.2; 10,2.; 11,2.6.7.10).

4.2.1 Research Unit 510

According to Schwartz (1992) the rain forest suddenly retreated around 3.000 to 2.500 BP and gave way to a relatively open vegetation only to reexpand after 2.000 BP (Maley and Brenac, 1998). Starting from these assumptions the Research Unit 510 explores whether the settling of the rain forest was linked to the supposed environmental change.

In early 2004 a survey was conducted in the region of southern littoral and southwestern inland Cameroon down to the Gabonesian and Equatorial Guinean border. This land-based reconnaissance profited greatly from previous work of R. Oslisly and P. Nlend Nlend who had followed up on road building activities of the Transafricana route from Ebolowa to Ambam and beyond (Oslisly, 2001). They had also surveyed the coastal area from Kribi to Campo (Figure 1) as well as from Campo to Ambam (Oslisly, 2002) (Figure 1). Oslisly discovered a large number of sites with pit complexes, two of which, Bwambé-Sommet (designation 'BWS') and Akonétye (designation 'AKO'), in the vicinity of Ambam and Kribi, respectively, were excavated by the archaeological and archaeobotanical groups of the Research Unit and will be described here later.

The survey in 2004 yielded also some tuyère fragments on the road Sangmélima–Djoum (Figure 1). They were dated between 2.208 and 2.054 BP (Table 1). Further

Table 1. Radiocarbon data mentioned in the text (calibrated with OxCal v.3.10 (2005)).

Site	Year	Feature	Lab. No.	Date	Cal. Age (2 sigma)
Ekak	1997	EKK 97/2	KI-4608	160 ± 30 BP	1.660–1.890 AD (77,3%)
					1.910–1.960 AD (18,1%)
Malapá	1997	MLP 97/2	KI-4614	1.700 ± 60 BP	AD 210–470 (89,7%)
					AD 480–540 (4,7%)
					AD 170–200 (1,1%)
Mfomakap	1997	MFK 97/2	KI-4611	2.220 ± 30 BP	380–200 BC
	1997	MFK 97/3	KIA-8456	2.217 ± 25 BP	380–200 BC
Mouanko-Epolo	1997	MOU 97/1	KIA-8457	1.783 ± 33 BP	AD 130–340
	1997	MOU 97/1	KIA-8458	3.860 ± 29 BP	2.470–2.270 BC (84,8%)
					2.260– 200 BC (10,6%)
Lobethal	1998	LBT 98/2	KIA-12948	157 ± 30 BP	1.660–1.890 AD (77,5%)
					1.910–1.960 AD (17,9%)
	1998	LBT 98/2	KIA-12947	8612 ± 50 BP	7.750–7.550 BC
	1998	LBT 98/3	KIA-12506	1.963 ± 36 BP	50 BC–AD 130
	1998	LBT 98/3	KIA-24745	1.911 ± 32 BP	AD 10–180 (92,9%)
					AD 190–220 (2,5%)
	1998	LBT 98/3	KIA-24746	1.842 ± 28 BP	AD 80–240
	1998	LBT 98/3	KIA-12946	8913 ± 49 BP	8.260–7.910 BC
	1998	LBT 98/4	KIA-12945	1.800 ± 27 BP	AD 130–260 (84,6%)
					AD 280–330 (10,8%)
	1998	LBT 98/4	KIA-12944	1.816 ± 34 BP	AD 120–260 (87,9%)
					AD 280–330 (5,9%)
					AD 80–110 (1,6%)
	1998	LBT 98/5	KIA-12507	805 ± 29 BP	AD 1.175–1.275
	1998	LBT 98/8	KIA-12943	631 ± 24 BP	AD 1.340–1.400 (56,5%)
					AD 1.280–1.330 (38,9%)
	1998	LBT 98/9	KIA-12949	1.888 ± 29 BP	AD 50–220
	1999	LBT 99/3	KIA-12940	1.799 ± 27 BP	AD 130–260 (83,7%)
					AD 280–330 (11,7%)
	1999	LBT 99/3	KIA-12939	2.146 ± 26 BP	230–90 BC (70,1%)
					360–290 BC (25,3%)
	1999	LBT 99/4	KIA-24743	1.860 ± 28 BP	AD 80–230
Lobethal	1999	LBT 99/4	KIA-24744	1.797 ± 28 BP	AD 130–260 (81,4%)
					AD 280–330 (14,0%)
	1999	LBT 99/4	KIA-12938	1.753 ± 25 BP	AD 220–390
	1999	LBT 99/5	KIA-12937	2.162 ± 35 BP	370–90 BC
	1999	LBT 99/5	KIA-24748	156 ± 23 BP	1.720–1.820 AD (49,7%)
					1.910–1.960 AD (18,6%)
					1.660–1.710 AD (16,2%)
					1.830–1.880 AD (10,9%)

Site	Year	Feature	Lab. No.	Date	Cal. Age (2 sigma)
Yatou	1999	YAT 99/1	KIA-12942	1.974 ± 26 BP	40 BC–AD 80
	1999	YAT 99/1	KIA-12941	1.227 ± 36 BP	AD 680–890
	1999	YAT 99/4	KIA-24751	1.974 ± 26 BP	40 BC–AD 80
	1999	YAT 99/4	KIA-24749	1.985 ± 26 BP	50 BC–AD 70
	1999	YAT 99/4	KIA-24750	1.964 ± 28 BP	40 BC–AD 90 (94,3%)
					AD 100–120 (1,1%)
Akonétye	2004	AKO 04/2	KIA-24729	1.770 ± 27 BP	AD 130–350
	2004	AKO 04/2	KIA-24730	1.685 ± 25 BP	AD 320–420 (81,8%)
					AD 250–300 (13,6%)
	2004	AKO 04/3	KIA-24739	1.755 ± 29 BP	AD 210–390
Bwambé Beach	2004	BWB 04/0	KIA-24738	1.485 ± 29 BP	AD 530–650
	2004	BWB 04/1	KIA-24731	1.610 ± 26 BP	AD 400–540
Kpwé-Monekpwé	2004	KPW 04/1	KIA-24740	2.208 ± 26 BP	370–200 BC
	2004	KPW 04/2	KIA-24741	2.054 ± 19 BP	120 BC–AD 10 (87,1%)
					160–130 BC (8,3%)
	2004	KPW 04/3	KIA-24742	2.131 ± 28 BP	210–50 BC (85,2%)
					350–310 BC (10,2%)
Kribi	2004	KRB 04/1	KIA-24734	1.514 ±21 BP	AD 530–610 (86,5%)
					AD 440–490 8,9%)
	2004	KRB 04/1	KIA-24735	1.460 ± 25 BP	AD 560–650
	2004	KRB 04/2	KIA-24736	2.146 ± 31 BP	240–50 BC (69,3%)
					360–280 BC (26,1%)
	2004	KRB 04/4	KIA-24737	2.215 ± 32 BP	380–200 BC
	2004	KRB 04/7	KIA-24732	1.736 ± 28 BP	AD 230–390
	2004	KRB 04/7	KIA-24733	1.760 ± 24 BP	AD 210–380

reconnaissance and ensuing test excavations in Kribi and its surroundings revealed about 15 structures, dating from 2.215 to 1.514 BP (Table 1). For one of those features, a cultural layer in the road bank on the site of Catholic Mission of Kribi ('KRB'), two radiocarbon data of 1.760 and 1.736 BP (Table 1) were taken, since they were associated with a richly decorated flat based ceramic vessel (Figure 3).

A pit in Bwambé Beach ('BWB'), dated from 1.610 to 1.485 BP (Table 1) also displayed a characteristic flat-based pottery with comb-stamped, zigzag as well as incised patterns. Sherds from the surface could be refitted with those of the excavated 2 m deep structure. Open bowls as well as globular pots with short, partially outturned, comb-stamped rims and a very coarse temper are common. The site is of great importance since archaeological work so far mainly focussed on earlier ceramics. These structures exemplarily highlight some of the cultural activities in the region of coastal Cameroon around the end of the first millennium BC and the first centuries of the first millennium AD.

Having been able to locate several sites with complex structures, systematic field work was begun in November 2004 at Bwambé-Sommet. Three of over thirty pits of depths between 2,00 m to 3,00 m located by Oslisly and his team were excavated in a one

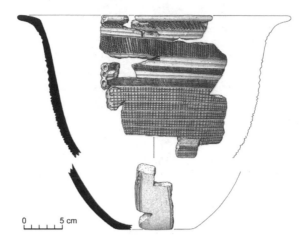

0   5 cm

Figure 3. Large, highly decorated ceramic vessel from KRB 04/7.

month campaign. Generally pits are regarded as refuse dumps, storage containers or traps (Mbida Mindzie, 1995/96, 44–47). Judging from our excavation the 'refuse pit assumption' is certainly acceptable as a hypothesis as far as secondary utilisation is concerned since we recovered large accumulations of ceramic sherds and charred macrobotanical remains from them. However, several uncommon features, we discovered at Bwambé-Sommet, suggest other possibilities as well.

The edges of pits 2 and 3 (BWS 04/2 and BWS 04/3) were lined with numerous ceramic sherds and stones. It is difficult to imagine any functional aspect with regard to storage or discard of refuse. The third pit we excavated there (BWS 04/1) had a similar fill but lacked this feature. The time invested in creating this lining must have been considerable.

Moreover, one of the pits mentioned (BWS 04/3) yielded several ceramic vessels which probably were smashed using large stones in the process of the refilling. Although it cannot be proven definitely that the vessels were not already broken at the time of deposition, it seems likely that the stones lying now interspersed with the vessels had been used to destroy them. This must have happened after the lining of the pit since pottery and stones belong to the last phase of the pit.

Having both flat and rounded bases many pots are comparatively thick. They are decorated by incision and impression. Horizontal bands, herringbone and rocked zigzag motifs are most prominent.

As to stone tools, a quartzite pounding or upper grinding stone (Figure 4,1) and several other grinding stones were found in BWS 04/1 and BWS 04/3 (Figure 4,3.6.7, Figure 5,1). This is not surprising since seeds of pearl millet (*Pennisetum glaucum*, a cereal adapted to a dry climate) are associated with these pits. For a detailed discussion of the archaeological as well as archaeobotanical evidence and its impact of these results on the debate of the settling of the rain forest see Eggert *et al.* (in press).

The pottery of all pits from Bwambé-Sommet is very homogeneous. It consists primarily of globular vessels with short, outturned and for the most part grooved rims.

Another stone discovered in BWS 04/1 has several roundly shaped cavities (Figure 5,4), as if the stone had been drilled with a stick. Furthermore a sandstone was found which exhibits several lateral grooves (Figure 4,5). It appears feasible that the stone has been used as an arrow shaft smoothener since incidentally, two quartz arrowheads (Figure 4,2.3) were discovered on the surface in close proximity of this pit. Needless to say, however, that

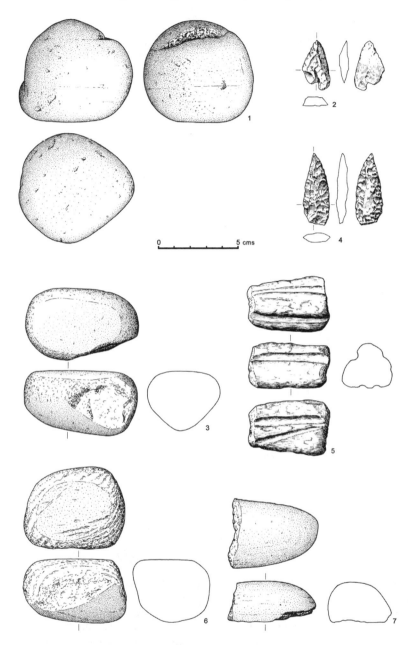

Figure 4. Stone artefacts from Bwambé-Sommet.
1.3.6 BWS 04/1; 2 BWS 04/101; 4 BWS 04/102; 5.7 BWS 04/3.

they are undated. An utilization as an arrow shaft smoothener seems further implausible, since Zimmermann (1995) argues that shaft smootheners consist of two pieces with a flat, regular surface. Other authors have labeled similar objects as grooved stones ('Rillensteine'; Hahn, 1991; Rupp, 2005), grinding stones with groove (Schleifsteine mit Rille; in: Bolus,

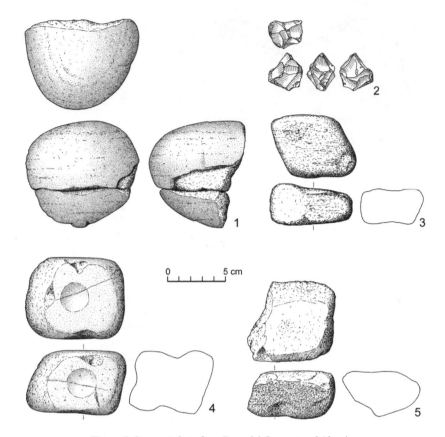

Figure 5. Stone artefacts from Bwambé-Sommet and Akonétye.
1 BWS 04/3; 2.4 BWS 04/1; 3.5 AKO 05/6.

in prep.) or polissoirs (de Beaune, 2000). Comparable artefacts have been found in north-eastern Nigeria (see Rupp, 2005, 202, Figure 102, p. 269 Tafel 17,2, p. 270 Tafel 19,1.2) and in the Lopé Faunal Reserve, Gabon (Assoko Ndong, 2000/01, 136, Planche 2,5, p. 163, Planche 8,7, p. 225, Planche 17,1.2.3, p. 425, Planche 47,5; Assoko Ndong, 2002, Figure 3,6.7).

Interpretations for grooved stones range from arrow (shaft) smootheners (Deacon and Deacon, 1999, 157, 158; Inskeep, 1987, 143, 144 and Plate 10; Hahn, 1991, 244), bead polishers (Stow, 1905, 6667; de Beaune, 2000; 105 and Planche VIII,2) to ropemaker gauges (Assoko Ndong, 2000/01, 134; Rupp 2005, 208-212) which are used to knot several strands of rope into one (for an explanation see Boisseau and Soleilhavoup, 1992; Rupp, 2005, 208-212). Other purposes of utilization, such as smoothing of bones, antler, specifically harpoons, arrowheads (Stow, 1905; 66, 67) and awls are also conceivable. That flint-like materials were present at the time is exemplified by a core (Figure 5,2) found in BWS 04/1. The radiocarbon data taken from samples of pearl millet seeds and oil palm (*Elaeïs guineensis*) of the pits coincide within the 2.380 to 2.200 BP time bracket (Eggert *et al.,* 2006, Table 4).

In January 2005 the south-eastern part of Cameroon was surveyed since it was reported that ceramic finds had been made in the area of the Chutes Chollêt on the Ngoko River as well as at Boulou ('BOL') and Ndangayé ('NDA') in the Lobéke National Park. Both areas border on the region which Eggert had already surveyed in 1987. The sites of Boulou and Ndangayé are located in immediate proximity to so-called 'enclosed' savannas. Being of limited extension they are to be found within the rain forest. At Boulou marsh-like conditions prevail, while Ndangayé is situated directly on the Lobéke River. The pottery of both sites is very coarse, intensely fired and sparsely decorated, although rim sherds with irregular circumferential impressions occur regularly (Figure 6,9 and see figure 6,8 for similar pottery found in Mokounounou). The bases are flat and extremely thick. Considering the closeness of salt containing soils of the savannas and the abundance of thick-based crude and broken ceramics (Figure 6,6.7) it seems likely that this pottery had been used in the process of salt-production by means of fire-induced evaporation. Ceramic vessels similar to those from Boulou and Ndangayé have been described by M. Hees in the context of salt-production (Hees, 2006, 391, Figure 6). Since only surface finds have been collected and thus no samples for ^{14}C treatment were taken the age of the sites cannot be fixed yet.

Apart from a few sherds, tuyère fragments and miniature bowls (Figure 6,4) in the vicinity of Moloundou ('MLU'), of which similar ones had also been found in the survey of the region by Eggert in 1997 in Nkomo ('NKN') and Mokounounou ('MKN') (Figure 2) (Figure 6,1.2.3.5), nothing of interest was discovered during the Ngoko river survey from Mambéle to Chutes Chollêt.

February and March of 2005 were spent with excavations at Akonétye and small-scale surveys in the surroundings of Ambam. Several features with ceramics and an abundance of charcoal could be located at Minyin ('MIY'), Abang Minko'o ('ABM') and Mekomo-Abang ('MEA') (Figure 2).

The site of Akonétye is located about 20 km to the north of Ambam (Figure 2). Eight features, consisting of five pits, two graves and a V-shaped ditch, were excavated. While all features had been damaged by road building activities related to the Transafricana route, they provided the archaeological team with a large amount of ceramics and several iron implements. Two decorated hoe-like iron objects and several iron bracelets were found in a partly destroyed structure (AKO 05/2), while one axe, three spearpoints, two ankle bracelets and a multitude of arm bracelets as well as one at the time of the excavation unidentifiable object, all made of iron, were found in AKO 05/6. Furthermore one smoothed egg-shaped stone was found. In addition, several intact and richly decorated ceramic vessels were discovered in these structures. On the basis of the content of the two features mentioned, their arrangement, orientation as well as their size they are interpreted as graves. It should be stressed, however, that no bones or teeth were found which is, at least as far as bones are concerned, rather characteristic for highly acid rain forest soils. These specific soil conditions also have a very negative effect on the surface and, consequently, on the preservation of the decoration of the ceramics. In some cases the original surface has been completely destroyed.

The other pits at Akonétye which were not visible but below 0,5 m reached a depth of up to 4,5 m (in case of structure AKO 05/5) and exhibited some interesting features. One iron knife (Figure 7), several other small iron artifacts as well as a multitude of ceramic sherds were recovered from the excavated structures.

Though differing in size, the decoration of the pottery from these structures is very homogeneous, displaying mainly rocked zigzag, incised and stamped patterns. Most of the decoration runs from the rim down to the lower part of the vessel, stopping just a few centimeters above the base. The pottery is exclusively flat-based with the better part consisting of ovally to roundly shaped pots. This and the homogeneity of the radiocarbon data (between 1.815 and 1.692 BP, Eggert *et al.*, 2006, Table 4) point to a relatively close temporal proximity of the various structures. A multitude of stone tools, mainly consisting of sandstone or gneiss were found. Some of them exhibit shallow depressions. It appears as if the stones had been

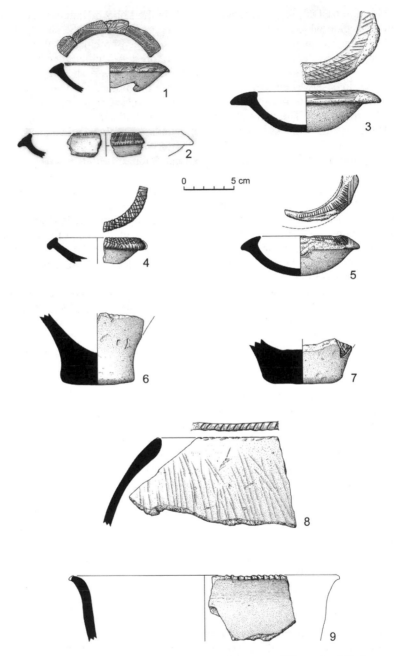

Figure 6. Pottery from Nkomo, Moloundou, Mokounounou, Ndangayé and Boulou.
1.2 NKN 97/101, 3.5.8 MKN 97/1, 4 MLU 05/103, 6.9 NDA 05/103, 7 BOL 05/103.

drilled for a purpose which remains unknown, since there are no complete holes to make these stones suitable as, e.g., weighing stones for digging sticks (Figure 5,3.5). An utilization as a pestle for food, such as nuts, vegetables and grains (de Beaune, 2000, 77–78 and Planche VI,1) or other materials such as ochre is conceivable, but cannot be proven.

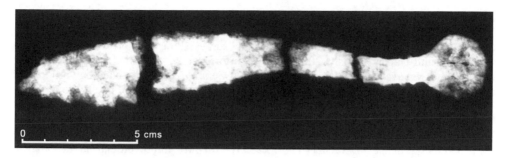

Figure 7. X-ray photography of iron knife discovered in AKO 05/3.

One feature of pit AKO 05/4 deserves special attention as the pit had been covered with a layer of ceramics from several broken pots.

It is evident that only the surface has been scratched of what lies beyond the excavated site of Akonétye since what has been found so far has only been exposed by chance in the process of road construction. The existence of a V-shaped, 2 m deep ditch clearly indicates larger settlement activities in the area. Thus further fieldwork, not only at this site, is necessary to specify some of the fundamental conditions of settlement and life at the beginning of the first millennium. For the moment, however, it is necessary to concentrate on basics, i.e. establishing a firm sequence of relative and absolute time, before in-depth, in the sense of intra-site, analysis can begin.

The site of Minyin (designation 'MIY'), located during a survey in 2005 about 10 km south of Ambam, has at least six features within a length of 500 m. Only three of them were excavated in 2006. Contrary to the sites dug in 1998/99, 2004 and 2005, only a small number of ceramics could be retrieved, albeit the considerable depth of the pits of up to 3 m. As yet the pottery has not been analyzed in detail. It does not resemble, however, to that of the nearby sites of Akonétye and Abang Minko'o. Nonetheless, the site of Minyin is interesting in that each pit contained a number of iron objects and tuyère fragments. Also exceptional is the high number of slag as well as large lateritic and quartzitic stones. As a first hand guess we might interpret these structures as refuse pits in the context of smithing activities. The radiocarbon data fall between 2.001 and 1.739 BP (Eggert *et al.,* 2006, Table 4).

The second site excavated in 2006 is Abang Minko'o (designation 'ABM'), situated 5 km north of the Gabonesian border and 20 km south of Ambam. Also discovered in 2005, at least 11 features were visible in the road bank. Only three of them, two pits and a small ditch, were excavated. These features held only a small quantity of eroded and highly fragmented, although very characteristic, ceramics. This pottery with its rocked zigzag, herringbone and impressed decoration resembles that of Bwambé-Sommet, a site which lies at a distance of about 150 km. Short, grooved and outturned rims as well as flat bases characterize the ceramics of both sites. Thus, it is not surprising that the radiocarbon data both fall into the last half of the first millennium BC (Eggert *et al.,* 2006, Table 4). Furthermore it seems to be very significant indeed that pearl millet was found at Abang Minko'o as well.

4.2.2 Other research

Apart from the research reported here, most archaeological work in the north-western part of Central Africa began only in the 1980s (de Maret, 1986; Clist, 1986; 1987; Essomba, 1989). Research intensified in the 1990s (Essomba, 1992; Mbida Mindzie, 1995/96) but only from the late 1990s onwards a more systematic exploration of archaeological sites is being conducted. In this context, the efforts of R. Oslisly should be emphasized, since he accompanied greater road building activities and surveyed the rain forest area in southern Cameroon and Gabon (Oslisly, 2000; 2001). Unfortunately, a detailed publication of the results is not yet available. Generally, research on Central African Archaeology progressed markedly in the last half decade or so (see Clist, 2006; Oslisly, 2006). The doctoral theses of Assoko Ndong (2000/01) and Clist (2004/05) as well as evidence on the cultivation of banana (Mbida *et al.*, 2000; 2001; 2005) may also serve as very impressive examples. Further, the archaeological transect effected in the context of construction of the trans-Cameroonian pipeline (Lavachery *et al.*, 2005) not only improved our archaeological knowledge significantly but also showed the potential of areas still untouched.

4.3 DISCUSSION

As has been demonstrated here, the joint effort of Archaeology and Archaeobotany yielded some very important results concerning the later prehistory of South-West Cameroon. For the first time it was shown by the Research Unit 510 that people at Bwambé-Sommet and Abang Minko'o disposed of pearl millet at the time around 2.200 BP. This may be archaeologically attested by the findings of upper grinding stones in the pits of Bwambé-Sommet, although stone might have been used to crush other (plant) materials. Analysis of macrobotanical remains and pollen profiles indicate that people at these sites lived in a rain forest habitat. Climatic conditions, however, must have been drier than today to make cultivation of pearl millet possible (for a detailed discussion see Eggert *et al.*, 2006). The existence of pearl millet clearly points to some kind of cultural connection between the savanna and the rain forest since botanists agree that the origin of pearl millet and its use as staple food is connected with the savanna (Tostain, 1998; Kahlheber and Neumann, in press).

Also, it is important to mention that arrowheads, quartz and flint artifacts as well as grooved stones were partly found in context with the pit features. Since we lack bone material we are largely ignorant of an important part of the nutrition basis of the people in question. Nevertheless, we may assume that, as today, they supplemented their diet by hunting. In this context they will have used not only iron but stone implements as well.

For Central African Archaeology the discovery of richly furnished graves in the rain forest at Akonétye around 1.800 BP is very important indeed. The associated iron grave goods inform us that this metal was being produced at that time. That iron ore was being locally reduced even earlier is suggested by slag and tuyère fragments at the site of Minyin.

It needs not be stressed that we need much more research on the later prehistory of Central Africa before we will be able to securely link different regions. This implies the establishment of detailed regional ceramic sequences as well as a tight temporal resolution in terms of ^{14}C data especially for regions 'that appear to be in [...] urgent need of enhanced chronological control' (Wotzka, 2006, 274). Ceramics play an important role in the establishment of regional chronologies, since it is not only highly variable as far as shape and decoration are concerned but most resistant to the natural processes of destruction occuring in tropical soils as well. Very often nothing but pottery withstands aggressive soils. The similarity of pottery from different sites (e. g. from Bwambé-Sommet and Abang Minko'o) or similar inventories from Gabon and Cameroon (as described in Eggert *et al.,* in press—see Mbida Mindzie 1995/96; 2002; Assoko Ndong 2000/01; 2002; Clist 2004/05; 2006) attest to the success of this procedure the establishment of regional sequences. Generally, research should be conducted by joint approaches, be it archaeologists, archaeobotanists, zoologists, geographers or ethnographers. This will, of course, only be possible under exceptional circumstances and in so far it remains an ideal which can be realized only partly.

From the archaeological work conducted so far, it is relatively certain that most of the sites described here were once situated in rain forest territory. Nevertheless, we need to keep in mind the importance of a more or less direct connection of the rain forest and the savanna. To understand the process of the settling of the rain forest the archaeological structures of the surrounding savanna need to be scrutinized and compared in the same manner which we applied to those in the forest. Only then we will be able to create the basis for the kind of fine-meshed grid that will allow us to meaningfully interpret and interconnect localities within and without the rain forest.

ACKNOWLEDGEMENTS

I am most grateful to Manfred K.H. Eggert not only for his help and suggestions, but also his relentless criticism. He also kindly brushed up my English. Further thanks are due to our collaborators from Frankfurt University Alexa Höhn, Stefanie Kahlheber, Katharina Neumann, Astrid Schweizer and Barbara Eichhorn from the archaeobotanical sections of the project, as well as Mark Sangen and Joachim Eisenberg for their help in the field and fruitful discussions. The large amount of fieldwork would not have been possible without our numerous collaborators from Cameroon and Germany. These are Liane Giemsch, Silja Meyer, François Ngouoh, Olivier Nkokonda, Pascal Nlend Nlend, Anselme Ossima Ossima, Andreas Sattler, Jules Tsague and Andreas Willmy, to name but a few. I take the opportunity to thank the Deutsche Forschungsgemeinschaft for financing the project, the archaeological part of which is directed by M. K. H. Eggert. We are obliged to M. Heinze of the GTZ (Gesellschaft für Technische Zusammenarbeit) at Yokadouma, who drew our attention to some sites and finds in the eastern part of Cameroon. He assisted our work in every possible way. We would also like to thank the guides of the WWF (World Wildlife Fund) of the Lobéke Reserve for their assistance.

REFERENCES

Assoko Ndong, A., 2000/01, *Archéologie du peuplement holocène de la reserve de faune de la Lopé, Gabon*. Vol. 2, Unpublished Ph.D. dissertation. Université Libre de Bruxelles, Academic Year 2000–2001.

Assoko Ndong, A., 2002, Synthèse des données archéologiques récentes sur le peuplement à la Holocène de la réserve de faune de la Lopé, Gabon. *L'Anthropologie*, **106 (1)**, pp. 135–158.

Atlas Cameroun, 1979, *Atlas de la République Unie du Cameroun*. Les atlas de Jeune Afrique, edited by Laclavère, G., (Paris: Editions Jeune Afrique).

Boisseau, P. and Soleilhavoup, F., 1992, Pierres à rainures du Sahara. Paléotechnologie des cordes, des peaux et des cuirs. *L'Anthropologie*, **96 (4)**, pp. 797–806.

Bolus, M., in prep., Schleifsteine mit Rille (Pfeilschaftglätter). In *Steinartefakte—vom Altpaläolithikum bis in die Neuzeit*, edited by Floss, H., (in prep.).

Clist, B.-O., 1986, Le néolithique en Afrique centrale: Etat de la question et perspective d'avenir. *L'Anthropologie*, **90 (2)**, pp. 217–232.

Clist, B.-O., 1987, Early bantu settlements in west-central Africa: a review of recent research. *Current Anthropology*, 28 **(3)**, pp. 380–382.

Clist, B.-O., 2004/05, Des premièrs villages aux premiers Européens autour de l'Estuaire du Gabon: Quatre millénaries d'interactions entre l'homme et son milieu, Vol. 2. Unpublished Ph.D. dissertation. Université Libre de Bruxelles, Academic Year 2004–2005.

Clist, B.-O., 2006, Mais où se sont taillées nos pierres en Afrique Centrale entre 7.000 et 2.000 bp?. In *Grundlegungen. Beiträge zur europäischen und afrikanischen Archäologie für Manfred K.H. Eggert*, edited by Wotzka, H.-P., (Tübingen: Francke Attempto Verlag), pp. 291–302.

de Beaune, S. A., 2000, Pour une Archéologie du geste. Broyer, moudre, piler des premier chasseurs aux premiers agriculteurs (Paris: CNRS Editions).

de Maret, P., 1986, The Ngovo Group: An industry with polished stone tools and pottery in Lower Zaïre. *African Archaeological Review*, 4, pp. 103–133.

Eggert, M.K.H., 1984, Imbonga und Lingonda: Zur frühesten Besiedlung des äquatorialen Regenwaldes. In *Beiträge zur Allgemeinen und Vergleichenden Archäologie*, 6, pp. 247–288.

Eggert, M.K.H., 1987, Imbonga and Batalimo: ceramic evidence for early settlement of the equatorial rain forest. *African Archaeological Review*, 5, pp. 129–145.

Eggert, M.K.H., 1992, The Central African rain forest: Historical speculation and archaeological facts. *World Archaeology*, 24, pp. 1–24.

Eggert, M.K.H., 1993, Central Africa and the Archaeology of the equatorial rain forest: Reflections on some major topics. In *The Archaeology of Africa: Food, Metals and Towns*. One World Archaeology, 20, edited by Shaw, T., Sinclair, P., Andah, B. and Okpogo, A., (London, New York: Routledge), pp. 289–329.

Eggert, M.K.H., 1994/95, Pots, farming and analogy: early ceramics in the Equatorial Forest, *Azania*, 29–30, pp. 332–338.

Eggert, M.K.H., 2002, Southern Cameroun and the settlement of the equatorial rain forest: Early ceramics from fieldwork in 1997 and 1998–99. In *Tides of the Desert/Gezeiten der Wüste: Beiträge zu Archäologie und Umweltgeschichte Afrikas zu Ehren von Rudolph Kuper*, edited by Lenssen-Erz, T., Tegtmeier, U., Kröpelin S. *et al.*, (Köln: Africa Praehistorica, 14), pp. 507–522.

Eggert, M.K.H., Höhn, A., Kahlheber, S., Meister, C., Neumann, K. and Schweizer, A., in press, Pits, Graves and Grains: Archaeological and Archaeobotanical Research in Southern Cameroun. *Journal of African Archaeology*, 4 (2).

Essomba, J.-M., 1989, Dix ans de recherche archéologiques au Cameroun méridional (1979–1989). *Nsi*, 6, pp. 33–57.

Essomba, J.-M., (ed.) 1992, *L'archéologie au Cameroun: Actes du premier Colloque international de Yaoundé (6–9 janvier 1986)*, (Paris: Karthala).

Hahn, J., 1991, Erkennen und Bestimmen von Stein- und Knochenartefakten. Einführung in die Artefaktmorphologie. In *Archaeologica Venatoria*, 10, edited by Müller-Beck, H.-J., (Tübingen: Verlag Archaeologica Venatoria).

Inskeep, R. R., 1987, Nelson Bay Cave, Cape Province, South Africa. The Holocene Levels, (Oxford: BAR International Series 357).

Kahlheber, S. and Neumann, K., in press, The development of plant cultivation in semi-arid West Africa. In Rethinking agriculture: archaeological and ethnoarchaeological perspectives. In *One World Archaeology* edited by Denham, T.P., Iriarte, J. and Vrydaghs, L., (London: University College Press).

Lavachery, P., MacEachern, S., Bouimon, T., Gouem, B.G., Kinyock, P., Mbairo, J. and Nkokonda, O., 2005, Komé to Ebomé: Archaeological research for the Chad Export Project, 1999–2003. *Journal of African Archaeology*, 3, pp. 175–193.

Maley, J. and Brenac, P., 1998, Vegetation dynamics, palaeoenvironments and climatic change in the forests of western Cameroon during the last 28.000 years B.P. *Review of Palaeobotany and Palynology*, 99, pp. 157–187.

Mbida, C.M., 1995/96, *L'émergence de communautés villageoises au Cameroun méridional: Etude archéologique des sites de Nkang et de Ndindan*. Vol. 2, Unpublished Ph.D. dissertation. Université Libre de Bruxelles, Academic Year 1995–1996.

Mbida, C.M., 2002, Ndindan: Synthèse archéologique d'un site datant de trois millénaires à Yaoundé. *L'Anthropologie*, **106 (1)**, 2002, pp. 159–172.

Mbida, C.M., Van Neer, W., Doutrelepont, H. and Vrydaghs, L., 2000, Evidence for banana cultivation and animal husbandry during the first millennium BC in the forest of southern Cameroon. *Journal of Archaeological Science*, **27**, pp. 151–162.

Mbida, C.M., Doutrelepont, H., Vrydaghs, L., Swennen, R.L., Beeckman, H., de Langhe, E. and de Maret, P. 2001, First archaeological evidence of banana cultivation in central Africa during the third millennium before present. *Vegetation History and Archaeobotany*, **10**, 1–6.

Mbida, C.M., Doutrelepont, H., Vrydaghs, L., Swennen, R., Swennen, R., Beeckman, H., De Langhe, E. and de Maret, P., 2005, The initial history of bananas in Africa: A reply to Jan Vansina, Azania, 2003. *Azania*, **40**, pp. 128–135.

Oslisly, R., 1994/95, The Middle Ogooué Valley: Cultural changes and Palaeoclimatic implications of the last four millennia. In *The Growth of Farming Communities in Africa from the Equator Southwards*, Azania 29/30, 1994/95, edited by Sutton, J.E.G., (Nairobi: British Institute in Eastern Africa) pp. 324–331.

Oslisly, R., 2001, Archéologie et Paléoenvironment dans L'uto de Campo-Ma'an. Etat des connaissances. Unpublished survey report April 2001.

Oslisly, R., 2002, Premiers Résultats de la Mission de Prospection Archéologique sur l'axe routier Ebolowa-Ambam. Unpublisehd survey report February 2002.

Oslisly, R., 2006, Les traditions culturelles de L'Holocène sur le littoral du Cameroun entre Kribi et Campo. In *Grundlegungen. Beiträge zur europäischen und afrikanischen Archäologie für Manfred K.H. Eggert*, edited by Wotzka, H.-P., (Tübingen: Francke Attempto Verlag), pp. 303–317.

Oslisly, R. and Peyrot, B., 1992, L'arrivée des premiers métallurgistes sur l'Ogooué, Gabon. *African Archaeological Review*, **10**, pp. 129–138.

Rupp, N., 2005, Land ohne Steine. Die Rohmaterialversorgung in Nordost-Nigeria von der Endsteinzeit bis zur Eisenzeit. http://publikationen.ub.uni-frankfurt.de/volltexte/2005/2355/

Schwartz, D., 1992, Assèchement climatique vers 3000 B.P. et expansion Bantu en Afrique centrale atlantique: quelques réflexions. *Bulletin de la Société Géologique de France*, **163**, pp. 353–361.

Stow, G. W., 1905, *The native races of South Africa*, edited by Theal, T.M., (London: Swan Sonnenschein).

Tostain, S., 1998, Le mil, une longue histoire: hypothèses sur sa domestication et ses migrations. In *Plantes et paysages d'Afrique. Une histoire à explorer*, edited by Chastanet, M., (Paris: Karthala), pp. 461–490.

Williams-Schmid, M.R., 2001, *Keramikführende Befunde aus Mouanko-Lobethal, Province du Littoral, Kamerun*. Unpublished M.A. thesis, Eberhard-Karls-Universität Tübingen, 2001.

Wotzka, H.-P., 1995, Studien zur Archäologie des zentralafrikanischen Regenwaldes: Die Keramik des inneren Zaïre-Beckens und ihre Stellung im Kontext der Bantu-Expansion. In *Africa Praehistorica, 6*, (Köln: Heinrich-Barth-Institut).

Wotzka, H.-P., 2006, Records of activity: radiocarbon and the structure of Iron Age settlement in Central Africa. In *Grundlegungen. Beiträge zur europäischen und afrikanischen Archäologie für Manfred K.H. Eggert*, edited by Wotzka, H.-P., (Tübingen: Francke Attempto Verlag), pp. 303–317.

Zimmermann, A., 1995, Austauschsysteme von Silexartefakten in der Bandkeramik Mitteleuropas. In *Universitätsforschungen zur Prähistorischen Archäologie*, **26**, (Bonn: Habelt).

CHAPTER 5

The Batié palaeopodzol and its palaeoclimatic and environmental significance

Mesmin Tchindjang, Samuel Aime Abossolo, Joseph Armathée Amougou, Jude Mphoweh Nzembayie and Kah Elvis Fang
Department of Geography, University of Yaoundé I, Cameroon

ABSTRACT: This paper focuses on Late Holocene palaeoenvironmental changes in north western Cameroon by studying a podzolic soil profile within alluvial sediments in a mountainous area (Batié). A sand pit offered an excellent site and gave a good perspective for obtaining an insight into the palaeoenvironmental history. Samples of podzolic as well as of alluvial soils were collected from the quarry. Aside from the stratigraphical description of the profile, chemical and granulometric data were determined. It was evidenced that there was a strong modification of the environmental conditions and especially of vegetation coverage at about 3.000 years BP. The studied profile section located on a fluvial terrace 6 m high, was probably accumulated as a consequence of strong '*savannization*' and deforestation processes interconnected by significantly drier climatic conditions in the Late Holocene. This seems to be confirmed by a strengthened fluvial morphodynamic activity with spatial shifts of channels of the Latse and the Che Ngwen Rivers. The Batié study at the outskirts of the Bamileke land suggests that palaeopodzol formation resulted from strong bioclimatic changes interfered with human activities.

5.1 INTRODUCTION AND OBJECTIVES

Podzols are found in many parts of the world, however, they had been first described in temperate regions. Scientific works during the colonial period also revealed their existence in lower latitudes. The extension of scientific research to the rest of the world after World War II led to their recognition also in tropical regions. Research on podzols is often geographically classified into two main domains: mountains or uplands and lowlands podzols.

In many tropical regions, podzols often constitute remnants of the Quaternary era. Their occurrence depends on many factors amongst which are topography, climate and vegetation. The profile studied in Batié can be described as a mountain podzol (1.200 m asl.). It obviously shows a succession of morphoclimatic sequences from the past to the present. A detail of the vertical section and sedimentological parameters justifies their occurence in excellent drainage conditions at the foot of the tectonically caused Batié escarpment. The vertical section—about 6 m high—was accessible because of sand exploitation by the local population.

The recent vegetation at the site consists of degraded sub-mountainous forest (Letouzey, 1985) that is characterised by the presence of *Elaeis guineensis* which marks an ecological limit with a coverage rate ranging from 20 to 50%. The main objectives and questions that arise from this profile are focussing on the climate controlled morphodynamic landscape evolution and soil-climate interrelationships. Is podzol formation still an ongoing recent or a fossil process? It seems that the present climate does no longer permit such an evolution, since the humus strata is quite thick (50 cm). Due to the absence of empirical data, one may suggest that the mechanism seems to be quite ancient compared to the present day ones. This can be noticed through observations of the deepness of horizons and of the morphoscopy which differs from that of the sub dry climate. It is also of an interest to know, if the Batié site reflects general trends in overall former landscape and climate dynamics or if it simply has to be considered as a singletherefore regional proxy data signal.

5.2 STATE OF THE ART

Within the tropics the first observations on podzols were made by explorers like Deccair (1904), Sphence (1908) and Ranann (1911) who described these soils as white sands. It was a simple description based only on color. However, more recent descriptions—already before World War II—were based on their pedological nature (i.e. horizons, texture, structure, morphological aspects, climate and vegetation). Dieb and Hackensberg (1926) carried out studies on black and hardened horizons while Babet (1933) compares the spodic horizons of the Congo to that of the Parisian Basin. The first time the russian language based, scientific term 'podzol' was used in 1936 by Hardon (1936). Other authors like Richards (1941) Riquier (1948), Leneuf (1956), Bocquier and Boissezon (1959), Aubert (1965) and Klinge (1965) also used the term. It is worth mentioning here that the ideas of these authors were obtained from the work of Schwartz (1988). More recent publications, however, focused on the dynamics of minerals (organic matter, titanium, iron and aluminium) within the podzol's horizons (Turenne, 1975; Flexor, 1975).

In Cameroon just as in other tropical countries, evidence of former climatic and environmental changes were often reconstructed from lacustrine sediments and a major period to take notice of this is 3.000 years BP as a turning point in climate conditions. Studies on this topic have been introduced by Servant (1973) and Maley (1981) in the Lake Chad basin, by Brenac (1987) by sediment cores from seven West Cameroon lakes (Barombi Mbo, Barombi Kotto, Debunscha, Disoni, Ejagham, Bémé, Asom); by Giresse *et al.* (1994) in Lake Barombi Mbo, by Zogning *et al.* (1997) in Lake Njupi, and Nguetsop *et al.* (2004) in Lake Ossa, for quoting just a few of them.

Furthermore, Kadomura (1982) proposed that 'in the intertropical regions of Africa, the effects of both, climatic changes and human activities on the evolution of the physical environment are rather greater than generally expected'. He concluded in 1994 from radiocarbon data and stratigraphic sequences of slope and valley deposits that the environmental history of the western Cameroon landscape can be summarised into four periods. From 28.740–10.000 BP: cooler and drier climate characterized by the spread of the grassland with montane tree taxa *(Podocarpus latifolius* or *milanjanus)*. From 10.000–3.300 BP: return of warmer and wetter climate and development of submontane rain forest dominated by *Syzygium staudtii*. About 3.100 BP: strong aridity and cooling suggested by the appearance of sudanic species as *Olea hoechstetteri*. This period marks the beginning of the savannization by the spreading of a savanna woodland which is possibly linked to the influence of humans. From 3.000–2.600 BP: the beginning of grassland landscape extension with a maximum around 1.700–1.600 BP.

Other works on podzolisation linked to environmental change in the tropical regions are those of Flexor *et al.* (1975) in Brazil, Turenne (1977) in French Guyana and Schwartz (1985–1990) on the Bateke sands in Congo Brazzaville. The last author established the fact that the podzols developed on the Bateke sands were characterized by spodic thick strata indurated in an ortstein horizon. These horizons accumulated important quantities of humic substances. From physical and chemical analysis Schwartz (1988) made ^{14}C datations obtaining ages for a period from 40.000 to 30.000 BP. By this it could be evidenced that podzolisation in the tropics is an important indicator for Late Quaternary climatic and environmental changes.

Up to now, there was no description of a podzol in the Cameroonian mountain area. However, on a worldwide perspective, some scientists have distinguished two types of podzols in tropical regions which are: mountain or upland podzols (zonal soils) by Klinge (1968); they are similar to podzols in temperate regions, and lowland podzols found on sandy plains (Table 1a, b). In South America, it is estimated that lowland podzols cover a surface of about 0,6 Mio ha, meanwhile, upland podzols cover about 0,004 Mio. ha as shown in table 1b. In Cameroon there are no details existing on the distribution of podzols.

5.3 GEOGRAPHICAL SETTING

Batié district occupies an area of 64 km² in the High Plateau Division. It stretches from 5°15 to 5°20 N and 10°16 to 10°21 E in the direction of the Cameroon Volcanic Line (Figure 1). It forms a granitic section of the Bamileke plateau. The physical milieu is made up of a series of tectonised granite plateaus (1.400 m asl.); deeply dissected by valleys and covered by a basaltic trapp. The Bamileke plateau is a constellation of various size and slope ranges of massifs (hillocks) originating from different rock substrata such as: Pou

Table 1a. Regional extension of tropical areas and share of humid tropical forest climate in millions of hectares according to Klinge (1968).

	Central and South America and the Caribbean's	South Sahara, Africa and Madagascar	South East Asia	Australia and New Zealand	**Total**
Surface area of tropical zones	560	2.170	400	345	4.475
Tropical zones of forest humid climate	680	230	240	2	1.152
Total	2.240	2.400	640	347	5.627

Table 1b. Distribution of tropical lowland podzols in millions of hectares according to Klinge (1968).

	Central and South America and the Caribbean's	South Sahara, Africa and Madagascar	South East Asia	Australia and New Zealand	**Total**
Podzols under forest humid climate	0,6	0,052	1,25	—	1,4
Podzols of other climatic region	0,004	0,003	0,2	2,5	2,7
Sub tropical region	1	0	0	2,1	3,1
Total	1,604	0,055	1,45	4,6	7,2

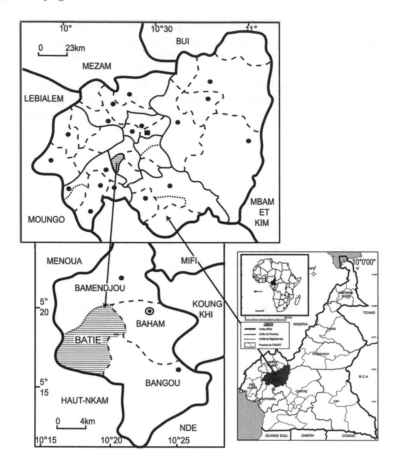

Figure 1. Location map of Batié, NW Cameroon (source: Tchindjang, 1996).

granite mountain (1.724 m asl.), Bani granite massif (1.921 m asl.), Bana androgenic massif (2.097 m asl.), Bangou volcanic massif (1.924 m asl.), and Bamboutos caldera mountains (2.740 m asl.). A great fault scarp surrounds the edges of this horst. The west comprises of a tectonic escarpment of 700 to 1.000 m asl. that overhangs the coastal region. Eastwards of the Bamileke plateau, on the Nun plain, there is a fault scarp of 200–300 m height. Northwards, the plateau terminates at the foothills of the Bamenda Highlands (1.800–3.011 m asl.) which project above the Ndop basin. A fault scarp more than 500 m high separates this basin.

5.3.1 Geology

The Batié granite mountain is formed by a series of hilly and faulted blocks as shown by the cross section, locally called Kong (Mount), which stretches from the south west to the north east (Figure 2). The Pan African granite of the Batié massif (Talla, 1995) is situated at the heart of the vast Batié-Dchang anticlinorium (Dumort, 1968). From a petrographic and mineralogical viewpoint, the Batié granite is classified among the confined syntectonic or syncinematic granites (Dumort, 1968). The base rock is a monzonitic granite with two micas of porphyric texture presenting a structure called 'horse teeth' due to the high development of orthose (7–10 cm). Their main components are quartz, micas, and

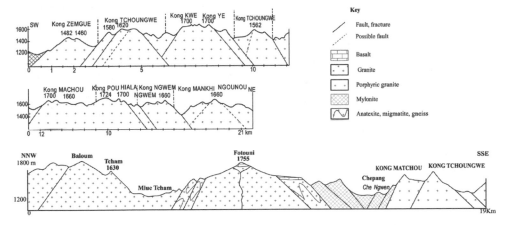

Figure 2. Geological cross section of wider Batié study area (source: Tchindjang, 1996).

amphibole. This porphyric granite is a faulted rock made up of amphibolite fragments that represent a trenching rock. At the edge of the escarpment the presence of mylonites (Figure 2) can be noticed. It is worth mentioning here that mylonites were formed during the Pan African event meanwhile block faulting into horsts and grabens occurred during the Cainozoic. These were caused from the scrubbing of the granite. The granite transits at the threshold of the massif with migmatites, anatexites and fine granite.

5.3.2 Geomorphology and tectonics

Geomorphological features of the Batié granite are closely linked to lithology and tectonics during the Cainozoic of the area. The main forms observed consist of poly-convex or multi-convex hills or croups and large knolls in oval shapes. Poly-convex landscapes lie between 1.400 and 1.600 m asl. and their slopes vary between 40 to 80%. The mechanical debiting in lumps as well as various boulders of this granite, tors and castle kopje landforms, the weathering that is responsible for the coarse sand and scouring surfaces, are in direct link with the lithology and tectonics. At the lower part of these plateaus, (that is, after the main tectonic escarpment), many small hills have a 'banana shaped' morphology of 5–10% of slope gradient. Their bases are made up of flat and smooth colluvial basements (glacis).

A tectonic map shows the main escarpment as well as the main strike directions SW–NE that organise the shifting of the area and the massif (Figure 3).

The escarpment forms the watershed for the complete region. It is situated at the turning point of three fracture systems. The SW–NE system which orientates the rivers Mlue Tcham, Che Ngwen and Lache water courses; the N–S direction with slight variation to NNW–SSE, and the NW–SE direction. The main faults (strike 48–58° E) created a tectonic corridor Mlue Tcham–Che Ngwen (of more than 40 km wide) as seen in figure 3. This tectonic 'corridor' appears like a drain pipe dominated by small banana hills shape morphology and glacis.

5.3.3 Drainage

The hydrological network (Figure 4a–b) shows that the streams are almost rectilinear and tectonic structures influence their dominante run-off direction. Given that a high drainage

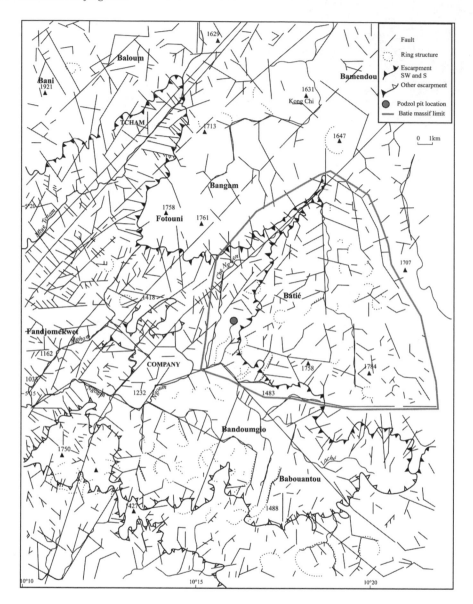

Figure 3. Batié tectonic map.

density favours the formation of podzols, Batié offers adequate conditions for its formation because it has a high drainage density of 4–5 streams per km². The hydrological network of the Batié massif has a dendritic oval form with a radial tendency (Figure 4b). The Venn fracture diagram shows the preponderance of Precambrian features (Strike 11–15° and 20–60° E) which underlines the shifting of the Batié granite mountain complex. The second group (strike 30° and 48–50° E) is essentially made up of faults formed during the Cainozoic which led to the block faulting of the Batié granite mountains into horsts and grabens.

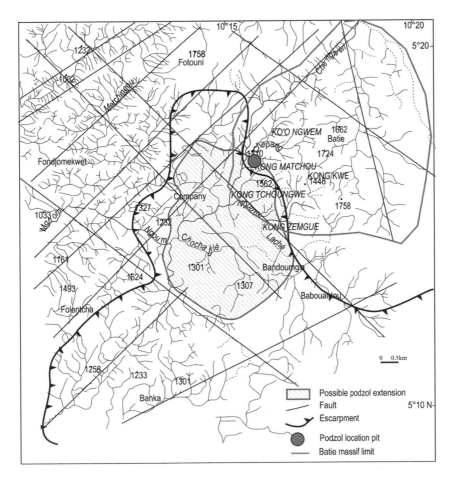

Figure 4a. Batié hydrological network.

The last group (strike 104°, 110° and 140° E) has some effect on the shield nature of the Batié massif that determined the circular structures due to vertical movements (Figure 4b).

5.3.4 Soils

There are three main soil types in the region: typical ferrallitic soils originating from granite, migmatite or mylonite, hydromorphic soils and less developed soils.

Ferrallitic soils have a very low organic content, estimated at less than 50%. They generally have three horizons, but can degenerate into 4, 5 or 6 horizons: horizon A, sandy clayish strata with organic matter and a grey (7,5R 6/0) or a dark brown color (2,5YR 4/2), horizon B_1, sandy clayish strata with several large grains (blocks), constituted of hardened limestone and quartz pebbles with a dark red (10R 3/6) to brownish red color, horizon B_2, red (10R 5/8) in color, and often dominated by a stone line and horizon C, parent rock in decomposition (saprolite).

Concerning this first group of highly homogeneous ferrallitic soils (otherwise referred as ferralsols), a granulometric analysis shows a domination in large particles (gravels and large

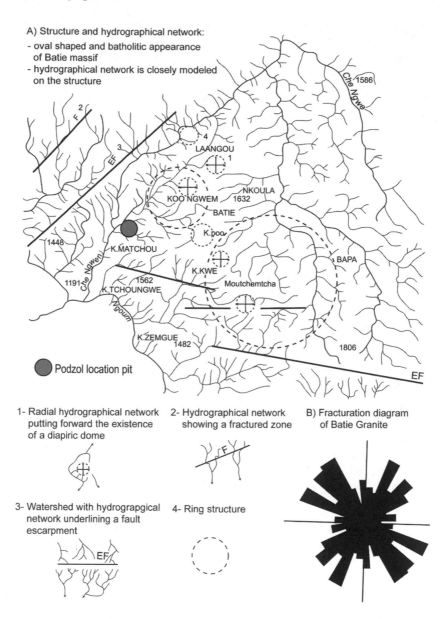

Figure 4b. Batié oval shaped hydrological network.

grained sand) which constitute 80%, meanwhile fine particles (averagely grained sand, fine sand and silt) make up of 20%. The second types of soils (hydromorphic soils) are formed due to high saturation of water. They are mainly found along water courses, marshes and/ or intermontane basins. They are made up of alluvial and/or colluvial deposits. Generally, hydromorphic soil strata in this region hardly reach over 200 cm in thickness. Less developed soils are mainly regosols, lithosols and fluvisols. Their distribution is haphazard within the whole region on slopes, valleys and hilltops. Their thickness can vary from 50 cm to 200 cm.

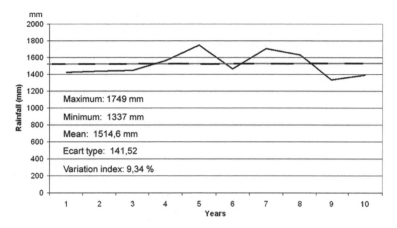

Figure 5a. Batié irregular annual rainfall of the last ten years.

It is worth mentioning that podzols are much more complex soils and differ considerably from the three main soil types described above. It is not uncommon to find some characteristics of the other three soils types within podzols.

The Batié massif is made up of red ferrallitic soils on coarse textured granite (sandy gravel to gravelo–sandy) and yellow ferrallitic soils on migmatite and mylonite ranging from sandy texture to sandy clayish silt (limon). At ground level, the passage from unsaturated ferrallitic soils to hydromorphic soils or podzols with ortstein formation, which may suggest that pedogenesis is still going on.

5.3.5 Climate

Batié's climate reveals a great paradox if compared to that of the rest of the Bamileke highlands. The mean annual rainfall is about 1.515 mm and the variation index is high: 9,34 mm (Figure 5a).

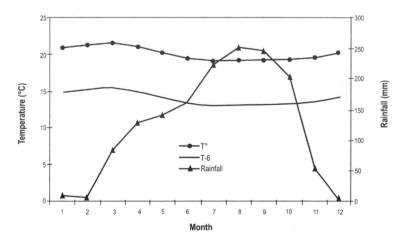

Figure 5b. Batié hydrothermical abacus.

The hydro-thermal abacus, built through rainfall and temperature data, shows a concentration of humid periods between June and November (Figure 5b). There are three dry months (December–February), four semi-dry months (November, March, April and May) and five humid months (June–October). Conclusively it can be affirmed that Batié has a semi dry climate.

5.3.6 Vegetation and land use changes

Batié has experienced drastic environmental changes through deforestation to a large extent caused by the action of humans. The former Bamileke territory (including Batié) was covered by a thick montane forest (1.800–2.800 m asl.), a submontane forest (800–1.800 m asl.) and also an atlantic semi deciduous forest. This primary atlantic forest, with Sterculaceae and Ulmaceae species is found in the Bamileke meridional appendage in contact with the cameroonian sedimentary littoral basin. This humid, semi-humid and densely forested plateau carries a diversified faunal species like elephants (*Loxodonta africana*), gorilla (*Gorilla gorilla*), chimpanzees (*Pan troglodyte*), lion (*Panthera leo*), panther (*Panthera pardus*), buffalo (*Syncerus cafer*) and others. Later, hedges were planted with vegetal species carefully selected by man for food and fodder (avocado, *Persea americana*, mango,

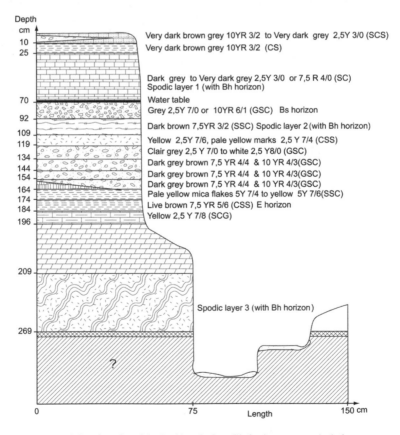

Figure 6. Stratigraphy of the Batié podzol profile horizons respectively layers.

Mangifera indica, banana, *Musa sapientium*); for religious or cultural purpose (tree of peace, *costus afer*, kola, *Cola verticillata*, etc). Inside the hedges crops were cultivated (beans, cassava, cocoyam). Raffia palms which were often tapped grew in river valleys. In any case, between 1900 and 1950, fences were established in the major parts of the Bamileke and Batié area giving such a beautiful and organized landscape that was qualified as a domesticated landscape by Letouzey (1985). This fenced landscape caused mainly the retreat of the original montane vegetation. The main para-climax vegetation (Tchindjang, 1996) is a degraded sub-montane forest that has been transformed into high savannas or peri-forestal savannas at the edges. Also prairies can be found on the hilltops.

Recent evolution is characterized by forest species re-conquering savanna environment like *Elaeis guineensis* that dominates the landscape at 1.200 m asl. These transformations by humans go a long way to positively influence and trigger the formation of podzols.

5.4 METHODS

Digging of the Batié sands by the local population rendered possible the sampling of the profile. Between 1998 and 2006 more than eight field trips were organized in order to collect samples from the spodic stratigraphy of the podzolic profile; field descriptions and several lab analyses were made for each stratum. Soil colors were determined by the Munsell soil color charts. Collected samples were taken to the Geomorphology and Chemistry Labs of Yaounde I University for further analyses. Samples were weighted and treated by stabilised oxygenated water (30 volumes). This oxygenated water contains about 9% in weight of hydrogen peroxide (H_2O_2) and it is susceptible to releasing about 30 times of its volume in oxygen. It was used for: assessing the rate of organic matter—it bleached the sample and separated the organic matter from the sand. The organic matter floated meanwhile the sand remained at the bottom; measuring the amount of sand in each sample—after four days, the specimen was completely cleaned of its organic content and sandy remnant was dried and weighted; analysing the morphoscopic structure for establishing granulometric curves—this was effected by sieving of the sand through sieves of different mesh dimensions.

From fifteen samples collected at a depth of 210 cm, eight were crushed and powdered and subsequently taken to the Laboratory of Inorganic Chemistry in the Faculty of Science at Yaounde I University. The following components were determined: Al_2O_3 by complexometry and colorimetry, Mg and Ca by AAS (atomic absorption spectrometer), K and Na by photometry and flame test and Fe_2O_3 by titrimetry or colorimetry.

5.5 RESULTS AND INTERPRETATION

5.5.1 Spatial and stratigraphic analysis

The Batié soil profile (Table 2, figure 6) is located at the footslope of the south west trending escarpement of the central Bamileke land (1.200 m asl.). The total area covered by similar podzol-like features was measured in the field up to 15.330 m². However, it is estimated that its distribution could be much more extend in the region.

The profile is characterized by its thickness of above 6 m and the composition of its different strata. From top to bottom, there are 15 strata from an observed section of almost 3 m depth. Table 2 gives the detailed description of these strata. Within the profile at a depth of 70–92 cm, there is a reddish coloured liquid originating from the leaching of upper soil horizons. There is a high concentration of iron (44–68%) within the layers mostly

Table 2. Description of the Batié podzol profile characteristics.

Depth in cm	General soil characteristics and observations
0–10	Very dark brownish grey 10YR 3/2 to very dark grey 2,5Y 3/0 (62,8% clay 23,74% sand).
10–25	Very dark brownish grey (10YR 3/2) to white (2,5Y 8/0), clayish and silty, with sand (80,79%) and some clay (10,36%).
25–70	Dark to very dark grey; silty and clayey. Contains 47,10% of clay and 37,7% of sand. The end of the strata marks the recent water table. A reddish liquid can be observed at the lower part of the strata. This layer is build up of anmoor soils and dead leaves. Aside of rotten leaves a tree trunk was found inside this horizon. The rusty liquid at the end of the strata is proof of the presence of precipitated iron. Analysis revealed that this psammitic ferrallitic soil layer contains 51% of iron and 21% of aluminium (Bh horizon).
70–92	Bs horizon, grayish stratum (25Y 7/0 or 10YR 6/1). Some gravels, sandy to clayey texture; 30,10% sand, 46,48% clay. Particles are polyhedral, coarse and large.
92–109	Dark brown (7,5YR 5/2) and brown (2,5Y 7/0), 71,17% sand, 18,37% clay. Silty to sandy, partly clayey.
109–119	Bh horizon, dark brown (7,5YR 5/2) to yellowish (2,5Y 7/6) and olive yellow (2,5Y 6/6). Clayey to sandy–silty texture; 23,39% sand, 57,49% clay. This layer gets quickly hardened when exposed to air. Contains 60% of organic matter with 44,82% of iron and 37,92% of aluminium.
119–134	Clear grey (2,5Y 7/0) to whitish (2,5Y 8/0) soil. Texture of gravels, sand and clay; 67,6% sand and 16,29% clay (E-Horizon).
134–144	Dark greyish brown (7,5YR 4/4 and 10YR 4/3) to white (2,5Y 8/0). Gravelly, sandy to clayey; sand (76,57%), clay (11,73%) and organic matter (11,19%).
144–154	Dark greyish brown (7,5YR 4/4) to clear brownish white (10YR 6/4). Gravelly, sandy to clayey; sand (85,89%), clay (7,81%).
154–164	Grey (5Y 6/1) to clear grey (5Y 7/1) colored stratum. Gravelly, sandy to clayey; sand (76,92%), clay (10,42%).
164–174	Yellow (5Y 71/6) to clear brownish white (10YR 6/4) layer. Silty, sandy to clayey texture; sand (85,89%), clay (7,81%), 'Illuvial horizon'
174–184	Grey (5Y 6/1) to clear grey (5Y 7/1). Clayey–Silty–Sandy texture: sand (76,92%), clay (10,42%) and organic matter (14,85%).
184–196	White (5Y 8/2), pale yellow (2,5Y 714) and yellow (5Y 716). Sandy–clayey–gravelly texture; sand (42,73%), clay (34,38%).
196–209	Bh Horizon, shiny brown (7,5Y 5/8) to reddish yellow (7,15Y 618) stratum. Ortstein layer with sighns of an anmooric to oligotroph soil. Texture: sand (56,70%), clay (25,71%).
>209	Bh Horizon, dark brown (7,5YR 5/2) and Yellow (2,5Y 7/6). Ortstein horizon with 82,48% of sand and 7,69% of clay. This spodic horizon probably indicates the first humid deposit that has been hardened while in contact with air (70,03% organic matter, 53,86% iron and 30,52% of aluminium.

constituted of organic matter. This gives evidence for the illuviation process which causes the accumulation and development of an iron pan.

In general, the texture of this probably multiphased podzol is coarse, sandy and angular that resulted from the granite weathering (grus formation). The sand samples analysed for the different strata revealed a relative homogeneity within the profile. There is also the observation of a remarkable superimposition from bottom to top: texture sizes varying from top (40 μ to 4.000 μ) to bottom (63 μ to 10.000 μ) as to be seen in figure 7.

Figure 7 equally shows three optimal troughs (lowest points): 75 cm depth, 109 cm and 184 cm on quartiles Q1, Q2, Q3, and Q4.

Starting from the bottom of the profile i.e. 184 cm depth, a first trough can be noticed. This point which contains organic matter (70,03%) can be considered as the first period of deposition of organic matter (subsequently sequestrated as carbon). This alluvial deposit reflects a still humid period which could be estimated at 3.300 BP (Hypothesis). The second trough is located at a depth of 109 cm. It contains 56–60% of organic matter. It can be considered as the second period of organic deposition perhaps linked to deforestation. This period is no longer as humid as the first one and could be eventually estimated at 2.100 BP. It's worth mentioning that this other alluvial deposit contains some large particles of coarse sand. The last trough located at a depth of 75 cm is another stratum of great organic deposition (89–95%). It is the last humid period marked by deforestation and carbon sequestration. It is much more humid than the second period but less than the first one. It's a much more recent period (1.700–1.500 BP?) dominated by a fine alluvial deposit. The end of the strata marks the recent water table. Rotten leaves and a tree trunk were found in this horizon.

On the other hand, three peaks can be observed in figure 7 at 15 cm, 92 cm and between 134 to 174 cm depth which generally show larger sand grains. Each of these peaks

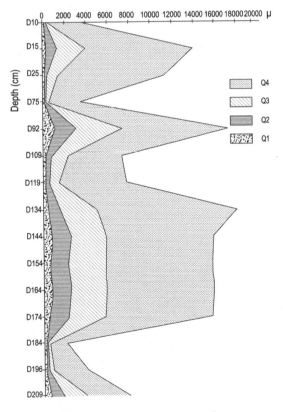

Figure 7. Quartiles of podzol samples.

are interpreted to correspond with a dryer climatic period. A bottom to top interpretation begins with the first peak (134–174 cm). It is the largest peak which corresponds to a severe dry period (aridification around 3.100–2.600 BP) marked by colluvial deposition. Fluvial system mostly seems to have dried out at that time. The second peak in 92 cm depth could correspond to a brief, but severe dry period (2.000 years BP) with comparable changes. The last peak around 15 cm depth may correspond to a slightly longer and dryer period around 1.000 BP. However, there is a restructuring of the drainage system marked by the fluvial re-cutting of the river bed. All these three peaks give some evidence of the beginning and the evolution of the savannization processes linked to man's influence as well as climate changes.

The morphoscopic analysis shows many angular sandy grains of 315 μ with debris of micas, weathered debris, heavy minerals, and quartz as well as quartzite debris.

The granulometric curves show three main types (Figure 8). Parabolic and sigmoid well sorted features (P1, P4, P13, P14). Parabolic or sigmoid curves represent areas of huge organic matter accumulation. It corresponds to sandy silty clayish material. In these strata leaves, tree trunks, lignin and roots were observed; giving possible evidence for striking palaeoecological changes. This material was transported by fluvial processes over relatively short distances with an origin from the nearby Batié plateau. This could support the idea that severe deforestation formerly occurred on this plateau. There are two partly degraded parabolic curves: P3 and P7. Linear worst sorted curves (P2, 6, 8, 10). These curves represent material deposited by colluvial processes on slopes. The materials are regarded to be characteristic for dryer periods with more open vegetation coverage. Hyperbolic relatively sorted facies (P5, 9, 11, 12). These reflect deposition by decantation in a relatively paludal milieu. Both colluvial and alluvial processes occurred within these layers. These curves are a transition between the parabolic sigmoid and linear curves. In the same line, there is a kind of transitional landscape evolution between dryer (colluvial) and wetter (alluvial) climatic periods.

5.5.2 Palaeoenvironmental approach

Organic matter and traces of fossil vegetation

The presence of the spodic soil seems to be interconnected with the lateral transfer of solid matter. Traces of roots and three tree trunks, indicators of tree trunks and dead leaves within

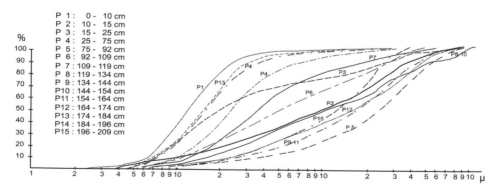

Figure 8. Granulometric curves of Batié podzol sand (Tchindjang, fieldwork, 2006).

the organic layers, acajou colour and black leaves (lignin) point to the former presence of an ombrophilous forest or mountain mesophile forest. Slope colluvia were deposited locally. The evolution towards compaction and spodic material is linked to the drying up of the region.

The observation of a hard shell originates from two factors: high organic matter content ranging from 11–90% (Figure 10); it shows an inverse relationship between the proportion of sand and that of organic matter content within the profile. The vegetation observed is of the hygrophilous type.

It is important to note that the thickness of the organic layer as shown by the profile (50–70 cm) gives further evidence to the relative abundance of trees roots and trunks. This is proof of carbon sequestration by a thick forest whose destruction seems to be closely linked to human activities due to the settlement of the Bamileke people around 3.000 BP in this region. It's worth mentioning that climatic changes as proved by Kadomura (1994) are contributing factors to these changes mainly during dry periods as shown in the profile.

Water table

There are three water tables located below the three troughs earlier mentioned. The water table close to 70 cm depth shows seasonal variations of about 1 m in the dry season, but remain permanent in the rainy season. Its level never attains upper horizons in the dry season. It is a hanging water table that percolates through the non compacted spodic strata.

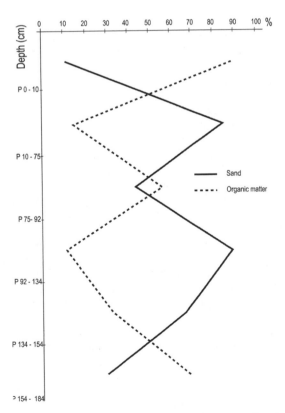

Figure 9. Inverse relationship between organic matter and sand.

The water table is linked to a non hardened base level. Since it is located on the riverside, this podzol profile can be considered as a hydromorphic, multilayered palaeopodzol. The variation of water level is evidently linked to the intensity and spatial distribution of rainfall that favours lateral contributions. The perennial high humidity induces suffusion in lower soil horizon as can be seen on the last spodic horizon (see figure 6). This has caused a retreat of the slope by about 1 m within several years only.

Historical aspects of podzol evolution

Podzolisation seems to be an 'old' phenomenon because of two reasons: there is no precise limit between the water tables and the spodic strata. The present hydromorphic situation seems not to have caused the formation and compaction of the ortstein layer. Some non podzolised layers are hydromorphic, meanwhile others are not. The river shapes the last spodic layer for more than 40 cm. The spodic process is equally a relatively 'old' phenomenon and the presence of hanging water table could have played a role for this. The presence of a fossil flora (roots and tree trunks) apparently homogenous and more specially lignin traces incompletely carbonised give a further probable evidence for the old nature of podzol formation in this area. This fossil flora and the unusual abundance of organic matter within the profile could be regarded as are preliminary proof that the Batié sub-montane forest sequestered huge quantities of carbon.

5.6 DISCUSSION

In a recent review of podzolisation processes, Lundstrom *et al.* (2000), described podzols as soils which develop in a humid climate, under vegetation that favors the development of a 'Moor' humus layer. Batié, as explained before, has a humid climate that favours the development of a permanent forest vegetation. Contrary to the study carried out by Lundstrom *et al.* (2000) which shows that podzols have a horizon depleted in Fe and Al compared to the parent rock; the Batié podzol shows a reverse tendency. Thus its parent rock is a calco-alkaline granite rich in silica (>70%) and aluminium (14,90%) and relatively poor in iron (<5%), meanwhile the eluvial E-horizon and illuvial B-horizon, enriched in organic matter (11–15%), contains Al and Fe in great proportions (iron from 28,26 to 68,36%; aluminium from 22,06 to 43,12%).

The ortstein layer shows two main horizons of humic matter accumulation: one horizon is dark in colour, with organic cement which engulfs quartz grains. This horizon is hardened, relatively crumbly at dry state. Unlike the relatively high calcium values of the Bateke plateau podzol in Congo Brazzaville, that of the Batié podzol is low in calcium. This could be an indicator of a relatively active biological milieu. It is however not the case due to the hardened nature of the spodic horizon. This even hinders the formation of termite mounds in the area. Other components can be seen on figure 10 based on chemical analyses. The proportion of sodium is relatively high. The main elements having very high values are: iron, aluminium and potassium. Sodium, magnesium and silica have very low values (Figure 10).

Consequently the low C/N relationship further confirms a low biological activity and the aging of organic matter in spodic strata. This phenomenon is accompanied by a reduction of nitrogen which is evacuated by the water table in the form of soluble compounds. The hardening of spodic strata requires drying. This could probably originate from two sources. Firstly, the amelioration of external drainage through the reinforcement of the base level and, secondly, a dry climate transiting to an arid climate and ends by the reinstallation of a forest vegetation.

5.7 CONCLUSIONS

This study deals with the pedological nature of Batié. Based on a multidisciplinary approach, combining pedology, sedimentology and chemical analysis, the study proposed a periodical functioning of this soil complex in relation with geomorphologic and environmental factors responsible for a particular hydrological pattern. There is a strong support of the idea that oxidoreduction processes in depth took place during the wet periods. Considerations can be made on the existence of a high drainage density which signifies preferential paths for podzol formation during these wet periods. This favoured the transport of organic acid Al rich-solutions from the surface organic litter to the lower horizons where they had been accumulated in granitic sands.

Actually, the Batié palaeopodzol does not correspond with recent alluvial levels. It rather reflects the base level of the river with reduced solid load. This solid load diminishes during humid climatic periods of exceptional high annual rainfall (1.800–2.000 mm). In this region with signs of tectonic subsidence, one can understand the dynamism of soils with neo-tectonics that functions greatly at the edges of the escarpment. This study has collected some evidence that the Batié palaeopodzol is a relic soil. The morphological and floristic environment that led to the formation of a podzol in Batié is different from that of

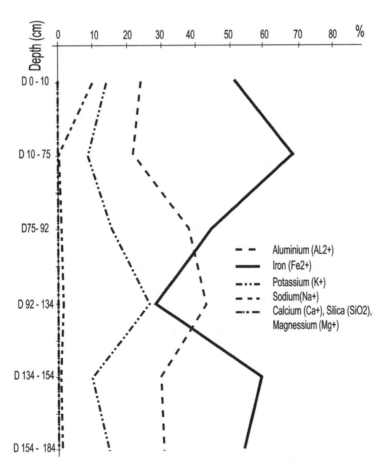

Figure 10. Chemical composition of Batié podzol.

the present. Thus, it is possible to search systematically similar formations at the edges of all the tectonic escarpements surrounding the Bamileke highlands because of comparable drainage conditions and processes. Such research will enable a better correlation with the proxy-data signals around the Barombi-Mbo Lake or elsewhere.

ACKNOWLEDGEMENTS

Special thanks go to Dr. Kecha Mbadcam Joseph of the Laboratory of Inorganic Chemistry at the Faculty of Science, University of Yaoundé I.

REFERENCES

Babet V., 1933, Exploration de la partie méridionale des plateaux Batéké. *Bulletin Service des Mines A.E.F.*, 1947, **3**, pp. 21–56.

Bocquier G. and Boissezon P., 1959, *Note relative à quelques observations pédologiques effectuées sur le plateau Batéké (région du Pool, Rép. du Congo).* ORSTOM Brazzaville, p. 19.

Brenac, P., 1988, Evolution de la végétation et du climat dans l'Ouest Cameroun entre 25.000 et 11.000 BP. *Travaux de la Section Scientifique et Technique de l'Institut Français de Pondichéry*, pp. 91–103.

Dumort, J.C., 1968, *Notice explicative de la feuille de Douala Ouest au 1/500 000.* BRGM, Direction des Mines et de la Géologie, Yaoundé, p. 69.

Flexor, J.M., De Oliveira, J.J., Rapaire, J.L. and Sieffermann, G., 1975, La dégradation des illites en montmorillonite dans l'alios de podzols tropicaux humo-ferrugineux du reconcavo bahianais et du Para. *Cahiers ORSTOM, série Pédologie*, Vol. 13, **1**, pp. 41–48.

Giresse, P., Maley, J. and Brenac, P., 1994, Late Quaternary palaeoenvironments in the Lake Barombi Mbo (West Cameroon) deduced from pollen and carbon isotopes of organic matter. *Palaeo 3*, **107**, pp. 65–78.

Kadomura, H. and Hori, N., 1990, Environmental implications of slope deposits in humid tropical Africa: Evidence from southern Cameroon and western Kenya. *Geographical reports* (Tokyo Metropolitan University), **25**, pp. 213–236.

Kadomura, H. and Kiyonaga, J., 1994, Origin of Grassfields landscape in the West Cameroon Highlands. In *Savannization processes in Tropical Africa 2*, (Tokyo Metropoliytan University, Department of geography), pp. 47–85.

Kadomura, H., 1995, Palaeoecological and paleoohydrological changes in the humid tropics dutring the last 20.000 years, with references to Equatorial Africa. In *Global Continental Palaeohydrology,* edited by K.J. Gregory, L. Starkel and V.R. Baker. (London: John Wiley and sons Ltd.), pp. 177–202.

Klinge, H., 1965, Podzol soils in the Amazon Basin. *Journal of Soil Science*, **16**, pp. 96–103.

Klinge, H., 1968, *Report on tropical podzols*, (Rome: Food and Agriculture Organization, IV), p. 88.

Klinge, H., 1973, Root mass estimation in lowland tropical forest of central Amazonia, Brazil. I. Fine root masses of a pale yellow latosol and a giant humus podzol. *Tropical Ecology*, **14**, pp. 29–38.

Leneuf, N. and Ochs, R., 1956, *Les sols podzoliques du cordon littoral en basse Côte d'Ivoire*, (Paris: Congrès de la science du sol), pp. 529–532.

Letouzey, R., 1985, *Notice de la carte phytogéographique du Cameroun au 1/500 000,* (Toulouse: Institut de la Carte Internationale de la végétation; Yaounde: Institut de recherches agronomiques).

Lunstrom, U.S., Van Breemen, N. and Bain, D., 2000, The podzolisation process. A review. *Geoderma*, **94**, pp. 91–107.

Maley, J., 1981, Etudes palynologiques dans le basin du Tchad et paléoclimatologie de l'Afrique Nord tropicale de 30 000 ans à l'époque actuelle. *Travaux et Documents de l' ORSTOM*, p. 129.

Munsell Color C0., 1992, Munsell soil color charts. Revised edition. Newburgh, New York, USA.

Nguetsop, V.F., Servant-Vildary, S. and Servant, M., 2004, Late Holocene climatic changes in West Africa, a high resolution diatom record from equatorial Cameroon. *Quaternary Science Reviews*, **23**, pp. 591–609.

Richards, P.W., 1941, Lowland tropical podzols and their vegetation. *Nature,* **148 (3774)**, pp. 129–131.

Schwartz, D., 1985, *Histoire d'un paysage: le lousseke. Paléoenvironnement Quaternaire et podzolisation sur sables Batéké (quarante derniers millénaires, région de Brazzaville, R.P. du Congo)*. Thèse Doctorat, Université Nancy I, p. 211.

Schwartz, D., Delibrias, G., Guillet, B. and Lanfranchi, R., 1985, Datations par le ^{14}C d'alios humiques: âge njilien (40.000–30.000 BP) de la podzolisation sur sables Batéké (République Populaire du Congo). *Comptes rendus de l'Académie des Sciences*, série 2, Vol. 300, **17**, pp. 891–894.

Schwartz, D., Guillet, B., Villemin, G. and Toutain, F., 1986, Les alios humiques des podzols tropicaux du Congo: Constituants, micro et ultra structure. *Cahiers ORSTOM, série Pédologie*, Vol. 36, **2**, pp. 179–198.

Schwartz, D., 1988, Histoire d'un paysage: le Lousséké. Paléoenvironnements Quaternaires et podzolisation sur sables Batéké. *ORSTOM, Collection Etudes et Thèses*, pp. 1–285.

Talla, V., 1995, *Le massif granitique panafricain de Batié Ouest Cameroun: Pétrologie-Pétrostructurale-Géochimie*. Thèse 3ème Cycle. Yaoundé, Université de Yaoundé I, p. 144.

Tchindjang, M., 1996, *Le bamiléké central et ses bordures: morphologie régionale et dynamique des versants. Etude géomorphologique*. Thèse de Doctorat, Université de Paris 7 Denis Diderot, p. 867.

Turenne, J.F., 1977, Modes d'humification et différenciation podzolique dans deux toposéquences guyanaises. *Mémoire ORSTOM*, **84**, p. 173.

Zogning, A., Giresse, P., Maley, J. and Gadel, F., 1997, The Late Holocene palaeoenvironment in the Lake Njupi area, West Cameroon: implications regarding the history of Lake Nyos. *Journal of African Earth Science*, **24/3**, pp. 285–300.

CHAPTER 6

New evidence on palaeoenvironmental conditions in SW Cameroon since the Late Pleistocene derived from alluvial sediments of the Ntem River

Mark Sangen

Institute of Physical Geography, Johann Wolfgang Goethe University, Frankfurt am Main, Germany

ABSTRACT: An interior delta in the lower course of the Ntem River near the sub-prefecture Ma'an was identified after interpretation of satellite images, topographical maps of SW Cameroon and geological as well as hydrological references and a reconnaissance fieldtrip to the study area. Here neotectonic processes have initiated the establishment of a 'sediment trap' (step fault), which in combination with environmental changes strongly generated the fluvial morphology. It transitionally led to temporary lacustrine and palustrine conditions in parts of this river section. Inside the interior delta an anastomosing multi-branched river system has developed, which contains 'stillwater locations', periodically inundated sections, islands and rapids. Following geomorphological, physio-geographical and sedimentological research approaches, the alluvial plain has been prospected and studied extensively. 91 hand-corings, including three NE–SW transects, were carried out on river benches, levees, cut-off and periodical branches, islands as well as terraces throughout the entire alluvial plain and have unveiled multi-layered, sandy to clayey alluvia reaching up to 440 cm depth (Figure 1). At many locations, fossil organic horizons and palaeosurfaces were discovered, containing valuable palaeoenvironmental proxy data. At these sites, through additional detailed stratigraphical analysis (close-meshed hand-coring and exposure digging) a comprehensive insight into the stratification (lamination) of the alluvia could be gained, clarifying processes and conditions that prevailed in the catchment area during the period of their deposition.

32 Radiocarbon data of macro-rests (leafs, wood), charcoal and organic sediment sampled from these horizons provided ages between 48.230 ± 6.411 and 217 ± 46 years BP (not calibrated). This constitutes the importance of the alluvia as an additional, innovative palaeoarchive for proxy data contributing to the reconstruction of palaeoenvironment and palaeoclimate in western Equatorial Africa. The further examination of the alluvia will not only provide additional information on the dynamics of vegetation, climate and hydrology (esp. fluvial morphology) in SW Cameroon since the 'First Millennium BC Crisis' (around 3.000 years BP), the main focus of the DFG-research project, but also on conditions prevailing since the Late Pleistocene, during the Last Glacial Maximum (~18.000 years BP), the Younger Dryas impact (~11.000 years BP) and the 'Humid African Period' (~9.000–6.000 years BP). $\delta^{13}C$-values ($-31,4$ to $-26,4$‰) evidence that at the particular drilling sites rain forest has prevailed during the corresponding time period (rain forest refuge theory). The sampled macro-rests all indicate rain forest dominated ecosystems, which were able to persist in fluvial habitats, even during arid periods.

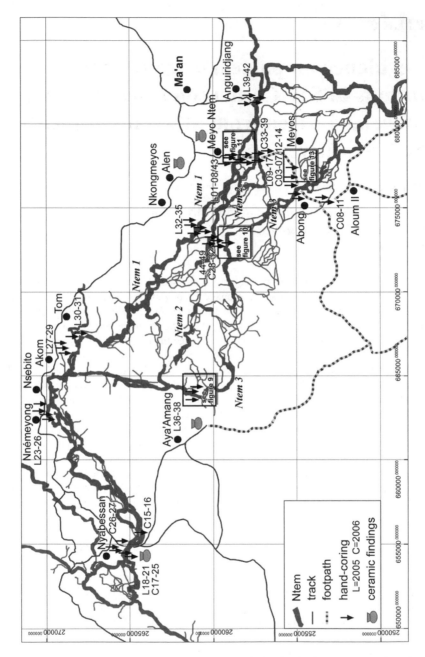

Figure 1. Map of the Ntem interior delta study area (210 km²) showing the three main Ntem branches, hand-coring sites, locations where ceramic was found, infrastructure and other sights of interest, mentioned in the text.

6.1 INTRODUCTION

The ReSaKo (Regenwald-Savannen-Kontakt)-project (sub-project J. Runge, Ru 555 14-1) investigates climatic, ecological and cultural changes in the tropical rain forest of SW Cameroon on interdisciplinary basis with Archaeobotanists (K. Neumann and A. Schweizer, University of Frankfurt) and Archaeologists (M.K.H. Eggert and C. Meister, University of Tübingen) in the framework of the DFG Research Unit 510 of the German Research Foundation (Deutsche Forschungsgemeinschaft, DFG). Evidence for interconnections between climatic fluctuations and modifications of the fluvial system Ntem and its sensitive eco-zones during the Late Pleistocene and Holocene are examined by methods of geomorphological, physio-geographical and sedimentological research. Climate fluctuations have strongly affected the hydrological cycle and composition of vegetation habitats (rain forest-savanna fringe), especially in Equatorial Africa's high sensitive and species-rich eco-systems (Gasse, 2000; Runge 2001 and 2002; Thomas, 2004). Alluvial sediments of the Ntem River's floodplain (Figure 1) serve as innovative palaeoenvironmental proxy data archives and the tentative interpretation of their multi-layered stratigraphical composition allows statements on processes and conditions during the time of their deposition. Besides, anthropogenic influence on landscape evolution in the study area will be recognized.

6.2 STATE OF THE ART

Cores of the deep-sea fans of the Niger (Zabel *et al.*, 2001; Lézine and Cazet, 2005) and Congo River (Marret *et al.*, 1998; Holvoeth *et al.*, 2005; Marret *et al.*, 2006), as well as lacustrine sediments of several lakes in West- and Central Africa (among others Barombi Mbo: Giresse *et al.*, 1994; Ossa: Nguetsop *et al.*, 2004; Reynaud-Farrera *et al.*, 1996; Wirrmann and Bertaux, 2001; Njupi: Zogning *et al.*, 1997; Bambili: Stager and Anfang-Sutter, 1999; Sinnda: Vincens *et al.*, 1998; Kitina: Elenga *et al.*, 1996; Bosumtwi: Talbot *et al.*, 1984 and Lake Tchad: Maley, 1981; Servant, 1983) have provided the most high-resolution and oldest proxy data for palaeoenvironmental, palaeoclimatic and palaeohydrological research in tropical West and Central Africa. These findings were mainly based on palynological studies initiated by ECOFIT (ÉCOsystèmes et Paléoécosystèmes des Forêts Intertropicales) in 1992. It focused on Holocene climate fluctuations and corresponding vegetation composition changes as well as rain forest–savanna margin dynamics (Servant and Servant-Vildary, 2000). In recent times, additional results derived from diatoms, phytoliths, stable organic carbon isotopic composition values ($\delta^{13}C$) and mineralogical data (reaching back to ~700.000 years BP) have been added to the research on palaeoclimate and palaeoenvironment of tropical Africa (Abrantes, 2003; Barker *et al.*, 2004). They substantiate changes in solar radiation (Milankovitch cycles: eccentricity ~100 ka, obliquity ~41 ka and precession ~21 ka) as the most logical reason for climatic fluctuations next to changes in ice-sheet extent (Heinrich events). These modifications were linked with changes of the sea surface temperatures (SST's) in the Atlantic Ocean (~2–6°C), south- and northward migration of the Inter-Tropical Convergence Zone (ITCZ), El- Niño events and non-linear feedbacks between atmosphere and ocean systems. According to the ECOFIT studies, this consequently provoked changes in energy- and moisture-fluxes in the Gulf of Guinea and strengthening/weakening of the African monsoon (Abrantes, 2003; Maley, 1997; Marchant and Hooghiemstra, 2004; Marret *et al.*, 2006; Nguetsop *et al.*, 2004). De Menocal *et al.* (2000) postulated a close coherence between the thermohaline ocean circulation and the intensity of the African monsoon. Mutations within this circulation system (e.g. occasional interruption of the up-welling of cold ocean waters in the Gulf of Guinea) can induce rapid and far reaching local climate changes. As assumed by Maley (1997), modifications of the Hadley and Walker cells trigger changes in energy and moisture fluxes between ocean, atmosphere and land surface. Also evidence for changes in seasonality (longer dry and shorter rainy seasons) in West Equatorial Africa were verified (Servant and Servant-Vildary, 2000).

The Holocene 'African Humid Period' (De Menocal, 2000) in West and Central Africa (maximum phase ~9.000–6.000 BP) already started around 14.000 years BP in equatorial regions and was assumedly interrupted by aridity during the Younger Dryas (~11.000 BP) and further, shorter arid phases (9.600–9.400 BP and 8.400–8.000 BP) before it was terminated around 4.000 BP (Barker *et al.*, 2004; Lézine and Cazet, 2005). It is assumed that during this period the solar radiation reached a higher value (~4–5%, corresponding to a total of ~470 Wm^{-2}) and thereby induced an increase of temperature (around up to 5°C) and probably ~30–50% higher monsoonal precipitation (Thomas, 2000; Barker *et al.*, 2004). Holvoeth *et al.* (2005) gave evidence for fluctuating influx of terrestrial organic material into the oceans by rivers, changing vegetation composition as well as shifts in the hydrological cycle. During pluvial phases the input of terrestrial organic material into the Niger and Congo fans was immense higher. Barker *et al.* (2004), Gasse (2005), Kadomura (1995), Thomas (2004), Lezine and Cazet (2005) and Marret *et al.* (2006) among many others (esp. Maley) provide most detailed descriptions of palaeoenvironmental conditions in Equatorial Africa (esp. Cameroonian and Congolian regions) for different time-scales and based on various evidences. Hall *et al.* (1985), Preuss (1986), Runge (1992, 1996) and Thomas and Thorp (1980, 1995) were among the first scientists using fluvial-morphological and geomorphological approaches in the context of environmental changes in western and Equatorial Africa, apart of the early studies on the sedimentation dynamics of the Niger (Pastouret *et al.*, 1978) and Congo Rivers (Giresse *et al.*, 1982). Giresse *et al.* (2005) and Viers *et al.* (2000) have recently measured sediment fluxes in the Sanaga and Nyong Rivers. They proved that during the late Holocene, climate related phases of higher (2.800–2.000 BP) and lower (5.000–3.000 BP) sediment fluxes occurred in the Sanaga catchment.

Kadomura and Hori (1990) also found evidence for increased anthropogenic activities in the rain forest ecosystem around 3.000 years BP. This time period is also believed to be the onset of the hypothetical 'Bantu Migration' (Phillipson, 1980; Schwartz, 1992) within a time period defined as 'First Millennium BC Crisis'. The expansion of human activities might have been induced by regression of tropical rain forest and widespread savannization, favoured by changes in climatic and hydrological conditions, which led to the partial breakdown of the rain forest between 3.000 and 2.500 BP and its replacement by pioneer formations and savannas (Maley, 2002; Runge, 2002 and Vincens *et al.*, 1999). Settling and sedentariness in the tropical rain forest of Central Africa may additionally have caused major impacts on the distribution of the rain forest-savanna margin, at least since the initiation of metallurgy around 2.600 years BP (Schwartz, 1992).

Nevertheless, the majority of conclusions concerning the palaeoenvironmental conditions in Central Africa are based on palynological and lacustrine archives. It is however essential to integrate palaeohydrological investigations into the reconstruction of environmental changes because they offer indicators for realized hydrological changes in the past (Baker, 2000). Although still quite incipient and incomplete, Latrubesse (2003) and other scientists have already offered exemplary spectacular and detailed findings for the understanding of hydrological and environmental changes in tropical central South America. Similar studies and results are also necessary in tropical Central Africa to contribute to an improved understanding of environmental change in this region.

6.3 STUDY AREA: NTEM INTERIOR DELTA

6.3.1 Physiogeographical settings

The Ntem River is a perennial river, 460 km long, with a catchment area of 31.000 km², which stretches over the countries of Equatorial Guinea, Gabon and (mainly) SW Cameroon.

Around 100 km upstream the Ntem River's mouth into the Atlantic Ocean near Campo, an interior delta was formed by neotectonical activities (cp. J. Eisenberg, this issue). This section of the Ntem, which is located in the department Ambam near the sub-prefecture of Ma'an in the district 'Vallée du Ntem', shows an anastomosing, multi-channelled character with three main branches (Ntem 1–3, from NE to SW). It stretches from 2°14'N and 10°39'E to the NW, where it is limited by the waterfalls of Menvé'élé (2°24'N, 10°23'E) near the village Nyabessan. The surface of the alluvial basin (~210 km²) has an SE–NW expansion of ~35 km and a maximum width (NE–SW) of 10 km (Figure 1). The northern section (Ntem 1) is marked by several rapids formed by outcropping basement (charnockites, gneisses and granites), whereas in the southern part (Ntem 3) many periodically inundated river sections occur, where thick fine-grained alluvia have been accumulated. At Meyo Ntem and bellow the waterfalls of Menvé'élé, thick (1–2 m) layers of ferruginized conglomerates (Ntem gravels, cemented by iron and manganese oxides) were observed, which indicate environmental modifications since their initial formation.

In the upper river course (from the source near Oyem in Gabon, ~700 m a.s.l., to the Gabon/Cameroon border at ~580 m a.s.l.) the Ntem River drains a cratonic peneplain surface with low inclination (2‰). From the Cameroon/Gabon border to Ma'an (518 m a.s.l.) the inclination decreases to 0,27‰ and the Ntem River affiliates its major tributaries (Kom, Kie, Kye, Mboro, Mgoro, Nlobo, Nye, Rio Bolo, Rio Guoro and Woro). Inside the interior delta, the Ntem confluences with the Mvila, Ndjo'o and Biwomé and inclination increases (4,5‰) to reach it's maximum of 5‰ behind the border of the interior delta (waterfalls of Menvé'élé, ~384 m a.s.l.). Here the Ntem River crosses an up to 80 m deep V-shaped valley, descending 200 m over a distance of 40 km (Figure 2).

The longitudinal profile of the Ntem's section which is described as interior delta shows the positions of outcropping basement but also plain stretches where multi-layered alluvial sediments have been deposited (Figure 3). The basement outcrops may have forced

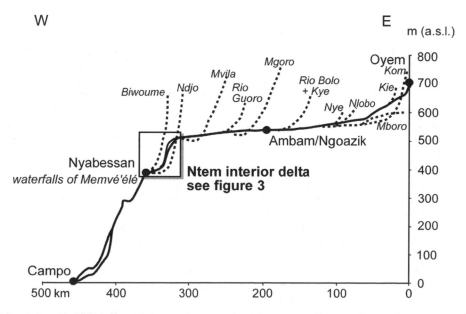

Figure 2. Longitudinal profile of the Ntem River from its source in Oyem (Gabon) to the mouth near Campo on the Atlantic coast. The figure shows major tributaries and locations of the gauging stations Ambam/Ngoazik and Nyabessan as well as the interior delta and the waterfalls of Menvé'élé (modified after Olivry 1986).

m (a.s.l.)

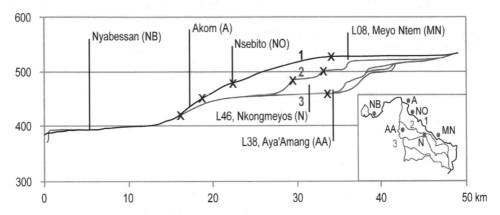

Figure 3. Longitudinal profile of the Ntem's multi-channelled interior delta showing positions of basement outcrops (X), rapids (Akom and Nsebito) and hand-corings mentioned in the text and in figure 8. Based on SRTM DEM data (obtained from the GLCF server of the University of Maryland).

the river to create nickpoints in its recent anastomosing to anabranching pattern by avulsion. Initially these locations would have served as sediment traps.

6.3.2 Climate and hydrology

The climate ('Equatorial Guinea Climate') in the catchment area is tropical to semi-humid with a mean annual temperature almost constantly around 25°C and a short (April–June) as well as a long (September–November) rainy season, corresponding to the seasonal shifting of the meteorological equator (ITCZ) and the African monsoon. Rain fall in the catchment area ranges between 1.500 mm on the plateau near Oyem and 3.000 mm on the coast near Campo. According to Olivry (1986) the mean annual rain fall averages 1.695 mm in the entire Ntem catchment and 1.675 mm at the meteorological station Nyabessan (2°24'N, 10°24'E; 385 m a.s.l.). Figure 4 shows mean monthly rain fall and temperature values for Ambam (2°23'N, 11°16'E) and mean monthly discharges of the Ntem at Ngoazik (~12 km S of Ambam) during the period 1954–1979. Annual discharges usually show two maxima in May and especially in November, associated with the rainy seasons.

Sometimes this for the Gulf of Guinea typical rain fall and river discharge regime is interrupted and rain fall as well as discharge in the short rainy season increases. Maley (1997) associates this with periodical ENSO-events, which modify moisture fluxes between ocean and land surfaces in the Gulf of Guinea.

At Ngoazik (2° 18'N, 11° 18'E; 535 m a.s.l.), 150 km upstream of Nyabessan, discharge valued ~260 m³/s between 1954 and 1991 (catchment area: 18.100 km²). The discharge shows high fluctuations and some kind of periodicity (3–4 years) in which dry as well as wet hydrological years occurred. Additionally, a slight wetter phase in the 1960s and a drier in the 1970's can be recognized (Figure 5). Mean monthly discharges of the Ntem at Nyabessan, with a catchment area of 26.350 km², range between 100 and 1.600 m³/s and show high annual variability as well as discharge values in general. Between 1978 and 1991, mean annual discharges at Nyabessan averaged ~382 m³/s.

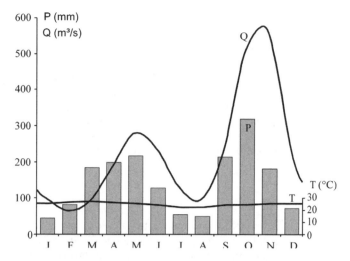

Figure 4. Mean monthly temperatures and rain fall at the meteorological station Ambam (2°23'N, 11°16'E; 561 m a.s.l.) and mean monthly discharges at the gauging station Ngoazik (2° 16' N, 11° 19' E; 535 m a.s.l.), located ~12 km S of Ambam, for the period 1954–1979.

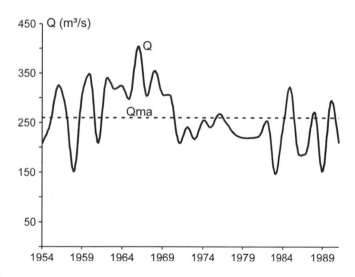

Figure 5. Fluctuating mean annual discharges of the Ntem at Ngoazik during the period 1954–1991, with an average discharge (Qma) of ~260 m³/s (modified after Olivry 1986, new data provided by J.-C. Ntonga, Centre de Recherches Hydrologiques, Institut de Recherches Géologiques et Minières (CRH-IRGM), Yaoundé).

The present river's suspended load is characterised by high content in organic material and minerals. The high outflow of organic suspended load into the Gulf of Guinea can be very good recognized on satellite images of the Ntem's fan near Campo. According to the dense vegetation cover in the river basin, erosion is nowadays very low where human activity is reduced. Coarse sands and pebbles which have been deposited in the initiative phase of the river basin evolution in a probably much less forested environment (braided river character), are presently only part of the sedimentary load during torrential flood events.

6.3.3 Vegetation

After Letouzey (1985) and Tchouto Mbatchou (2004) the potential vegetation in the study area can be described as tropical lowland evergreen rain forest (mainly Caesalpiniaceae) and mixed, semi-evergreen and semi-deciduous forest. On floodplains swamp forest with *Rapphia* spp., *Uapaca guineensis* and *Gilbertiodendron dewrei* occurs. Actually in most of the study area (Ntem interior delta) secondary forests of different ages occur because human impact (e.g. shifting cultivation, cacao cropping (*Theobroma cacao*)) has intensely modified the natural vegetation. Dominating species are *Ceiba pentandra*, *Elaeis guineensis*, *Musanga cecropioides* and *Terminalia superba* (K. Neumann, personal communication).

6.3.4 Soils

According to Segalen (1967) three major characteristic soils prevail in SW Cameroon: Yellow to brown tropical ferralitic (clayey) soils ('sols ferrallitiques jaunes sur les roches acides (gneiss)'), gneiss and granites covered by red to brown ferralitic soils ('sols ferrallitiques rouges sur les roches acides') and alluvial soils ('sols alluviaux') in floodplains and valleys.

6.4 METHODS

After the interpretation of satellite imageries and topographical as well as hydrological data a reconnaissance field trip was made in 2004 (February–March), where the existence of an interior delta in the lower catchment area of the Ntem River could be proved (Runge *et al.*, 2005). This was followed by field work in the dry seasons of 2005 (January–March) and 2006 (February). During this physio-geographical, geomorphological and sedimentological field work, closer observations and 91 hand-corings were undertaken in the floodplain. The multi-layered alluvial sediments, which could primarily be found at stillwater locations (periodically inundated cut-off channels, terraces (rainy/dry season), levees), were recovered and mostly sampled until maximum depths of 440 cm. The sediments were sampled with Edelman-corer in 20 cm layers. At some locations, a thin percussion-probe (3 cm diameter, 50 cm length) was used when ground water level was reached and recovery with Edelman-corer was not possible.

After texture and soil colour were determined in the field, the samples were evaluated at the laboratory of the Institute of Physical Geography in Frankfurt. Here pH (solved in 0,1 n KCl, after Meiwes *et al.*, 1984), grain sizes (after Köhn), soil colour (wet and dry, after Munsell), organic material and carbon contents (with LECO EC-12), nitrogen (after Kjeldahl) and dithionite as well as oxalate solvable quantum of iron and manganese (after Mehra and Jackson, 1960; with AAS Perkin Elmer Analyst 300) were analysed.

During this work, samples of vegetation macro-rests, fossil organic horizons with charcoal and palaeosoils were collected for ^{14}C (AMS) radiocarbon dating and determination of stable organic carbon isotopic composition values (δ^{13}C). The arising data provides preliminary (maximum) ages of the alluvial sediments as well as evidence on former vegetation composition (C3/C4 species; Runge, 2002). In total 32 samples from different locations and depths were extracted and analysed by the Friedrich-Alexander-University of Erlangen-Nuremberg.

During the investigations inside the interior delta also ceramic fragments have been found which could indicate abandoned human settlements in the tropical rain forest margin (Figure 1).

6.5 ALLUVIAL SEDIMENTS OF THE NTEM INTERIOR DELTA

During fieldwork in 2005 and 2006, 91 up to 440 cm deep hand-corings were carried out in the Ntem interior delta between the villages Ma'an and Nyabessan (Figure 1). At most locations multi-layered alluvial sediments have been found in the floodplain, between a recent lower (dry season) and an upper (rainy season) terrace and in periodically inundated river sections.

On the most northern branch of the Ntem (Ntem 1), between Anguiridjang and Akom, suitable coring sites were rare and mainly sandy sediments were recovered overlying basement rock. Here the river bed is generally formed by basement and at many locations rapids have developed. At some places on the rapids gravels were found, mostly contracted to conglomerates by iron and manganese and coated by a black (iron-manganese) patina.

To get an overview of the stratigraphy of the alluvial sediments inside the Ntem's interior delta, three N–S-transects were set up through the interior delta following existing small pathways through the rain forest (Meyo Ntem, Nkongmeyos, Nyabessan).

Additionally, selective hand-corings were made across the river course (N- and S-bench at each particular location) at many sites inside the interior delta. On the basis of these samples, three sites with different time slots were selected for detailed stratigraphical analysis during fieldwork in 2006. In order to achieve a particularized, comprehensive insight and documentation of stratigraphy and sedimentation structures of some of the drilling sites examined in 2005 (Meyos, Nkongmeyos, Nyabessan), additional corings were carried out. Also, several exposures were dug along natural levees, flood banks and nickpoints within the bore-hole transects (Figure 6).

The first transect between Meyo Ntem (Ntem 1) and Aloum II (behind Ntem 3) contains 39 bore holes (at locations Aloum II (5 bore holes), Meyos (12) and Meyos II (7)) over a distance of about 8,5 km. The next one was made near Nkongmeyos (16 bore holes) with a length of 5 km and the third one at Nyabessan (26 bore holes) with a length of around 2 km. In contrast to the locations in the middle part of the interior delta (Akom, Nkongmeyos, Nnémeyong and Tom), where relatively thin layers of sandy sediments (sand >80–90%) over basement occur, corings at Abong, Anguiridjang, Aya'Amang, Meyos and Nyabessan reached the greatest depths with 3–4 m.

In many profiles, coarse-grained sandy sediments were found in the deepest layers, indicating palaeochannels (fluvial strata) and/or the transition zone towards the saprolite and basement. From depths of 2–3 m upward, in most profiles coarse-grained sediments are replaced by fine-grained silty, clayey and sandy sediments. In some of the profiles these changes in texture are very abrupt. Besides Anguiridjang, at all these locations more or less fine-grained sediments (silt + clay >60%) occurred in the upper layers, indicating channel-fill and floodplain overbank deposits.

The coarse-grained sandy sediments at the base of most profiles are sometimes associated with organic material, alternating in very thin layers. Occasionally these organic horizons are overlying the saprolitised basement or laterite. At the locations Meyos (between cores L16 and 17), Nkongmeyos (L44-46) and Nyabessan (L18, 19 and 22) this kind of fossil organic sediments occur in continuous horizontal layers. This has been proved during detailed investigations in 2006. The colours of the alluvial sediments are very similar at all investigated locations. The fossil organic horizons (20–80 cm thick), which occur in depths between 100 and 360 cm, are marked by brown to dark brown and black colours (2,5Y, 7.5YR or 10YR 5/1; 10YR 3/1), which change to lighter, greyer colours when exposed to the surface (oxidation). They mostly lie beneath the groundwater level and are thereby well preserved. At the base of this layer concentrated organic macro-rests occur and the often muddy sediments have putrefied character and mouldy smell, indicating swampy environment (gyttja) (Figure 7).

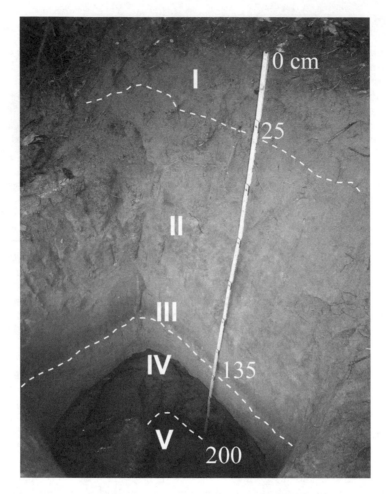

Figure 6. Exposure site at Meyo Ntem. The stratigraphic work was made near the cores L05 and L08 on the S-bench of Ntem 1 and evidenced the existence of a fossil organic horizon/palaeosurface, which was also found during hand-coring of L05 and L08 in 2005, several metres next to and behind this exposure site. V: fossil organic horizon, IV: layer showing reduction processes, III: thin, concentrated iron film, II: layer showing oxidation processes and I: humus top layer.

The fossil organic horizons are covered by a layer with characteristics of reduction processes, which is of clayey or loamy facies and occurs in grey colours (2,5Y 7/2 or 8/2; 10YR 8/1). It is covered by a clayey to loamy layer which is marked by oxidation processes. This layer is often more than 100 cm thick and shows beige to brown colour with orange (2,5Y 7/3, 7/4 or 8/2) to dark orange (10YR 7/8) oxidation spots. Both facies units are divided by a concentrated iron film, marking the groundwater level. Usually a thin (0–20 cm) brownish (10YR 5/2, 6/2 or 7/2) humus layer forms the top of the profile. Sediment organic matter (SOM), total carbon (C(tot.)) and nitrogen (N) contents reach maxima in the deepest facies units, i.e. in the fossil organic horizons and show again increasing regime towards the upper layers after reaching a minimum above these horizons. In fossil organic horizons they reach extreme maxima (L17 Meyos: SOM 17,07%, C(tot.) 9,9%, N 0,134% in 280–300 cm; L19 Nyabessan: SOM 16,9%, C(tot.) 9,8%, N 0,3% in 380–400 cm). These sandy to clayey

Figure 7. Concentrated fossil organic macro-rests occurred below 260 cm depth (dotted white line) at coring site C08. This material was widespread found in the hand-corings carried out at the site Meyos (see figure 13). Similar macro-rests, which were recovered from 280–300 cm at the coring site L17, were dated to a [14]C (AMS) age of 14.263 ± 126 years BP (not calibrated).

layers are also marked by dark colours: 2,5YR 3/1, 5YR 2/1, 7.5YR 3/2, 7.5YR 5/1 or 10YR 3/2, 4/1 and 5/1. Here, especially where (concentrated) fossil organic material occurs, pH is very low, between 3 and 4. The lowest pH values reach 2,28 (L17, Meyos), 2,70 (exposure Nyabessan) and 2,99 (L37, Aya'Amang). Contents of Fe and pH (always between 4 and 6) are low in the deepest facies units and increase towards the top of the profiles to reach one or more maxima and afterwards decrease again. C/N-ratio reaches high values in the fossil organic horizons, afterwards remaining low and stable towards the top. Figure 8 shows major characteristics of selected alluvial sediment profiles.

At the site Aya'Amang, hand-coring of L37 was undertaken on an upper terrace, some 50 m from the river bed on the E-bench.

From 300 cm onwards, a fossil organic horizon (min. 80 cm thickness) was found, where SOM, C(tot.), N and C/N rapidly increase in coarse to fine-grained sandy (~80%) texture, while Fe(tot.) and pH reach lowest values. Macro-rests from 320–340 cm were dated to a not calibrated [14]C (AMS) age of 30.675 ± 770 years BP. Cores L36 (3 m from river branch on E-bench) and L38 (5 m from river branch on W-bench) provided similar results, although fossil organic horizons already occurred in 140 cm (L36) and 200 cm (L38) with Middle Holocene ages of 4.341 ± 60 and 5.306 ± 64 years BP, respectively. The layers above the fossil horizons are characterized by sandy to clayey, fine-grained sediments, indicating floodplain overbank deposits. Silty and clayey facies decrease with growing depth together with SOM, C(tot.) and N contents, reaching minima in ~300 cm. Total iron content (Fe(tot.)) and pH show slightly increasing tendencies, with iron content maxima in 90 and 170 cm and pH maximum in 280 cm, whereas C/N-ratio remains stable.

At Nkongmeyos, across the Ntem branches 1 and 2, basement crops out and rapids are frequent. Nevertheless, on the N-bench of Ntem 3 a periodically (during rainy season) inundated upper terrace was sampled and stratigraphically studied. Another fossil organic horizon embedded in coarse- to fine-grained sandy sediments was identified in core L46 between 140 and 85 cm.

Organic macro-rests from this horizon provided [14]C (AMS) radiocarbon ages (not calibrated) of 8.291 ± 89 (in 80–100 cm) and 10.871 ± 99 years BP (120–140 cm). Sediment characteristics of L46 prove the existence of a palaeosurface. The Fe(tot.) maximum in ~70 cm marks the stratification shift and the high C/N-ratios, as well as SOM and C(tot.) in the deepest layers are evidencing indicators for a buried fossil organic horizon, which is again characterized by low pH. While SOM, C(tot.) and C/N show decreasing tendency towards the surface, N and pH values generally increase. Recent fine-grained overbank deposits are overlying these sandy sediments. During stratigraphical investigations in 2006,

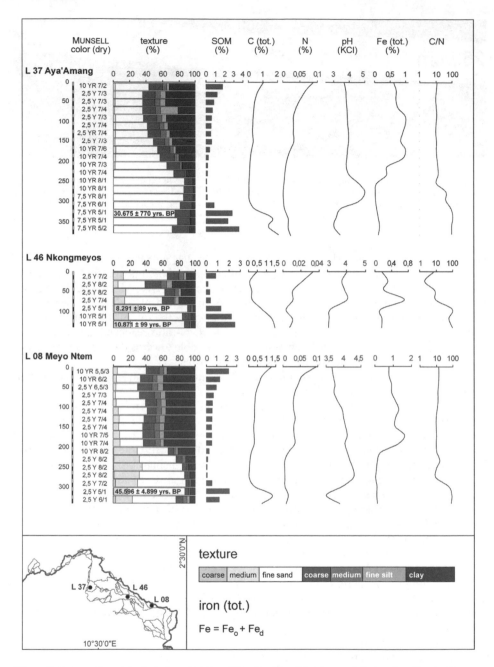

Figure 8. Some sedimentological characteristics of the alluvial sediments sampled at coring sites Aya'Amang (L37), Nkongmeyos (L46) and Meyo Ntem (L08) with appendant ^{14}C-data.

the existence of an extended fossil organic horizon (max. 50 cm thickness) below ~85 cm could be proved at this site in cores C29–32.

Rapids and outcropping basement partially covered with pebbles in the river bed prevail close behind the site L08, Meyo Ntem. This core was taken from an upper terrace,

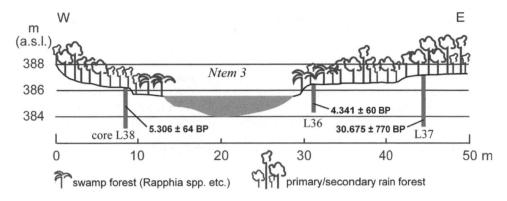

Figure 9. Sketch showing physio-geographical information (geomorphology, vegetation) and hand-corings carried out at the location Aya'Amang as well as appendant ¹⁴C-data.

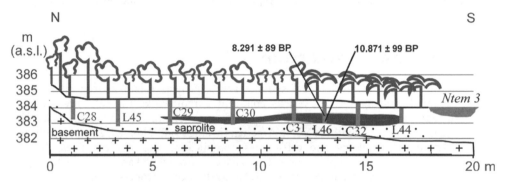

Figure 10. Sketch of location Nkongmeyos across the N-bench of Ntem 3, showing physio-geographical information (geomorphology, vegetation) and hand-corings as well as appendant ¹⁴C-data.

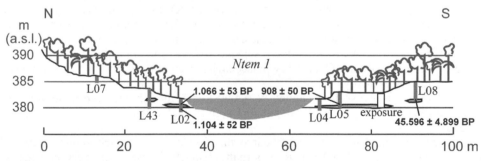

Figure 11. Sketch of the first part of transect Meyo Ntem-Aloum II on the N- and S-bench of Ntem 1, showing physio-geographical information (geomorphology, vegetation), the exposure site described in figure 6 and hand-corings as well as appendant ¹⁴C-data.

marking the river bench in the rainy season, on the S-bench of Ntem 1, some 20 m from the water level in the dry season. After fossil organic sediments were found on the N-bench (L02 and 43) in 120 cm depth, a thin fossil organic horizon could be also verified on the S-bench. In profile L05, located 3 m from the river bed, it was found in 220–280 cm depth, in the exposure site below 200 cm (Figure 6) and in profile L08 in 290–340 cm.

Sediment analysis data of profile L08 shows almost similar dynamics to data analyzed from L37, with maxima in SOM, C (tot.) and C/N from the base of the profile until around 300 cm depth (total maxima in ~320 cm). Texture shows coarse- to fine-grained sands and again very low pH occurs, whereas Fe and N reach lowest values. From 280 cm upwards SOM, C(tot.) and N slightly increase to reach further maxima at the top of the profile. In contrast, pH continually drops until the top after reaching its maximum in 270 cm and C/N-ratio stays stable from 290 cm upwards. The Fe-maximum in ~180 cm marks the stratification shift between the lower coarse-grained and upper sandy to clayey fine-grained sediment layers. Macro-rests from the 320–340 cm layer gave a (not calibrated) ^{14}C AMS radiocarbon age of 45.596 ± 4.899 years BP. In L05 a quite young age of 908 ± 50 years BP was found. In 2006, near this site an exposure was dug, in order to achieve a comprehensive insight into the stratification of the alluvial sediments (Figure 6). The different stratigraphical layers and sediment units could be clearly verified and separated. Like in profiles L05 and 08, the deepest sediment units (from the base to ~200 cm depth) are coarse- to fine-grained sandy and of fluvial character towards the base. They are covered by silty to clayey fine-grained sediments representing channel-fill and overbank floodplain deposits.

6.6 RESULTS AND INTERPRETATION

The interpretation of stratigraphical sedimentary sequences found in tropical rivers is complex and fluvial models as well as conceptual frameworks for palaeohydrological research in tropical areas are still incipient and incomplete, although in the last years a lot of progress is made, especially in South America (Miall, 1996; Thomas, 2004; Latrubesse, 2005). Additional problems and difficulties complicating the studies of the fluvial-morphological, palaeogeographical as well as palaeoenvironmental and palaeohydrological history of the study area (Ntem interior delta) are the lack of appropriate geomorphological and topographical maps as well as aerial images of the study area. Along comes the inaccessibility of the sparse inhabited region, the lack of transport in the floodplain and the dense vegetation cover (tropical rain and swamp forest). Let alone the general problems interpreting fluvial deposits and connected forms in the tropics (e.g. interbedding of alluvial and colluvial sediments, inclusion/interstratification of laterite crusts and stonelines, hiatus problem, space-time correlation). Nevertheless, the alluvial sediments found in the interior delta of the Ntem River are an eminent suitable archive with appropriate proxy data for the reconstruction of Late Quaternary palaeoenvironmental conditions in the study area, covering the last 50.000 years.

According to the so far obtained ^{14}C (AMS) data, the inherited information covers Late Pleistocene, more precise Middle to the Upper Pleniglacial, as well as Holocene times (see table 1).

Baring in mind that the insights into the fluvial structure and sedimentary and depositional pattern of the Ntem River's multi-channelled system are very punctuated, interpolations and interpretations must be careful and speculative. Because of the fact, that access for over-viewing sedimentary structures for facies stratification analysis is very limited in the study area and sampled sediments were almost always disturbed, sedimentary units were mostly difficult to identify. Following the preliminary results and findings of the first field work campaigns and other studies from western Equatorial Africa, especially the ECOFIT programme (Servant and Servant-Vildary, 2000), some tentative interpretations and conclusions can be given:

Regarding the details mentioned before, it can be stated that in Middle to Upper Pleniglacial times mainly coarse-grained sandy sediments have been deposited on a lower

Table 1. 32 ^{14}C AMS radiocarbon ages (not calibrated) and corresponding δ^{13}C-values obtained from organic macro-rests, organic sediment as well as charcoal, sampled from fossil organic horizons and palaeosurfaces inside the interior delta during field work. The ages range between 48.230 ± 6.411 and 217 ± 46 years BP. Analysis and data were provided by the AMS-Laboratory of Friedrich-Alexander-University in Erlangen-Nuremberg.

core	name	depth (cm)	Munsell col. (dry)	org. mat.	^{14}C (not cal.)	δ^{13}C
L2 Meyo Ntem	Erl-8249	100–120	4/2 10 YR	wood, etc.	1.066 ± 53	−27,0
	Erl-8250	160–180	3/2 10 YR	wood, etc.	1.104 ± 52	−29,2
L5 Meyo Ntem	Erl-8251	220–240	3/2 7,5 YR	leafs, etc.	908 ± 50	−29,4
L8 Meyos Ntem 3	Erl-8252	300–320	3/2 2,5 Y	wood	45.596 ± 4.899	−31,4
L17 Meyos Ntem 3	Erl-8253	280–300	2/1 10 YR	wood	14.263 ± 126	−27,3
L14 Abong	Erl-8254	340–360	4/2 7,5 YR + 6/8 10 YR	wood	48.230 ± 6.411	−29,6
L32 Nkongmeyos	Erl-8255	60–80	6/2 10 YR	charcoal	217 ± 46	−28,0
L34 Nkongmeyos	Erl-8256	100–120	3/2 10 YR	wood	435 ± 51	−27,4
L49 Nkongmeyos	Erl-8257	180–200	6/2 2,5 Y + 4/1 2,5 Y	org. sediment	10.775 ± 144	−30,1
L46 Nkongmeyos	Erl-8599	80–100	5/2 2,5 Y + 3/1 10 YR	wood, etc.	8.291 ± 89	−27,5
	Erl-8258	120–140	4/2 2,5 Y	wood, etc.	10.871 ± 99	−28,3
L40 Anguiridjang	Erl-8259	360–380	3/1 10 YR	wood, etc.	2.339 ± 52	−27,9
L30 Tom	Erl-8260	140–160	5/2 10 YR	wood, etc.	1.381 ± 49	−28,2
L27 Akom	Erl-8261	100–120	4/1 10 YR	wood, leafs	443 ± 57	−26,4
	Erl-8262	160–180	4/2 10 YR	leafs	427 ± 52	−27,6
L36 Aya Amung	Erl-8263	140–160	3/2 10 YR	leafs	4.341 ± 60	−27,2
L37 Aya Amung	Erl-8264	320–340	6/2 2,5 Y + 4/1 2,5 Y	wood	30.675 ± 770	−30,8
L38 Aya Amung	Erl-8265	220–240	5/2 2,5 Y	wood	5.306 ± 64	−31,4
L24 Nnémeyong	Erl-8266	80–100	3/2 7,5 YR	wood	441 ± 46	−30,6
	Erl-8267	120–140	4/2 7,5 YR	wood	671 ± 52	−28,7
L25 Nnémeyong	Erl-8268	160–180	4/1 10 YR + 5/2 10 YR	wood	21.908 ± 302	−27,0
	Erl-8269	220–240	3/1 10 YR + 10/2 10 YR	wood	22.398 ± 316	−29,1
L18 Nyabessan	Erl-8270	140–160	4/1 2,5 Y	leafs, etc.	587 ± 64	−29,1
	Erl-8271	200–220	4/1 2,5 Y	leafs, etc.	2.337 ± 55	−31,9
L19 Nyabessan	Erl-8272	320–340	3/1 7,5 YR + 3/2 7,5 YR	wood, etc.	2.189 ± 52	−29,5
L22 Nyabessan	Erl-8273	140–160	5/1 2,5 Y	wood, etc.	3.894 ± 57	−28,1
C13 Meyos	Erl-9567	120–140	5Y 7/1 + 5Y 2,5/1	wood, etc.	14.020 ± 106	−28,1
	Erl-9571	250–263	5Y 2,5/2	org. sediment	18.372 ± 164	−27,6
Aloum II	Erl-9574	78–88	10YR 7/1 + 10YR 6/8	org. sediment	8402 ± 67	−30,6
C11 Aloum II	Erl-9575	120–140	10YR 6/2 + 10YR 4/2	wood, etc.	6979 ± 58	−30,1
C20 Nyabessan	Erl-9576	430–440	n.a.	org. sediment	2479 ± 43	−28,8
C27 Nyabessan	Erl-9577	270–280	n.a.	org. sediment	3829 ± 46	−28,5

terrace of the Ntem River with braided sandy character. This unit is intermixed with high content of organic material and macro rests. This sedimentary unit was found at the locations Abong L14: 48.230 ± 6.411 BP (340–360 cm), Meyo Ntem L08: 45.596 ± 4.899 BP (320–340 cm) and Aya'Amang L37: 30.675 ± 770 BP (320–340 cm). These are up to now the oldest sediments found in the Ntem River's interior delta. In this phase the river was morphogenetically more active, moving coarse sandy sediment and aggrading the fluvial system. These findings are conform with earlier results from Central Africa of

Kadomura (1995), Schwartz and Lanfranchi (1991) and Thomas and Thorp (1995), who describe cool but moist semi-arid to sub-humid conditions with rivers showing braided character until around 40.000 years BP (70.000–40.000 Maluékien, after Schwartz and Lanfranchi, 1991) followed by a relative sub-humid period until around 30.000 years BP (Njilien) with enhanced erosion and meandering rivers. It is remarkable, that at the base of all hand-corings no gravels or pebbles could be found. Only near rapids at Meyo Ntem and Nsebito as well as below the waterfalls of Menvé'élé layers of ferruginized conglomerates (Ntem gravels and pebbles, coated with iron and manganese oxides) were observed, which prove an arid climate of origin. It is assumed that the formation of these ferruginized gravel layers took place in an earlier, initial phase of river system development and was probably accompanied by neotectonic events.

The next sedimentary unit is comprised of medium to fine-grained sandy sediments, which are generally lacking organic material. Some organic material (wood etc.) found in the profile Nnémeyong (L25) provided ^{14}C-dates of 21.908 ± 302 (in 160–180 cm) and 22.398 ± 316 years BP (220–240 cm). This unit may represent a cooler and sub-humid period occurring until the onset of the LGM aridity around 20.000–18.000 years BP (Léopoldvillien, after Schwartz and Lanfranchi, 1991). The LGM was associated with striking forest regression (Maley and Brenac, 1998), but preserved forested zones (rain forest refuge theory; Maley, 1998 and Lezine, 2005) especially across river systems, and probably also across the interior delta. This period was marked by aggradation with erratic and inhibited discharge, but river bed erosion and avulsion processes during broader floodpeaks occurring with rainstorms. As ^{14}C-data is very scarse, this assumption is very tentative. As climate became more humid after the LGM with enhanced runoff and probably catastrophic stripping of sediments, hiata might widespread occur in the sedimentary record. Nevertheless, one profile representing this time period, more precise the terminating LGM, was found at Meyos.

Here a sandy to clayey palaeosurface is overlying coarse sandy sediments and saprolite. The palaeosurface has a swampy character and probably constitutes an abandoned palaeochannel which was transformed into a back-swamp during the LGM, when rain fall and river runoff was considerably reduced. The oldest organic material from location Meyos yielded a ^{14}C (AMS) radiocarbon age of 18.372 ± 164 years BP at the base of the swampy palaeosurface (Figure 13). The overlying muddy and clayey organic sediments indicate periodical inundation of the site and accumulation of overbank deposits.

After this, more humid conditions for the period between ~14.000 and 8.000 years BP. are assumed, with a possible interruption around ~11.000 (Younger Dryas). This might indicate the prematurely onset of the 'Humid African Period' (Barker et al., 2004; DeMenocal, 2000; Lézine and Cazet, 2005) in SW Cameroon. More likely, the swampy sediments at Meyos might be buried by deposits representing the Pleistocene-Holocene transition (~13.000–10.000 BP), when climate became unstable and high floods as well as frequent and increased storms occurred, which were also documented in the deposits of the Niger (Pastouret et al., 1978) and Congo Rivers (Giresse et al., 1982; Marret et al., 1998) and the record of several lakes (mentioned in part 2), where lake levels rapidly increased. During this period, hill-wash and colluviation might have been severe. Simultaneously, the older deposits of the alluvial plain were vertically and laterally excavated, ancient deposits largely eroded and new channels created by river shifting in a more anastomosing river pattern.

With the onset of the Holocene Pluvial around ~10.000 BP, these processes might have intensified, but with the following reforestation of the study area (since ~9.500 BP; Maley and Brenac, 1998) also terraces and river banks were stabilised. The transition of the Ntem River's pattern from braided to anabranching or anastomosing might more likely fall into this period. The stabilisation of the landscape and climate (towards recent seasonal conditions) might have promoted a less migrating river pattern carrying an abundant rate

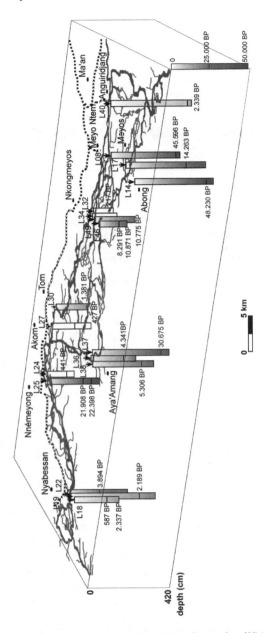

Figure 12. General overview of major interesting hand-corings and appendant ^{14}C AMS radiocarbon ages (not calibrated) carried out in the alluvia of the Ntem interior delta near Ma'an in 2005.

of fine-grained clayey, silty and sandy suspended load, which formed an upper floodplain in the Holocene. This sedimentary unit is widespread found in the interior delta. The fine-grained overbank deposits in the floodplain have evolved from lateral accretion, avulsion and crevasse processes during regular seasonal flooding in the rainy seasons. In this way, since ca. 10.000–8.000 BP many previously existing channels have become infilled with sediment and formerly active channels have been buried.

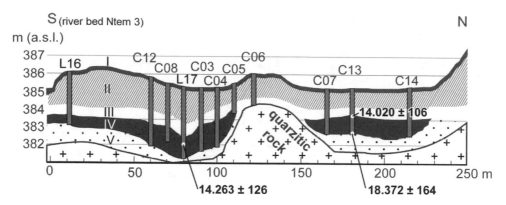

Figure 13. Results from stratigraphical field work in 2006 at the site Meyos. An extended palaeosurface (IV, min. age 14.020 ± 106 BP (120–140 cm, C13)) is overlying alternating thin layers of coarse sandy and organic sediments and basement (quarzitic rock) with a saprolitic transition layer (V). It is covered by layers with characters of reduction (III) and oxidation (II) processes and a thin humus top layer (I).

At locations where Early to Middle Holocene ^{14}C (AMS) dates were found in the alluvia (Aloum II: 8.402 ± 67 (in 78–88 cm) and 6.979 ± 58 BP (120–140 cm); Aya'Amang L36: 4.341 ± 60 BP (140–160 cm) and L38: 5.306 ± 64 BP (220–240 cm); Nkongmeyos L46: 10.871 ± 99 (120–140 cm) and 8.291 ± 89 BP (80–100 cm)), generally stable climatic and fluvial conditions in forested environment combined with accumulation of fine-grained clayey to sandy deposits are assumed. The organic material found at these locations was intermixed with thin fine-grained sandy layers. This period was marked by aggradation with multiple shifts from vertical to lateral accretion and initial floodplain construction in response to lower peak discharges.

Across the unified single Ntem channel near Nyabessan, a periodically inundated small-scale floodplain in swampy environment was discovered. These fine-grained clayey to loamy organic sediments of dark colour and up to 440 cm thickness indicate stillwater or even occasionally lacustrine conditions. They are most suitable for investigations and statements on the time interval quested by the interdisciplinary DFG-Research Unit 510. ^{14}C (AMS) radiocarbon ages yielded a maximum Late Holocene age of 3.894 ± 57 years BP (L22, 140–160 cm) and two ages falling into the period denominated as the 'First Millennium BC Crisis' (L19: 2.189 ± 52 (320–340 cm) and L18: 2.337 ± 55 years BP (200–220 cm)). First archaeobotanical results indicate that after 2.820 ± 70 years BP again drier conditions prevailed connected with a loss of primary and swampy species in the vegetation composition (A. Schweizer, personal communication). Textures of recovered alluvia (L18 and 19) from the same site show an increase of fine-grained sandy sediments (30–70%) for this time period. This time period might mark another drier climate impact, which is conform to results from Maley for the SW Cameroon region (Maley, 2002).

Although texture profiles might provide information on the evolution of processes at each particular coring site, their interpretation must remain tentative. In a further step eventual similarities among different sites might provide evidence for underlying causes leading to modified fluvial processes. Abrupt texture composition changes in the profiles can for example manifest transitions to modified fluvial behaviour, which can be induced by natural evolution, climatic changes, tectonic influences and manmade modifications. As the alluvia allow only inferences for several locations instead of the entire interior delta, the evolution of the whole multi-channelled river pattern inside the interior delta is still widely unclear and unknown.

Several stillwater locations were probably generated by remobilization of basement structures and successional formation of a step-fault inside the delta. Here multi-layered fine-grained alluvial sediments containing embedded palaeosurfaces have been identified as proxy data archives for palaeoenvironmental conditions in the Ntem catchment. Across the river bed of Ntem 1 (Anguiridjang, Nkongmeyos, Tom and Akom), which is widely fixed by basement structures, younger coarse- to fine-grained sandy alluvial sediments (up to 90% in the upper layers) indicate turbulent fluvial conditions. Processes like cut-and-fill cycles, channel incision, large-scale cross-bedding and slope erosion might have removed older sediments and replaced them by younger sandy sediments at these locations. The young age of these coarse sediments might also indicate the amplified influence of human activities inside the catchment area on the fluvial dynamics. The fact that across Ntem 1 the youngest and across Ntem 3 the oldest sediments occur, leads to another assumption and may be interpreted as a consequence of tectonic activity which forced the Ntem River to displace its main river bed northward successively.

Stable carbon isotopic values ($\delta^{13}C$) allow allocation of the dated organic remains/macro-rests to C3 (rain forest trees) or C4 (savanna grasses) dominated vegetation (species) units (cp. Runge, 2002). In the interior delta they range between –26,4 and –31,9‰ and all indicate that at the particular drilling sites rain forest (C3) has prevailed during the corresponding time period (rain forest refuge theory). This implies that across the sampled sites conditions were adequate for rain forest ecosystems to persist, even during periods before, during and after the Last Glacial Maximum. Palynologic research, which is carried out by Archaeobotanists from Frankfurt University (K. Neumann) will provide detailed vegetation reconstructions (pollen diagrams) from three sites inside the interior delta (Aya'Amang, Meyos and Nyabessan), where cores (percussion probe) were recovered in 2005. It is assumed that across the interior delta temporarily lacustrine (Nyabessan) and palustrine (Meyos, Nkongmeyos) conditions prevailed, which favoured the preservation of small rain forest refuge zones also during arid periods (LGM, Younger Dryas?, First Millennium BC Crisis).

The SW of Cameroon has been explicitly prospected by the sub-project ReSaKo of the DFG Research Unit 510 regarding the existence of alluvia suitable for palaeoenvironmental research. Starting from the assumption, that the rain forest-savanna margin has shifted southward during the last 4.000 years, special attention was drawn to the catchments recently covered by tropical rain forest. After primary the forming of the interior delta was supposed to be attributed to the Holocene, the fossil organic horizons however subsequently manifested a much longer time-scale, spanning from 48.230 ± 6.411 to 217 ± 46 years BP. The alluvial sediments inside the interior delta display a very high spatial and temporal resolution, which deserve a very local, small-scale approach concerning the interpretation of radiocarbon data and depositional as well as sedimentary patterns. The rain forest-savanna border probably never reached as far south as the Ntem River, because no evidence for savannization was found inside the interior delta's alluvia. The new findings will be correlated and compared with the results of ECOFIT (Servant and Servant-Vildary, 2000), especially concerning the assumed aridification around 3.000 years BP. They will also supplement the results derived from lacustrine and marine data archives.

Further-going research must lead to the interpretation of suitable information from the alluvial sediments, especially the origin of sandy sediments (aeolian/fluvial), in spite of all the complications and uncertainties coupled with research on tropical fluvial sediments (Thomas, 2004; Latrubesse, 2005). The influences of neotectonics and human interactions (Bantu question) on the evolution of the river system is still uncertain for the most part.

ACKNOWLEDGEMENTS

This research was realized by grants of the German Research Foundation (Deutsche Forschungsgemeinschaft, DFG) in Bonn, within the DFG Research Unit 510: 'Ecological and Cultural Change in West and Central Africa' (sub-project RU 555/14-3; Prof. J. Runge). I gratefully thank the Cameroonian colleagues from the 'Département de Géographie' at the 'Université de Yaoundé I' (Prof. Dr. M. Tsalefac, Dr. M. Tchindjang) for assisting and supporting fieldworks in 2004, 2005, and 2006, and Dr. B. Kankeu from the 'MINRESI (Ministère de la Recherche Scientifique et l'Innovation)'. Sincere thanks goes to the people from German Embassy, GTZ, SNV and WWF for their friendly support with formalities, information and logistics. I thank Mr. A. Scharf from Friedrich-Alexander-University in Erlangen-Nuremberg for providing the ^{14}C-data and Mr. J.-C. Ntonga ('Centre de Recherches Hydrologiques, IRGM') for hydrological data. Finally I thank all the friendly people from the region Ma'an for their unique support during field work and last but not least Prof. Dr. E. M. Latrubesse for several critical remarks and conceptual discussions on this text.

REFERENCES

Abrantes, F., 2003, A 340,000 year continental climate record from tropical Africa - news from opal phytoliths from the Equatorial Atlantic. *Earth and Planetary Science Letters*, **209 (1–2)**, pp. 165–179.

Baker, V.R., 2000, South American paleohydrology: future prospects and global perspective. *Quaternary International*, **72**, pp. 3–5.

Barker, P.A., Talbot, M.R., Sreet-Perrott, F.A., Marret, F., Scourse, J. and Odada, E.O., 2004, Late Quaternary climate variability in Intertropical Africa. In *Past Climate Variability through Europe and Africa*, edited by Battarbee, R.W., Gasse, F. and Stickley, C.E., (Dordrecht: Kluwer Academic Publishers), pp. 117–138.

deMenocal, P., Ortiz, J., Guilderson, T., Adkins, J., Sarntheim, M., Baker, L. and Yarusinsky, M., 2000, Abrupt onset and termination of the African Humid Period: rapid climate responses to gradual insolation forcing. *Quaternary Science Reviews*, **19**, pp. 347–361.

Elenga, H., Schwartz, D. and Vincens, A., 1993, Pollen evidence of late Quaternary vegetation and inferred climate changes in Congo. *Palaeogeography, Palaeoclimatology, Palaeoecology*, **109**, pp. 345–356.

Elenga, H., Schwartz, D., Vincens, A., Bertaux, J., de Namur, C., Martin, L., Wirrmann, D. and Servant, M. (1996): Diagramme pollinique holocène du lac Kitina (Congo): mise en évidence de changements paléobotaniques et paléoclimatiques dans le massif forestier du Mayombe. *Comptes Rendus de l'Académie des Sciences*, Séries IIa, **323**, pp. 403–410.

Gasse, F., 2000, Hydrological changes in the African tropics since the Last Glacial Maximum. *Quaternary Science Reviews,* **19**, pp. 189–211.

Gasse, F., 2005, Continental palaeohydrology and palaeoclimate during the Holocene. *Comptes Rendus Geoscience*, **337**, pp. 79–86.

Giresse, P., Bongo-Passi, G., Delibrias, G. and Du Plessy, J.C., 1982, La lithostratigraphie des sediments hemiplagiques du delta profond du fleuve Congo et ses indications sur les paléoclimats de la fin du Quaternaire. *Bulletin Société Géologie française, Série*, 24, pp. 803–815.

Giresse, P., Maley, J., and Brenac, P., 1994, Late Quaternary palaeoenvironments in the Lake Barombi Mbo (West Cameroon) deduced from pollen and carbon isotopes of organic matter. *Palaeogeography, Palaeoclimatology, Palaeoecology*, **107**, pp. 65–78.

Giresse, P., Maley, J., and Kossoni, A., 2005, Sedimentary environmental changes and millennial climatic variability in a tropical shallow lake (Lake Ossa, Cameroon) during the Holocene. *Palaeogeography, Palaeoclimatology, Palaeoecology*, **218**, pp. 257–285.

Hall, A.M., Thomas, M.F. and Thorp, M.B., 1985, Late Quaternary alluvial placer development in the humid tropics: the case of the Birim Diamond Placer, Ghana. *Journal geological Society London*, **142**, pp. 777–787.

Holbrook, J. and Schumm, S.A., 1999, Geomorphic and sedimentary response of rivers to tectonic deformation: a brief review and critique of a tool for recognizing subtle epeirogenic deformation in modern and ancient settings. *Tectophysics*, **305**, pp. 287–306.

Holvoeth, J., Kolonic, S. and Wagner, T., 2005, Soil organic matter as an important contributor to late Quaternary sediments of the tropical West African continental margin. *Geochimica et Cosmochimica Acta*, **69**, pp. 2031–2041.

Kadomura, H. and Nori, N., 1990, Environmental implications of slope deposits in humid tropical Africa: evidence from southern Cameroon and western Kenya. *Geographical Reports, Tokyo Metropolitan University*, **25**, pp. 213–236.

Kadomura, H., 1995, Palaeoecological and palaeohydrological changes in the humid tropics during the last 20.000 years, with reference to Equatorial Africa. In *Global Continental Palaeohydrology*, New York, edited by Gregory, K.J., Starkel, L., Baker, V.R. (New York: Wiley), pp. 177–202.

Latrubesse, E.M., 2003, The Late-Quaternary Palaeohydrology of Large South American Fluvial Systems. In *Palaeohydrology: Understanding Global Change*, New York, edited by Gregory, K.J. and Benito, G., pp. 193–212.

Latrubesse, E.M., Steveaux, J.C. and Sinha, R., 2005, Tropical Rivers. *Geomorphology*, **70**, pp. 187–206.

Letouzey, R., 1985, *Notice de la carte phytogéographique du Cameroun au 1/500 000*. Institut de la Carte Internationale de la Végétation, Toulouse and Institut Recherches Agronomique, Yaoundé.

Lézine, A.-M. and Cazet, J.P., 2005, High-resolution pollen record from core KW31, Gulf of Guinea, documents the history of the lowland forests of West Equatorial Africa since 40,000 yr ago. *Quaternary Research*, **64**, pp. 432–443.

Maley, J. (1981): *Etudes palynologiques dans le bassin du Tchad et paléoclimatologie de l'Afrique nord-tropicale de 30.000 ans à l'epoque actuelle*. ORSTOM, Trav. Doc., 1–129.

Maley, J., 1997, Middle to Late Holocene Changes in Tropical Africa and other Continents: Paleomonsoon and Sea Surface Temperature Variations. In *Third Millennium BC Climate Change and Old World Collapse*, Berlin, edited by Dalfes, H.N., Kukla, G. and Weiss, H., (Berlin: Springer Verlag), pp. 611–639.

Maley, J. and Brenac, P., 1998, Vegetation dynamics, Palaeoenvironments and Climatic changes in the Forests of West Cameroon during the last 28,000 years BP. *Review of Palaeobotany and Palynology*, **99**, pp. 157–187.

Maley, J., 2002, A Catastrophic Destruction of African Forests about 2,500 Years Ago Still Exerts a Major Influence on Present Vegetation Formations. *IDS Bulletin*, **33/I**, pp. 13–30.

Marchant, R. and Hooghiemstra, H., 2004, Rapid environmental change in African and South American tropics around 4000 years before present: a review. *Earth-Science Reviews*, **66**, pp. 217–260.

Marret, F., Scourse, J.D., Versteegh, G., Jansen, F.J.H. and Schneider, R., 1998, Integrated marine and terrestrial evidence for abrupt Congo River palaeodischarge fluctuations during the last deglaciation. *Quaternary Research*, **50**, pp. 34–45.

Marret, F., Scourse, J., Fred Jansen, J.H. and Schneider, R., 1999, Changements climatiques et paléocéanographiques en Afrique centrale atlantique au cours de la dernière déglaciation: contribution palynologique. *Earth and Planetary Sciences*, **329**, pp. 721–726.

Marret, F., Maley, J. and Scourse, J., 2006, Climatic instability in West Equatorial Africa during the Mid- and Late Holocene. *Quaternary International*, **150**, pp. 71–81.

Mehra O. P. and Jackson, M. L., 1960, Iron oxide removal from soils and clays by dithionitecitrate system buffered with sodium carbonate. *Clays and Clay Minerals*, **7**, pp. 317–327.

Meiwes, K.-J., König, N., Khanna, P. K., Prenzel, J. and Ulrich, B., 1984, Chemische Untersuchungsverfahren für Mineralböden, Auflagehumus und Wurzeln zur Charakterisierung und Bewertung der Versauerung in Waldböden. *Berichte des Forschungszentrums Waldökosysteme der Universität Göttingen*, **7**, pp. 1–67.

Nguetsop, V.F., Servant-Vildary, S. and Servant, M., 2004, Late Holocene climatic changes in West Africa, a high resolution diatom record from Equatorial Cameroon. *Quaternary Science Reviews*, **23**, pp. 591–609.

Olivry, J.C., 1986, *Fleuves et Rivières du Cameroun.* (Paris: Editions de'l ORSTOM).

Pastouret, L., Chamley, H., Delibrias, G., Duplessy, J.C. and Theide, J., 1978, Late Quaternary climatic changes in Western Tropical Africa deduced from deep-sea sedimentation off the Niger Delta. *Oceanologica Acta*, **1**, pp. 217–232.

Phillipson, D.W., 1980, L'expansion bantoue en Afrique orientale et méridionale: les témoignages de l'archéologie et de la linguistique. In *L'expansion* bantoue, Paris, edited by Bouquiaux, L., (Paris: SELAF), pp. 649–684.

Preuss, J., 1986, Jungpleistozäne Klimaänderungen im Kongo-Zaire-Becken. *Geowissenschaften in unserer Zeit*, Volume 4, **6**, pp. 177–187.

Reynaud-Farrera, I., Maley, J. and Wirrmann, D., 1996, Végétation et climat dans les forêts du Sud-Ouest Cameroun depuis 4770 ans B.P.: analyse pollinique des sédiments du Lac Ossa. *Comptes Rendus de l'Académie des Sciences*, Séries IIa, **322**, pp. 749–755.

Runge, J., 1992, Geomorphological observations concerning palaeoenvironmental conditions in eastern Zaire. *Zeitschrift für Geomorphologie N. F.*, Suppl.Bd. **91**, pp. 109–122.

Runge, J., 1996, Palaeoenvironmental interpretation of geomorphological and pedological studies in the rain forest 'core areas' of eastern Zaire (Central Africa). *South African Geographical Journal*, **78**, pp. 91–97.

Runge, J., 2001, *Landschaftsgenese und Paläoklima in Zentralafrika. Physiogeographische Untersuchungen zur Landschaftsentwicklung und klimagesteuerten quartären Vegetations- und Geomorphodynamik in Kongo/Zaire (Kivu, Kasai, Oberkongo) und der Zentralafrikanischen Republik (Mbomou)*, (Berlin, Stuttgart: Gebrüder Bornträger).

Runge, J., 2002, Holocene landscape history and palaeohydrology evidenced by stable carbon isotope ($\delta^{13}C$) analysis of alluvial sediments in the Mbari valley (5°N/23°E), Central African Republic. *Catena*, **48**, pp. 67–87.

Runge, J., Eisenberg, J. and Sangen, M., 2005, Ökologischer Wandel und kulturelle Umbrüche in West- und Zentralafrika–Prospektionsreise nach Südwestkamerun vom 05.03.–03.04.2004 im Rahmen der DFG-Forschergruppe 510: Teilprojekt, Regenwald-Savannen-Kontakt (ReSaKo). *Geoökodynamik*, **26**, pp. 135–154.

Schumm, S.A., 1977, *The Fluvial System*, (Chichester: Wiley).

Segalen, P., 1967, Les Sols et la géomorphologie du Cameroun. *Cahier ORSTOM, série Pédologique*, **2**, pp. 137–188.

Servant, M., 1983, *Séquences continentals et variations climatiques: evolution du bassin du Tchad au Cénozoique supérieur*, (Paris: ORSTOM).

Servant, M. and Servant-Vildary, S., 2000, *Dynamique à long terme des écosystèmes forestiers intertropicaux. Publications issues du Symposium international « Dynamique à long terme des écosystèmes forestiers intertropicaux », Paris, 20–22 mars 1996*, (Paris: UNESCO).

Stager, J.C. and Anfang-Sutter, R., 1999, Preliminary evidence of environmental changes at Lake Bambili (Cameroon, West-Africa) since 24.000 BP. *Journal of Paleolimnology*, **22**, pp. 319–330.

Schwartz, D., 1992, Assèchement climatique vers 3.000 B.P. et expansion Bantu en Afrique centrale atlantique: quelques réflexions. *Bulletin de la Société Géologique de France*, **3**, pp. 353–361.

Schwartz, D. and Lanfranchi, R., 1991, Les paysages de l'Afrique centrale pendant le Quaternaire. In *Aux origines de l'Afrique centrale*, Libreville, edited by Lanfranchi, R. and Clist, B. (Paris: Sépia), pp. 41–45.

Talbot, M.R., Livingstone, D.A., Palmer, P.G., Maley, J., Maleck, J.M., Delibrias, G. and Gulliksen, S., 1984, Preliminary results from sediment cores from Lake Bosumtwi, Ghana. *Palaeoecology of Africa*, **16**, pp. 173–192.

Tchouto Mbatchou, G. P., 2004, *Plant diversity in a tropical rain forest. Implications for biodiversity and conservation in Cameroon*. (Wageningen: University of Wageningen).

Thomas, M.F. and Thorp, M.B., 1980, Some aspects of geomorphological interpretation of Quaternary alluvial sediments in Sierra Leone. *Zeitschrift für Geomorphologie N. F.*, Suppl.Bd. **36**, pp. 140–161.

Thomas, M.F. and Thorp, M.B., 1995, Geomorphic response to rapid climatic and hydrologic change during the late Pleistocene and early Holocene in the humid and sub-humid tropics. *Quaternary Science Reviews*, **14**, pp. 193–207.

Thomas, M.F., 2000, Late Quaternary environmental changes and the alluvial record in humid tropical environments. *Quaternary International*, **72**, pp. 23–36.

Thomas, M.F., 2004, Landscape sensitivity to rapid environmental change—a Quaternary perspective with examples from tropical areas. *Catena*, **55**, pp. 107–124.

Vandenberghe, J., 2003, Climate forcing of fluvial system development: an evolution of ideas. *Quaternary Science Reviews*, **22**, pp. 2053–2060.

Viers, J., Dupre, B., Braun, J.-J., Deberdt, S., Angeletti, B., Ndam Ngoupayou, J. and Michard. A., 2000, Major and trace element abundances and strontium isotopes in the Nyong basin rivers Cameroon: constraints on chemical weathering processes and elements transport mechanisms in humid tropical environments. *Chemical Geology*, **169**, pp. 211–241.

Vincens, A., Schwartz, D., Bertaux, J., Elenga, H. and de Namur, C., 1998, Late Holocene climatic changes in western Equatorial Africa inferred from pollen lake Sinnda, Southern Congo. *Quaternary Research*, **50**, pp. 34–45.

Vincens, A., Schwartz, D., Elenga, H., Reynaud-Farrera, I., Alexandre, A., Bertaux, J., Mariotti, A., Martin, L., Meunier, J.-D., Nguetsop, F., Servant, M., Servant-Vildary, S. and Wirrmann, D., 1999, Forest response to climate changes in Atlantic Equatorial Africa during the last 4000 years BP and inheritance on the modern landscapes. *Journal of Biogeography*, **26**, pp. 879–885.

Wirrmann, D. and Bertaux, J., 2001, Late Holocene palaeoclimatic changes in Western Central Africa inferred from mineral abundance in dated sediments from Lake Ossa (Southwest Cameroon). *Quaternary Research*, **56**, pp. 275–287.

Zabel, M., Schneider, R., Wagner, T., Adegbie, A. T., De Vries, U. and Kolonic, S., 2001, Late Quaternary Climate Changes in Central Africa as inferred from Terrigenous Input in the Niger Fan. *Quaternary Research*, 56, pp. 207–217.

Zogning, A., Giresse, P., Maley, J. and Gadel, F., 1997, The Late Holocene palaeoenvironment in the Lake Njupi area, west Cameroon: implications regarding the history of Lake Nyos. *Journal of African Earth Science*, **24/3**, pp. 285–300.

CHAPTER 7

The evolution of the Holocene palaeoenvironment of the Adamawa region of Cameroon: evidence from sediments from two crater lakes near Ngaoundéré

Simon Ngos III
Department of Earth Sciences, University of Yaoundé I, Cameroon

Frank Sirocko and Rouwen Lehné
Institute for Geosciences, Gutenberg University, Mainz, Germany

Pierre Giresse
Laboratoire d'Etudes des Géo-Environnements Marins EA3678, Université de Perpignan, France

Michel Servant
Institut de recherche pour le Développement, Bondy Cedex, France

ABSTRACT: The palaeoenvironmental evolution of the Adamawa region in central Cameroon during the last 6.400 years is analysed from the study of two dated drill cores from two crater lakes: Lake Mbalang and Lake Tisong situated around Ngaoundéré town. These analyses concern the granulometry and mineralogy of sediments, the magnetic susceptibility, and the carbon content. The bulk sediment is a very fine organic clayey mud with generally more than 95% of its components below 50 μm diameter. Radiocarbon datations show that during the period considered, the sediment accumulation rate varies from one lake to the other. In Lake Mbalang, the sedimentation rate is nearly constant and close to the mean value of 0,93 mm/year while Lake Tisong shows a clear change in the rate of sediment supply which passes from a mean value of 3,4 mm/year between 3.780 and 2.606 cal. years BP down to 0,73 mm/year from 2.606 cal. years BP to the present. The major change observed in Lake Tisong is thought to have been caused by the geomorphological evolution around the lake due to explosive volcanism that has reduced the catchment area and consequently the sediment supply. Low magnetic susceptibility coupled with high organic carbon (OC) content are considered to be linked to the development and the erosion of a humic horizon at the surface of the catchment soils and is an indication of a wet period. On the other hand, high magnetic susceptibility values originate from the erosion of iron-rich sediments from denuded catchment soils on which the humic level is absent. This is confirmed by a low OC content and indicates dry periods. A synthesis of the results from these different analyses in Lake Mbalang revealed six successive sedimentation periods. Lake Tisong which covers a period less than 4.000 cal. years BP confirms the four recent phases. These phases are interpreted in terms of palaeoclimatic successions as follows:

The time from 6.400 to 5.000 cal. years BP appears as the wettest period, marked by the lowest magnetic susceptibility values and the highest OC content on M4. Fine laminations observed at the basal part of this core testify great depth of the lake at this time. From 5.000 to 4.000 cal. years BP follows a dry period that is not commonly recorded in equatorial latitudes. 4.000 to 2.800 cal. years BP marks the return of wetter conditions. 2.800 to 1.300–1.000 cal. years BP appears as the driest phase with the highest magnetic susceptibility values in both lakes and the lowest OC content in Lake Mbalang. The period from 1.300–1.000 cal. years BP to the present is considered slightly wetter than the precedent period, but with a deficit in rain fall.

This palaeoclimatic reconstitution shows two main differences with Lake Barombi-Mbo at the equatorial latitudes; the 5.000–4.000 dry phase present in the Adamawa is not recorded in the equatorial zone, and the 2.800 dry phase that ended around 2.000 cal. years BP in Lake Barombi-Mbo continues right to 1.300–1.000 cal. years BP in the Adamawa. In comparison to the northern region, high and low levels of the Lake Chad seem to correspond to wet and dry periods recorded in the Adamawa lakes. The only difference concerns the 2.800 dry phase that ended at 1.300–1.000 cal. years BP in the Adamawa whereas it continues to the present in Lake Chad. The general trend of the palaeoclimatic evolution from south to north during the second half of Holocene is marked by the northward appearance of a dry phase at 5.000–4.000 cal. years BP and a more and more northward extension of the 2.800 cal. years BP dry phase.

Considering geomorphological aspects of the area around the two lakes, Lake Mbalang looks more stable as attested by the absence of volcaniclastic layers in M4, as well as the constancy of the sediment supply rate in the lake. On the contrary, several volcaniclastic levels are present in the lake Tisong core, suggesting an explosive volcanic activity around the lake during this time. The sedimentation rate drops by a factor of more than four after 2.606 cal. years BP. These elements are not observed in Lake Mbalang some 20 km away from Lake Tisong implying that the geomorphological modification is localised just around this lake. The last volcanic episode recorded in the lake dated around 1.400 cal. years BP raises the problem of volcanic hazards around Ngaoundéré town and its more than 100.000 inhabitants.

7.1 INTRODUCTION

The continental sector of the Cameroon Volcanic Line (Mt. Cameroon to Adamawa plateau) is host to numerous lakes. Geographically, these lakes can be divided into three latitudinal groups: those located close to the coastal plain, those located on the western highlands and those on the Adamawa plateau in Central Cameroon.

This corresponds to a clear evolution of the vegetation and climate from the equatorial forest in the south to the Sudano-Guinean savanna in the Adamawa (Giresse et al., 1994a). One of the first limnological studies carried out on Cameroonian lakes was done by Kling in 1987.

After the two gas disasters of Lake Monoun (14. August 1984) and Lake Nyos (21. August 1986) that killed 37 and 1.746 people respectively (Sigvaldason, 1989; Tazieff, 1989), more and more interest became focused on crater lakes and marshes of Cameroon. The aim of these studies was on the one hand, to evaluate the different risks associated to each of these lakes and secondly to establish the palaeoenvironment of the Quaternary period. Till today amongst the lakes of the Cameroon Volcanic Line, only the southernmost group, have been sites of detailed studies. On the basis of palynological, sedimentological and geochemical analyses, Giresse et al. (1991), Giresse et al. (1994b), Maley and Brenac (1998) have been able to reconstruct the palaeoenvironment of the region from 24.000 BP. On the bases of palynology, diatom and sedimentological studies in Lake Ossa, Reynaud Farrera et al. (1996), Nguetsop (1998) and Wirrmann et al. (2001), Giresse et al. (2005) reconstructed the palaeoenvironment of this lake from 6.000 years BP to the present. Despite all these studies, much work still has to be done to understand the latitudinal movement of the forest and climate towards the north, especially in the Adamawa region were just a few works can be cited (Giresse *et al.*, 1994; Ngos III *et al.*, 2003). The Adamawa region has remained unexplored and makes correlation with results obtained from Lake Chad to those obtained from Lakes Barombi Mbo and Ossa difficult. The Ecosystems of Tropical Forest (ECOFIT) a joint programme by ORSTOM (IRD), CNRS and CIRAD (France) launched in the nineties, has the objective to analyse the forest–savanna movement within the last 10.000 years (this program was unfortunately stopped some years ago but we are still studying samples belonging to it). Lake sediments are one of the best preservers of the whole Holocene history. In this respect, we obtained two drilled cores from Lake Assom on the southern flanks of the Adamawa plateau, some 40 km north of the forest–savanna interphase. The first results obtained (Ngos III *et al.*, 2003) suggest very strongly the absence of the forest here since 4.500 years BP. Nevertheless, the lakes in the Adamawa plateau are the ideal sites for the study of Quaternary palaeoenvironment of the region.

The present work conducted within the ECOFIT program aims to reconstitute the palaeoenvironmental successions of the Adamawa plateau during the period covered by the drill cores: palaeoclimates, vegetation, geomorphology and sedimentation. The results obtained will be used to correlate with those of the southern lakes and the northern Lake Chad. Our work is based on the analysis of two six metres long drill cores from Lake Mbalang (core M4) and Lake Tisong (core T2) located around Ngaoundéré town.

7.2 LOCATION OF THE REGION AND GENERAL PRESENTATION OF THE TWO LAKES

The area of study is located between Latitude 6° and 8°N, and Longitude 11°30 and 15°45'E (Figure 1). Relatively, this area is composed of a high topographic unit (850–1.200 m) that separates the Benue plain (500–300 m) to the north from the Centre and South Cameroonian plateau (800–500 m) that declines gradually to the coastal plain.

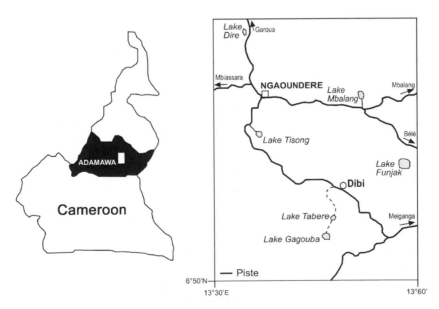

Figure 1. Map of the Adamawa region in Cameroon with location of the lakes around Ngaoundéré town.

Mbalang and Tisong are two crater lakes situated respectively 15 km eastward and 3 km southward from Ngaoundéré. The main characteristics of the two lakes studied are presented on table 1.

Lake Mbalang (7°19'N, 13°44'E) sits at 1.130 m asl. The area is about 50 ha with a maximum depth of 52 m. The general bathymetry describes an asymmetric bowl (Figure 2) with very steep slopes. The thermocline indicates a 5 °C decrease along a 20 m deep water column. According to this water stratification, column water stability appears "moderately stable" and the metalimnion is well developed (Kling, 1987). On the basis of radio-elements studies, it is assumed that the present catchment is moderate and that bioturbation processes are probable (Pourchet, 1988).

Lake Tisong (7°15'N, 13°35'E) is at an altitude of 1.160 m asl. Its area is only 8 ha and its maximum water depth is 48 m. A marked 6 °C thermocline is recorded through the upper 20 m, implying stratification and thermic stability of the waters (Figure 2). This lake is regarded as "moderately stable" and comprises an important hypolimnion (Kling, 1987). The bathymetry describes a symmetric bowl with marked steep slopes.

The radio-elements budget indicates high 135 Cs content and a high retention of the catchment. To the scale of the upper tenths centimetres, one notes the secular recurrence of successive small flood events (Pourchet, 1988).

7.3 CLIMATE AND VEGETATION

The Adamawa plateau is located in the altitudinal tropical climate which is a transitional between the Equatorial climate in the south and the Sahelian climate to the north.

There are two seasons: a rainy season (April to October) and a dry season (November to March) characterised by the harmattan winds from the Sahara desert that blow from the NE to the SW. Mean temperatures vary between 23 °C and 26 °C while mean annual precipitation is 1.500 mm (Figure 3).

Table 1. Presentation and main characteristics of Lake Mbalang and Lake Tisong.

Characteristics	Lake Mbalang	Lake Tisong
Location	7°19' N; 13°44' E	7°15' N; 13°35' E
Altitude	1.130 m	1.160 m
Maximum depth	52 m	48 m
Total area	50 ha	8 ha
Catchments	90 ha,	15 ha,
	100% savanna	100% savanna
Inlet	none	none
Exutory	present	absent
Specific erosion rate on the catchments	131,6 t. km²/year	—
Mean sedimentation rate	0,93 mm/year	1,58 mm/year
Soils on the catchments	Ferrallitic	Ferrallitic
Clay mineral present	7,24 Å; 13–14 Å;	7,17 to 7,19 Å;
	± gibbsite; ± siderite	allophane

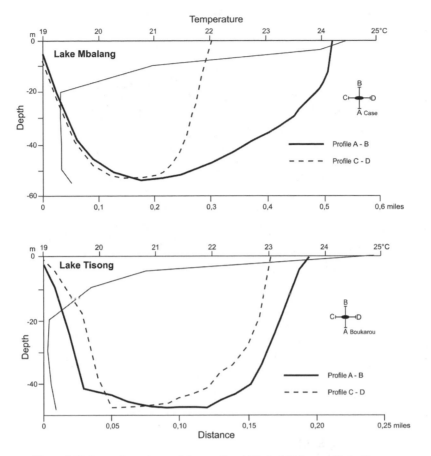

Figure 2. Bathymetric sections and thermoclines of Lake Mbalang and Lake Tisong
(after Pourchet *et al.*, 1988).

The vegetation map of Cameroon (Letouzey, 1968 and 1985) shows that the Adamawa plateau is covered by a Sudanese type woody savanna characterised by two main trees: *Daniella oliveri* (Caesapiniaceae) and *Lophira lanceolata* (Orcnacae). Northwards, the height and density of the vegetation drop gradually giving way to herbaceous savanna. The same vegetation is predominant in the southern part of the Benue plain. Siefferman (1966) was the first researcher to have brought out the palaeoclimate and palaeobotanical evolution of the region. He showed that between 10.000 and 4.500 BP there was a relatively humid climate of Sudanese-Guinean type, which evolved to a drier Sudanese type during the second half of the Holocene. In the southernmost part of the region, palynological and sedimentological studies in Lake Barombi Mbo (Maley and Brenac, 1998; Schwartz, 1992; Maley, 1992) have shown clear alternation of climate characterised by phases of forest extension or retreat during the last 25.000 years.

— 25.000 to 20.000 years BP forest incursion with the growth of a mountainous tree species called *Olea capensis*;
— 20.000 to 14.000 years BP opening of the forest leaving only islands of forest associated with the savanna;

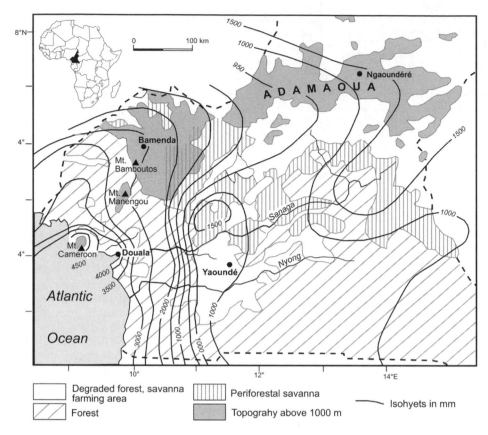

Figure 3. Vegetal cover and pluviometry; a = degraded forest, savanna, farming area; b = forest, c = periforestal savanna; d = topography above 1000 m asl; e = isohyets in mm. (Giresse *et al.*, 1994).

— 14.000 to 9.500 years BP another phase of forest extension;
— 9.500 to 3.000 years BP maximum forest density;
— 3.000 to 2.000 years BP sudden climatic change characterised by dryness with another phase of forest opening;
— 2.000 years BP up to now, new forest extension;

7.4 GEOLOGY

Adamawa is a mylonitic shear zone (N13 to N70) made of a conjugate system of faults going on to the Central African Republic and up to Eastern Africa on more than 2.000 km (Dorbath *et al.*, 1986). This zone underwent a dextral reactivation during the Pan-African orogeny. The structural evolution is generally raised by Cretaceous collapsing gaps as a consequence of the Pan-African reactivation, and in direct relation with the opening of the South Atlantic. Ring complexes are localised at the fault intersection zone. The three volcanic series (Gèze, 1943) generally described along the Cameroon volcanic line are present in the Adamawa plateau. The lower black series (upper Cretaceous to upper Eocene), which is composed of basalts and andesites, are conserved as yellowish clays. It is important to note that in the centre of the plateau, the immobilisation of aluminium during the Palaeogene period has led to very important bauxite accumulations in Minim Martap, Ngaoundal and Ngaoundourou. (Eno Belinga, 1972). The white medium series (end of Neogene) correspond to trachytes and phonolites lavas. The black upper series, mainly basaltic, are attributed to the quaternary volcanism.

Adamawa is covered with ferrallitic soils very rich in aluminium and iron oxides enveloping a more or less thick ferruginous crust. Halloysite is frequently observed in place of kaolinite. Even with 1.500 mm annual rainfall, the silica ratio today is not lowered enough to lead to the formation of gibbsite. Allophanes often evolve to metahalloysite (Giresse *et al.*, 1994).

7.5 MATERIALS, SAMPLING AND METHODS

The cores were drilled in the crater lakes using a compressed air corer built in Australia. This corer conceived to be used in lakes deeper than 10 m was able to drill 6 m long soft sediments in a plastic tube of 5,5 cm diameter. The only inconvenience was the loose of the 5–10 cm upper sediments of the core due to the release of the compressed air. The coring campaign was conducted in three lakes of the Ngaoundéré region. Several 5,5 to 6 m long cores were taken in lakes Mbalang, Tisong and Tabere (Ngaoundaba). This paper is based on the study of two cores drilled, one from the centre of Lake Mbalang (core M4), and the other from the centre of Lake Tisong (T2).

The 6 m long cores M4 (Mbalang) and T2 (Tisong) were opened at the Sedimentology Laboratory of the Institute of Geosciences of the University of Mainz (Germany). Before opening, the base of each core was sampled for radiocarbon dating at the University of Lyon I (France). Other AMS dating were done at the University of Kiel (Germany). After opening the two cores, a series of analyses were done either on the entire core or on samples from them. These are the Magnetic Susceptibility, the spectrography of near infrared bands (that are non exploited in this paper), carbon content analysis, thin sections for mineralogy, and preparation for pollen studies (results are still awaited). It is in Perpignan, France, that most of the thin sections were realised and studied; samples were washed every 20 cm on a 50 μm sieve to separate the fine (<50 μm) fraction from the coarser one (>50 μm). Coarser grains were then observed under binocular lenses and the whole sample or just the fine fraction analysed by XRD.

Selected samples were impregnated with resin and the microstructures studied in thin section.

7.5.1 Magnetic susceptibility

In order to describe the cores M4 and T2 concerning their composition, magnetic susceptibility has been specified by using the "MS2C Core Logging Sensor" from Bartington Instruments (see Dearing, 1994).

The measurement has been carried out in the SI-mode (Système International d'Unités) with a resolution of 2 mm. Results has been smoothed by a 100 point running mean (see figures 7 and 8).

7.5.2 Carbon content analysis

The total carbon content was analysed by combustion with a standard of 24,8% for 0,5 g of standard sample. 100 to 150 g of powder sample was weighed as the mean carbon content of the sample was estimated around 15% according to the results from other lakes of

Cameroon. Recalibration was done after every 40 measurements. To estimate the inorganic carbon content, and then deduce the total organic carbon, M4 samples were treated with dilute chlorhydric acid 0.1N and weighted. The estimated inorganic carbon content (mainly from siderite $FeCO_3$) was calculated on the basis of one tenth of the total lost weight, as Carbon represents 10% of the siderite molecule weight. M4 was chosen because its long period covered.

7.6 RESULTS

7.6.1 Description of the cores

The 6 m long core M4 (Lake Mbalang) (Figure 4) is made of dark grey mud showing some rare clear lamina at the base of the core, and two sand layers at 5,4 and 5,7 m deep. The whole material looks very clayey with abundant organic matter content.

Thin-section examinations indicate a brown organic matrix with few clayey components. Biogenic particles are composed of ubiquitous Spongiae spicules, some diatom cellular residues and numerous plant debris which are frequently ordered along several interlayered microlaminae. In some places, distribution of microcrystals is more or less aligned along the bedding. Siderite is ubiquitous, but only at small amounts. In a few places, nebula-like masses of very small (<5 μm) siderites prisms were observed. In some intervals, distribution of microcrystals is more or less aligned along the bedding. The siliciclastic phase is only a minor component of the sediment (except near 535, 550, and 570 cm). The small quartz grains are unworn and unsorted. They are scattered and rarely associated with reworked feldspars (mostly plagioclase) and some green minerals (mainly augite).

The T2 core (Lake Tisong) (Figure 3) is also a dark grey homogeneous fine mud having sandier lamina. Between 1,4 and 1,6 m deep a micro conglomerate layer from the reworking of coarse quartz grains and volcanic elements is observed. Other sand layers are observed at 4,45, 4,85 and 5,85 m deep. A piece of wood was found at 3,90 m deep.

The study of thin-sections allows us to note the widespread accumulation of Chlorophytae alga filaments. Clayey matter occurs as very small lensoid accumulations included in the algal microstructure. Fine grained black debris are scattered within the accumulation or are ordered along several interlayered microlaminae. Quartz grains are very scarce except in various pyroclastic accumulations where they are commonly associated with microlaths of plagioclase or more rarely grains of augite. This deposit displays disseminated yellow prisms of siderite on the order of 5 μm in diameter. A frequent habit is a very thin, less than one millimetre thick layer. However, at 583 cm deep, the thickness of the layer rises up to 3 mm.

7.6.2 Radiocarbon datations

The ages along the two cores appear on figure 3. From them, the calculated mean sedimentation rate on M4 is 0,93 mm/year and 1,58 mm/year on T2. The difference in the sedimentation rate can be explained on one hand by the ratio of catchments surface to total lake area that is lower in Mbalang than in Tisong, and on the other hand by the dip value of the area surrounding the lake greater in Tisong compared to Mbalang. The two parameters favour greater erosion on Lake Tisong catchments than in Lake Mbalang. A more detailed look shows a non similar evolution on the two lakes for the considered period.

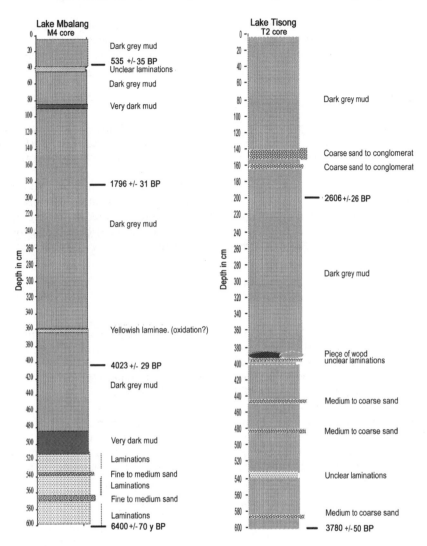

Figure 4. General presentation of M4 (Lake Mbalang) and T2 (Lake Tisong) cores.
The basal part of M4 is laminated and hosts some thin sand layers.
On T2, three volcanic rich layers are observed at 582–585, 160–162 and 140–150 cm depth.

On M4 the sedimentation rate does not vary enough around the mean value of 0,3 mm/year. The calculated value in mm/year are: 0,82 between 6.400 and 4.023 cal. years BP, 100 between 4.023 and 1.796 cal. years BP, 114 between 1.796 and 535 cal. years BP, and 0,71 after 535 cal. years BP. One can note a notable decrease of the sedimentation rate close to present.

On T2, It is observed a very pronounced lowering of the sedimentation rate by a factor of more than four after 2.606 cal. years BP. Between 3.780 and 2.606 cal. years BP the sedimentation rate value is 3,4 mm/year and just 0,76 mm/year from 2.606 to present.

The difference in the evolution of the sedimentation rate as observed in Lake Tisong is certainly due to some local geomorphological factors inherent to the immediate surrounding zone of this lake. It is possible that in Lake Tisong the volcanic deposits from neighbouring explosion craters as observed around 2.000 cal. years BP. have reduced the lake catchments, therefore orienting an important part of water flows and sediments outside the lake.

7.6.3 Granulometry

The vertical log obtained from the separation of the coarse fraction (>50 μm) from the fine one (< 50 μm) gives an idea of the granulometric evolution along the two cores (Figure 5). The sediment is mainly silty to clayey with generally more than 98% of the components below 50 μm.

Lake Mbalang presents two main sand layers at 80–100 cm and 560–580 cm deep. The basal part of the core (500–600 cm) presents a regular lamination partially masked due to the silty composition of the sediment. In Lake Tisong, the three main conglomeratic or coarse sand layers (140–152, 160–164 and 582–586 cm deep) correspond to clastic deposition from volcanic explosions. Two other centimetric sand layers are observed around 445 and 480 cm deep.

7.6.4 XRD analyses

They were conducted with the aim to better know the mineralogical composition of the whole sediment, and identify clay minerals and the volcanic elements. In both lakes, the uncristallised fraction close to opal is very important, due to the abundance of diatoms sponge spicules and phytolithes. The main mineralogical components of M4 sediments are: quartz, plagioclase feldspars, kaolinite and sometimes gibbsite. A semi-quantitative calculation of the ratio of quartz and plagioclase versus kaolinite and gibbsite gives a vertical evolution along M4 core (Figure 6). The alone peak appears around 1.300 ca. years BP. Quartz and plagioclase appear as the main mineralogical components of the coarse fraction. In T2, volcanic minerals like pyearoxenes and olivine (rare in M4) are concentrated along with quartz and plagioclase feldspars in the coarse fraction of some vocaniclastic layers.

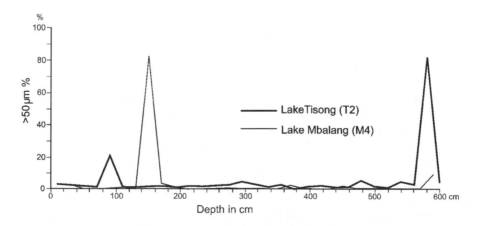

Figure 5. Granulometric logs of M4 and T2 from washing under a 50 μm sieve.

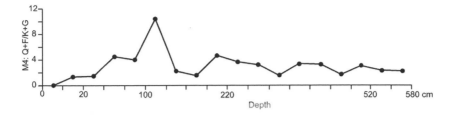

Figure 6. Plot of the ratio of quartz and plagioclase versus kaolinite and gibbsite (Q+F/K+G).

7.6.5 Magnetic Susceptibility

In regards to the composition of M4 and T2 sediments, ferromagnetic elements and ferromagnetic oxides are absent as well as hematite. The presence of goethite in these sediments is favoured by the oxidation of the uppermost sediments if the water column drops seriously (this is not the case for these deep crater lakes) or if the erosion affects denuded catchments ferrallitic soils without any humic horizon. A great amount of siderite can also raise the magnetic susceptibility values during periods of low detritic supply that favor chemical precipitations (Maher, 1986) (Figure 7).

The results obtained on the two lakes (Figure 5 and 6) are quite similar although Lake Mbalang shows higher values than Tisong in accordance with lower siderite quantities observed in Lake Tisong by XRD compared to Lake Mbalang.

In Lake Mbalang where no volcanic layer is recorded, the curve shows some evolutions and even alternations with low and higher values leading to distinguish six different sedimentation periods between 6.400 years BP and the present on M4 with the following values of magnetic susceptibility: from 6.400 to 5.000 cal. years BP 20–50; from 5.000 to 3.800 cal. years BP 50–85; from 3.800 to 2.800 cal. years BP 20–60; from 2.800 to 1.800 cal. years BP 40–110; from 1.800 to 1.000 cal. years BP 50–110; from 1.000 cal. years BP to present 50–90.

In Lake Tisong, three volcanic-rich layers have been identified at 1,40–1,52, 1,60–1,64 and 5,82–5,86 m deep; even if the presence of ferromagnesian minerals (olivine, augite, etc.) in these levels raises the magnetic susceptibility, it does not fundamentally modify the general trend as follows: from 3.780 to 2.700 cal. years BP 5–40; from 2.700–1.900 cal. years BP 20–60; from 1.900–1.300 cal. years BP 60–120; from 1.300 cal. years BP to present 10–50.

This general trend is comparable to what is observed in Lake Mbalang for the period between 3.780 cal. years BP and the present. If abstraction is made of the influence of volcanic layers, the magnetic susceptibility values are just lower in Tisong than in Mbalang.

7.6.6 Organic and inorganic carbon estimation

The evolution of M4 OC content with depth is presented on figure 8. These results show an average high total organic carbon content. The values lie between 5 and 18% on M4. It is observed a general decrease from the bottom to the top of the core. Compared to the general trend observed on M4, the OC values are low between 5.000 and 4.000 years BP and between 2.800 and 1.000 cal. years BP. The highest values are observed between 6.400 and 5.000 cal. years BP and between 4.000 and 2.800 cal. years BP. From 1.000 cal. years BP to the present the values rise again but smoothly. These contents are comparable to those obtained in some other crater lakes like Gagouba, Awing and Wum (Ngos III, 1992).

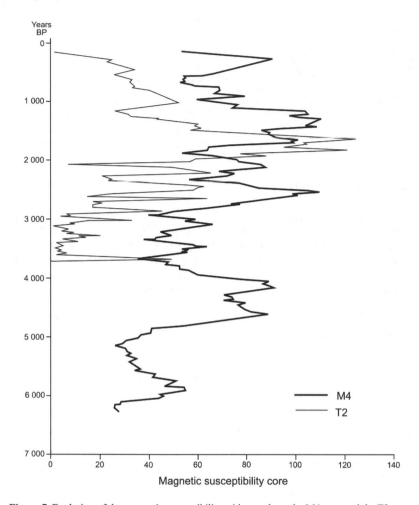

Figure 7. Evolution of the magnetic susceptibility with age along the M4 core and the T2 core.

The inorganic carbon fraction in M4 is less than 1% testifying low carbonate content in the sediments of this lake.

7.6.7 Palaeoclimatic interpretation

In his works in 1987, Kling has classified Lake Mbalang and Lake Tisong as "moderately stable" compared to "unstable" Lake Nyos and Lake Monoun, and "very stable" Lake Barombi Mbo. This means that palaeoenvironmental reconstitution can be made in these first two lakes that have not suffered important overturning events since a relatively long time. Lake Mbalang core M4 covering the longest period of sedimentation (6.400 cal years BP to the present) is considered here as the reference, while T2 is studied for comparison and confirmation.

In the two lakes, the OC content is negatively linked to the magnetic susceptibility (Table 2). The low values of OC are linked to the absence of humic cover on the surface

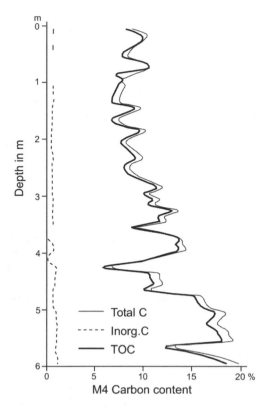

Figure 8. Plot of organic and inorganic carbon content along M4 core. The organic content shows a general decreasing trend from bottom to top, with marked deficit between 5.000 and 4.000, and between 2.800 and 1.000 cal. years BP.

of ferrallitic soils present on the catchments, favouring erosion of iron-rich sediments with high magnetic susceptibility (Dearing *et al.*, 1996). During these periods, siderite may form in great quantities as well as vivianite, which is always present in these lake sediments. High magnetic susceptibility values added to low OC content are therefore considered as a witness of drier periods, while low magnetic susceptibility values coupled with high OC content correspond to wetter periods.

M4 core shows fine laminations at its basal part, marked by algal microstructures from 6.400 to around 5.000 cal. years BP during this period, OC content reaches its highest values, coupled with the lowest magnetic susceptibility values. This period can be therefore considered as the wettest with a permanent great depth of the lake favouring laminations to take place inside the sediments.

On the contrary, highest magnetic susceptibility values and lowest OC contents are observed between 2.800 and 1.300 to 1.000 cal. years BP in M4. T2 shows the highest magnetic susceptibility values between 2.800 and 1.300 cal. years BP in accordance with M4 making this period the driest between 6.400 cal. years BP and the present. In Lake Assom just a few tens of km southward, this dry phase is recorded between 2.800 and 1.700–1.300 cal. years BP (Ngos III *et al.*, 2003), while in Lake Bosumtwi, Ghana, this phase started around 3.200 cal. years BP (Russel *et al.*, 2003). The general climatic evolution can be reconstituted as follows: from 6.400 to 5.000 cal. years BP, wettest climate inside the period

Table 2. Palaeoeclimatic reconstitution from Lake Mbalang and Lake Tisong since 6.400 cal years BP. Comparison of organic carbon content, M4 and magnetic susceptibility evolution along M4 and T2.

cal. years BP	M4 Magnetic susceptibility value	T2 Magnetic susceptibility value	OC content	Estimated climate	Volcanic-rich layers in Lake Tisong
6.400–5.000	20–50	—	20–15	Very wet	No
5.000–4.000	50–90	—	5–10	Dry	Yes
4.000–2.800	20–60	5–40	8–14	Wet	Important
2.800–1.300	40–110	20–60	7–9	Very dry	?
1.300–present	50–90	10–50	8–12	Less dry	?

of study, with catchments soils covered by a humic layer. From 5.000 to 4.000 cal. years BP, drier climate. From 4.000 to 2.800 cal. years BP, wet climate. From 2.800 to 1.300–1.000 cal. years BP, driest climate with denudation of soils on catchments. From 1.300–1.000 cal. years BP to present, wetter climate compared to the precedent interval.

The evolution described above here may be due to the solar oscillation coupled with blowing winds which regulate the climatic oscillations in this region.

7.6.8 Geomorphological evolution

Both Lake Mbalang and Lake Tisong have no river inlet recently. Sediment supply therefore is provided from smooth erosion on surrounding catchments. In these conditions, great variations in the sedimentation rate may not be observed at a millennium scale. This is the case of Lake Mbalang where the sedimentation rate is very constant in accordance with a stable environment around the lake. There is no volcaniclastic individualized layer in M4, witnessing on no explosive volcanism around the lake for the considered period. The presence of an exutory at the southern border of the lake limits the water depth variation below the present level. In Lake Tisong, is observed an important drop in the sedimentation rate after 2.606 cal. years BP from 3,4 to 0,73 mm/year. This cannot be explained by regional climatic variation because not recorded in Lake Mbalang. The most probable explanation comes from the presence of three volcanic-rich layers at 3.700, 1.800 and 1.600 cal. years BP which are clear indication of an explosive volcanic activity around the lake. The shift in the sedimentation rate can be explained by the modification of the catchments environment due to volcanic deposits that have reduced the area that supplies the lake in sediments. The existence of explosive volcanic activity at this area less than two millennia to the present constitutes a permanent hazard to neighbouring Ngaoundéré town with its more than 100.000 inhabitants. It must be possible from the surroundings to characterize the origin and the nature of this volcanism.

7.7 COMPARISON WITH SOUTHERN LAKE BAROMBI-MBO AND NORTHERN LAKE CHAD

One of the aims of this paper is to draw out a palaeoclimatic evolution the equatorial to the Sudanese latitude, taking profit of the transitional position of the Adamawa region that hosts the present lakes.

A comparison of the palaeoclimatic evolution reconstituted in Lake Barombi-Mbo by Giresse *et al.* (1994b) and Maley and Brenac (1998) amongst others shows some important evolution during the second half of Holocene. The period between 5.000 and 4.000 cal. years BP presents a clear tendency to dryness as compared to the end of the first half of Holocene. In Lake Barombi-Mbo, this period is considered as wet with the development of rain forest. The dry period recorded in Central Africa between 3.000–2.500 to 2.000 cal. years BP appears in the Adamawa to having last from 2.800 to close to 1.000 cal. years BP. The present humid period in the south is a slightly less dry episode compared to the preceding phase, with a persistence of a clear deficit in rain fall.

On the other hand, data on the Chad zone and Lake Chad are rare (Maley, 1972; Maley, 1981; Servant, 1973; Servant-Vildary, 1978; Durand et Mathieu, 1980). Although these authors have not accorded their views on the palaeoclimatic evolution of this zone during Holocene, some concordant points can be noted. The highest lake levels are recorded between 7.200 and 4.800, and between 3.700 and 2.800 years BP. On the other hand, the lowest levels are recorded between 4.800 and 3.700, and after 2.800 years BP. Compared to our results, the evolution of the Chadian zone presents similarities to the record of dry and wet periods during the second half of Holocene with just little differences in the timing. The main difference is observed after 2.800 cal. years BP where Chad experiences a hard dry period till now while in the Adamawa region wetness has risen up since around 1.000 cal. years BP.

In any case the evolution from south to north is marked by two main factors: the extension northward of the dry phases, and the presence and/or the extension from the Adamawa to Chad of some dry periods (5.000–4.000 cal. years BP) that are not recorded in the southern region. Current works and analyses on diatoms, pollens and the interpretation of recorded reflectance spectrographs of near infrared bands will add key knowledge for the comprehension of the palaeoclimatic variation in the Adamawa during Holocene.

ACKNOWLEDGEMENTS

This work has been launched within the late ECOFIT program; thanks to the former team manager for financial support. Special thanks to the Sedimentological Lab in Mainz for having financed the AMS datations and for the collaboration to finalise this paper. Thank also to Giresse's collaborators for all the aid to realise some analyses. Finally, thanks to the Cameroonian Ministry of Higher Education for financial support.

REFERENCES

Dearing, J.A., 1994, *Environmental Magnetic Susceptibility: using the Barington MS2 System* (Kenilworth: Chi Publishing), p. 104.

Dorbath, C., Dorbath, L., Fairhead, J.D. and Stuart, G.W., 1986, A teleseismic delay time study across the central african shear zone in the Adamawa region of cameroon, West Africa. *Geophysics Journal Astronomical Society*, **86**, pp. 751–766.

Durand, A. and Mathieu, P., 1980, Le Quaternaire sur la rive sud du Tchad (République du Tchad), *Cahiers ORSTOM, Série Géologie*, **11**, 1.

Eno Belinga, S.M., 1972, *L'altération des roches basaltiques et le processus de bauxitisation dans l'Adamaoua (Cameroun)*. These de Doctorat, Universite de Paris VI, p. 571.

Gèze, B., 1943, Géographie et géologie du Cameroun occidental. *Mémoires Muséum National d'Histoire Naturelle*, **XVII**, p. 320.

Giresse, P., Maley, J. and Brenac, P., 1994b, Late quaternary palaeoenvironments in the Lake Barombi-Mbo (West Cameroon), deduced from pollen and carbon isotopes of organic matter. *Palaeogeography, Palaeoclimatology, Palaeoecology*, **107**, pp. 65–78.

Giresse, P., Maley, J. and Kelts, K., 1991, Sedimentation and palaeoenvironment in Crater Lake Barombi-Mbo, Cameroon, during the last 25000 years. *Sedimentary Geology*, **71**, pp. 151–175.

Giresse, P., Maley, J. and Kossoni, A. 2005, Sedimentary environmental changes and millennial climatic variability in a tropical shallow lake (Lake Ossa, Cameroon) during the Holocene. *Palaeogeography, Palaeoclimatology, Palaeoecology*, **218**, pp. 257–285.

Giresse, P., Ngos, S. and Pourchet, P., 1994, Procéssus séculaires et géochronologie au 210Pb des principaux lacs de la dorsale camerounaise. *Bulletin de la Société Géologique de France*, **165**, 4, pp. 363–380.

Kling, G.W., 1987, *Comparative limnology of lakes in Cameroon, West Africa*. PhD Thesis, Duke University, p. 482.

Letouzey, R., 1968, Etude phytogéographique du Cameroun. *Encyclopédie Biologique*, **49**, p. 508.

Letouzey, R., 1985, *Notice de la carte phytogéographique du Cameroun au 1/500.000*. Institut de la carte internationale de la végétation, Toulouse, France.

Maher, B.A., 1986, Characterisation of soils by mineral magnetic measurements. *Physics of the Earth and Planetary Interiors*, **42**, pp. 76–92.

Maley, J., 1981, Etudes palynologiques dans le bassin du Tchad et paléoclimatologie de l'Afrique nord tropicale de 30.000 ans à l'époque actuelle. *Travaux et Documents de l'ORSTOM*, pp. 129–586.

Maley, J., 1992, Mise en évidence d'une péjoration climatique entre ca 2500 et 2000 ans BP. En Afrique tropicale humide. *Bulletin de la Société Géologique de France*, **163**, pp. 363–365.

Maley, J., 1997, Middle to late Holocene changes in tropical Africa and other continents: Palaeomonsoon and sea surface temperature changes, in: (Eds.). *Third Millennium B.C. Climate Change and Old Word Collapse,* edited by Dalfes, N., Kukla, G. and Weiss, H. Proceedings of the NATO Advanced Research Workshop on Third Millennium BC Abrupt Climate Change I (Berlin: Springer Verlag), pp. 611–641.

Maley, J. and Brenac, P., 1998, Vegetation dynamics, palaeoenvironments and climatic changes in the forests of West Cameroon during the last 28.000 years. *Revue Palaeobotany & Palynology*, **99**, pp. 157–188.

Ngos III, S., 1991, Paléoenvironnements lacustres du Cameroun au Quaternaire récent. *Rapport ORSTOM*, Université de Yaoundé, p. 19.

Ngos III, S., Giresse, P. and Maley, J., 2003, Palaeoenvironments of lake Assom near Tibati (south Adamawa, Cameroon). What happened in Tibati around 1.700 years BP? *Journal of African Earth Sciences*, **37**, pp. 35–45.

Nguetsop, V.F., Servant-Vildary, S. and Servant, M., 2004, Late Holocene climatic changes in West Africa, a high resolution diatom record from equatorial Cameroon. *Quaternary Science Reviews*, **23**, pp. 591–609.

Pourchet, M., Pinglot, J.P. and Maley, J., 1988, Résultats des mesures radiochimiques sur les sédiments de quelques lacs camrounais. *Collection ORSTOM*, p. 12.

Reynaud-Farrera, I., Maley, J. and Wirrmann, D., 1996, Végétation et climat dans les forêts du Sud-ouest Cameroun depuis 4.770 BP: analyse pollinique des sédiments du lac Ossa. *Comptes Rendus de l'Académie des Sciences de Paris*, Série 2a, **322**, pp. 749–755.

Russell, J., Talbot, MR. and Haskell, BJ., 2003, Mid-Holocene climate change in Lake Bosumtwi, Ghana. *Quaternary Research*, **60**, pp. 133–141.

Schwartz, D., 1992, Assèchement climatique vers 3.000 BP. et expansion Bantu en Afrique centrale atlantique: quelques refléxions. *Bulletin de la Société Géologique de France*, **163**, pp. 353–361.

Servant, M., 1973, *Séquences continentales et variations climatiques : évolution du bassin du Tchad au cénozoïque supérieur*. Thèse Doctorat es Sciences, Université de Paris VI, p. 348.

Servant-Vildary, S., 1978, Etude des diatomées et paléoclimatologie du bassin tchadien au Cénozoïque supérieur. *Travaux et Documents de l'ORSTOM*, **84**, (2 tomes).

Sieffermann, G., 1966, Les sols de quelques régions volcaniques du Cameroun. *Mémoires ORSTOM*, **66**, p. 183.

Sigvaldason, G.E., 1989, International conference on Lake Nyos disaster. Conclusions and recommendations. *Journal of Volcanology and Geothermal Research*, **39**, pp. 97–108.

Tazieff, H., 1989, Mechanisms of the Nyos carbon dioxide disaster and of so-called phreatic ateam eruption. *Journal of Volcanology and Geothermal Research*, **39**, pp. 109–116.

Youta Happi, J., 1998, *Arbres contre graminées: la lente invasion de la savane par la forêt au centre Cameroun*. Thèse de Doctorat, Université Paris-Sorbonne.

Wirrmann, D., Bertaux, J. and Kossoni, A., 2001, Late Holocene paleoclimatic changes in Western central Africa inferred from mineral abundance in dated sediments from Lake Ossa (Southwest Cameroon). *Quaternary Research*, **56**, pp. 275–287.

Youta Happi, J., 1998, *Arbres contre graminées: la lente invasion de la savane par la forêt au centre-Cameroun*. Thesis, Université de Paris- Sorbonne.

CHAPTER 8

Palaeoenvironmental studies in the Ngotto Forest: alluvial sediments as indicators of recent and Holocene landscape evolution in the Central African Republic

Marion Neumer and Eva Becker

Institute of Physical Geography, Johann Wolfgang Goethe University, Frankfurt am Main, Germany

Jürgen Runge

Centre for Interdisciplinary Research on Africa (CIRA/ZIAF), Johann Wolfgang Goethe University, Frankfurt am Main, Germany

ABSTRACT: Valley sediments in the expanded floodplains of tropical river systems represent valuable archives for the palaeoenvironmental research. The investigation of sediment layers from numerous boring- and exposure-profiles on the sandstone plateau of Gadzi-Carnot results in a mid to young Holocene climate change at the landscape ecological sensitive vegetation transition between rain forest and savanna. The analyses of the alluvia of the Mbaéré river cut deeply into the plateau and the sediments of the alluvial fan of the Sadika, a tributary of the Mbaéré, accomplished with conventional landscape-relevant as well as geomorphologic and pedological field techniques, show that the recent interaction of the environment indicate relevant references for the understanding of the processes of landscape history. The comparison of sediments from both fluvial systems clarifies thereby above all the meaning of the small-scale variability of petrographic, relief-conditioned, hydrologic and anthropogenic parameters on the morphodynamical processes and the potential conservation of evaluable sediment bodies. Considering the small surface discharge on the sandstones of the plateau and the relative hydrologic favour of the Mbaéré valley it is to be assumed that after evaluation of the results of the plateau region altogether Holocene climatic change to more arid conditions led particularly to an expansion of savannas on the plateau heights and a retreat of the rain forest vegetation into the topographically lower sections of the vast valley plains. Their predominantly sandy fillings recently represent an enormous water reservoir, which supplies ideal conditions of growth for an extensive flood forest vegetation under present tropical–alternating–wet conditions. In the context of the discussion about high glacial refugial areas of the lowland rain forest the landscape dynamics postulated for the investigation area admit the hypothesis that the Mbaéré valley represents probably the northernmost "fluvial refugial" of the Congo catchment area within the meaning of the "core area" conceptions of Colyn *et al.* (1991) and Maley (1995).

8.1 INTRODUCTION

Stated aim of the research project in the "Forêt de Ngotto", supported by the part of the DFG (German Research Foundation), is to demarcate, sample and evaluate sediment archives in a fluvial system by means of a fluvialmorphologic-palaeohydrological approach, still used little in the tropics, in order to close an existing regional "gap" regarding terrestrial palaeoenvironmental data for the Central African area. Basic consideration is that the layered alluvial sediments conserved in positions of relative morphological stability allow conclusions of the depositional conditions. In addition to conventional analyses radiocarbon datings (^{14}C) were accomplished to determine the age of sediments and fossil A_h-horizons (humous soil horizon). $\delta^{13}C$ values, appraised by isotope-chemical analyses of carbon isotopes, admitted a physiognomic classification of the dated humus containing soil samples to C_3- (mainly trec -) or to C_4- (mainly grass -) dominated plant communities (Mariotti, 1991). The results were consulted for the reconstruction of subrecent vegetation conditions in the investigated area against the background of changed and changing climatic conditions.

8.2 STATE OF THE ART

For a long time the influence of the Pleistocene glaciations on the global climate and on the expansion of tropical rain forests and other humid tropical to alternating wet biomes was discussed controversially. Above all numerous findings, derived from analyses of pollen, diatoms, phytoliths, isotopes of organic carbon ($\delta^{13}C$) and mineralogical data, which go back to more than 700.000 BP, could deepen the insights into global interactions and their influence on the climatic and vegetation history of tropical Africa in the meantime (Abrantes, 2003; Barker *et al.*, 2004). According to that variations in solar radiation (Milankovich cycles) are among changes in the expansion of the ice sheets (Heinrich events) possible explanations for past climate changes.

Geoscientific and vegetational studies in the low latitudes of Africa during the last 25 years confirm by now the assumption that climatic events in the Congo basin were characterised by a cooling around 2 to 4°C as well as a reduction of precipitation of up to 50% opposite today during the high glacial (Partridge *et al.*, 1999). Studies relating to the Postglacial and the Holocene give also information about regional and temporally not always corresponding climatic fluctuations, which were differently strong. These climatic fluctuations resulted in repeated advances and retreats of the equatorial lowland rain forest in particular in the transition between rain forest and savanna (Kadomura, 1995; Gasse, 2000). Indicators to these climate changes accompanied by environmental changes on the African continent were derived so far particularly from pollen analyses in limnic deposits. For West and Central African lakes (Talbot *et al.*, 1984; Giresse *et al.*, 1994; Elenga *et al.*, 1996; Zogning *et al.*, 1997; Vincens *et al.*, 1998; Stager and Anfang-Sutter 1999; Nguetsop *et al.*, 2004) as well as for the rift lakes of the marginal mountainous regions of the eastern Congo basin (Kendall, 1969; Degens and Hecky, 1974; Tiercellin *et al.*, 1988; Roberts and Baker, 1993) adequate data are available. Evaluations of distal continental oceanic drilling cores from the alluvial fans of the Niger (Zabel *et al.*, 2001) and the Congo (Marret *et al.*, 1998) give information about the fluctuating sediment discharge from the large fluvial systems and proxy data like aeolian landforms under forest indicate formerly drier climatic conditions. Different dune forms, in addition to blown-out depressions, were already described and studied since the 1960s (amongst others Alexandre-Pyre,1971; de Dapper 1981 and 1985; Soyer, 1983; Thomas and Shaw 1991; Alexandre *et al.*, 1994).

Terrestrial palaeoenvironmental data for the low latitudes of Africa, as provided by lacustrine sediments and aeolian land forms, are distributed sparsely throughout the area, so that the dynamics of the ecological systems in Central Africa from the LGM until today still contain significant spatial and temporal gaps. For the region between Lake Barombi

Mbo surveyed in West Cameroon (Maley and Brenac, 1998) and the studied valley of the Mbari in the southeast of the CAR (Runge, 2002) so far no palaeoenvironmental data are available. Oscillating precipitations are bound to climatic radical changes, which result in modifications of the vegetation coverage and definitely must have had also profound effects on the performance of erosion and sedimentation of fluvial systems in the tropics. These spatially well distributed accumulations represent terrestrial archives, which can contribute to the reconstruction of changes in landscape history. With the indication of the complexity of the still insufficiently investigated recent fluvial morphodynamics in the tropics and due to the widespread assumption, that the intensive chemical weathering doesn't allow datings of organic matter, attempts to deal with this approach of research were omitted for a long time. The consequent absence of generally accepted conceptions for the analysis of alluvions in the low latitudes can particularly be explained by the intensive relocating processes and the high entry of colluvial material (see Fölster, 1983). Investigations of tropical valley sediments indicate meanwhile that there are locations, where datable sediment bodies containing information on landscape history are conserved. Indications of their existence are to be found in investigations accomplished in South America, Australia and Southeast Asia (among others Stevaux, 1994; Stevaux and Dos Santos, 1998; Nanson and Price, 1998; Nott and Price, 1999; Latrubesse and Franzinelli, 2002). For the tropical regions of Africa the data is comparatively poor as a result of general research deficits in the last 20 years due to political and infrastructural problems of many countries in this region (Preuss, 1986; Zeese, 1991; Thorp and Thomas, 1992; Alexandre *et al.*, 1994; Runge, 2001, 2002).

8.3 STUDY AREA: MBAÉRÉ VALLEY AND THE SADIKA ALLUVIAL FAN

8.3.1 Geological, geomorphological and pedological settings

The research area is the approx. 40 km long middle and lower course of the Mbaéré valley in the SW of the CAR (3°45'–4°N, 16°55'–17°25'E). The river drains the plateau of Gadzi-Carnot, which slants down SSE towards the Congo Basin over 272 km with an average incline of 1,6‰. The Mbaéré covers a floodplain of up to 5 km in width in this sector, into which a today obviously fossil alluvial fan of the left tributary Sadika extends. The geological bedrock of the plateau is set up by the Cretaceous Carnot formation. It is dominated by sandstones, in which conglomerate-like layers or lenticular inclusions and thin layers of (clay) marl are interbedded (Censier and Lang, 1999). This geological formation expands in the W and SW of the CAR over a surface area of 46.000 km². It has maximum heights of approx. 650 m asl within the range of the floodplain of the Mbaéré located around 415 m asl. The transitions from the plateau heights to the valley are accentuated by steep slopes particularly in the NE.

The alluvial plain of the Mbaéré, which drains the majority of the plateau together with the Bodingué, is a very young sedimentation area considering the geological past (Boulvert, 1996). Both rivers flow exclusively on the sandstones of the Carnot formation, whose erosion implicates the development of significant alluvial plains downstream.

The catchment area of the Mbaéré is dominated by leached sandy ferralitic soils of reddish color, which can be bleached on the interfluves superficially and partially eroded on the valley sides. At the upper valley slopes mineral-rich, but little developed soils (Boulvert, 1983) have evolved. Pedogenic crusts, as they are frequently found in the landscape of the CAR, which normally have a strong influence on the vegetation coverage and the geomorphologic processes, exist only isolated at the level of the prospected section of the Mbaéré.

Figure 1. Geomorphological units of the Central African Republic
(modified after Les éditions Jeune Afrique, 1984).

8.3.2 Recent climatic and hydrologic conditions

The recent climate on the plateau is associated with the equatorial-moist-tropical type (Boulvert, 1986). Two precipitation peaks appear, the first in May and the second with the precipitation maximum in August. A dry season of three months, usually from December to February, is characteristic. Data from 1994 to 2000 reveal an average annual amount of precipitation of 1.692 mm, which fall during 109 days at Ngotto station (north of the Mbaéré valley) (Yongo, 2003).

The river system on the plateau is developed marginally due to the high cleaving of the sandstones. In the centre of the complex exist regions with a total area of up to 400 km^2, in which no superficial drainage appear (Chatelain and Brugière, 1999). The rate of groundwater recharge is adjusted to the the lowest level of the drainage in this region and correspondingly high. The main rivers are primarily supplied by the groundwater body and just to some extent by their tributaries, parts of which only drain permanently. The Sadika is one of the perennial tributaries. During the dry season it has up to 3 m in width and about 50 cm in depth, but during the rainy season it bursts its banks and then floods spaciously the alluvial fan. The Mbaéré records the highest water levels in the months of August until January, while in May the lowest water level is observed. Reliable discharge data are only available for the Lobaye, into which the Mbaéré drains near the settlement of Lébé. The temporal offset of the Lobaye regime compared to other rivers draining in the Oubangui can be attributed to the special drainage conditions of the sandstone plateau (Callede *et al.*, 1998).

8.3.3 Recent vegetation patterns

The Mbaéré valley is located in the transition zone of perpetually humid rain forests and semi-humid tree and grass-savannas ("savanna woodland"). It is primarily a semi-deciduous forest ("forêt dense semi caducifoliée", Boulvert, 1986). It is characterised by its great biodiversity and a dominance of Meliaceae and Sapotaceae (Lejoly, 1995). Its extensive primary forest status refers to the fact that the plateau is comparatively hostile to settlement (<2,5 inhabitants/km^2) due to its poorly developed drainage net and the accordingly small superficial discharge. Forestry operations take place north of the Mbaéré in the "Forêt de Ngotto". The exploitation takes place sustainably and reforestation arrangements flank

Figure 2. Flood forest at the Mbaéré River, downstream view at Kpoka settlement.

the activities. Due to the absence of pedogenic crusts, tree savannas rich in understory are common in the range of the settlements emerging isolated on the plateau. The floodplain of the Mbaéré is characterised in its whole width by the existence of forest, which is periodically flooded during the time by the end of July to the end of January ("forêt inondable") and differs clearly in its composition from the primary forest of the plateau region.

Figure 3. Seasonally flooded grass vegetation on the Sadika alluvial fan in the front, in the background the gallery forest of the Sadika (left hand side) and the Mbaéré flood forest (right hand side).

Along the river the vegetation is composed of *Raphia* sp., *Ficus* sp., *Miragyna ciliata*, *Milletia* sp. and *Irvingia ethii*. The population of *Raphia* sp. and *Ficus* sp. decreases with the distance to the river. Among the herbaceous plants Marantaceae are widely spread (Barrière *et al.*, 2000).

The surface of the Sadika alluvial fan is seasonally flooded and is covered by grass vegetation, which is surrounded by a belt of grass-/ bush savanna. The course of the Sadika is bordered by a narrow strip of gallery forest, which corresponds in its composition to that of the primary forest.

8.4 METHODS

The field campaign in spring 2004 was divided into a first phase, in which the focus was on the logistic preparation of the fieldwork and the observation of the research area, and a second phase, in which many samples were taken upstream the Mbaéré valley, starting at the level of the alluvial fan of the Sadika. In the dry season 2005—apart from a shorter mission in the Mbaéré valley—the alluvial fan of the Sadika was investigated. The sampling in the Mbaéré valley was mainly carried out by hand-corings with the EIJKELKAMP© Edelman hand auger, that enables the sampling with a probe for sandy sediments. The sediments were sampled in layers of 20 cm. The profiles were sampled down to maximum depths between 80 and 320 cm depending on the location of the groundwater table. Apart from the drilling profiles (covering up to 200 cm) exposures were dug on the alluvial fan of the Sadika, in which the sampling took place related to the horizons.

The sediment samples were evaluated conventionally in the pedological laboratory of the University of Frankfurt. The following analyses were accomplished: determination of the soil color in moist and dry condition (Munsell Soil Color Chart), grain size analysis (after Köhn), pH value (solved in 0,1 n KCl solution, after Meiwes *et al.*, 1984), determination of the organic matter and total carbon (with the carbon catalyser LECO EC-12) and the total nitrogen (after Kjehldal) to calculate the C/N ratio.

Radiocarbon datings (^{14}C) to determine the age of organic sediments and fossil humus-horizons (partly charcoal) were carried out via the AMS-Radiocarbon-laboratory at the University of Erlangen and Beta Analytic, Inc. (Florida, USA). Within the scope of these datings the ^{13}C values—relevant for the reconstruction of subrecent vegetation cover—were determined by isotope-chemical analyses of carbon isotopes.

8.5 REGIONAL INVESTIGATIONS

8.5.1 Sampling of the Mbaéré valley and the Sadika alluvial fan

A reference drilling (G2) within the valley between Grima and Bambio, still water-logged at the end of the dry season, reached only one meter in depth and therefore was not analysable adequately. Due to this general view emerged from these and other corings it has to be assumed that under constant influence of water in combination with the coarse-pored sands no interpretable traces regarding landscape history have been preserved in the zones of the valley permanently affected by water. At sites where lower terrace-like alluvial bodies, cut by the river are accessible, sampling near the river is possible at the end of the dry season. These locations, which are characterised by distinctly drier soil conditions and another vegetation, composition, are located in the proximity of Kpoka (K2, K3, K4) and near Grima (G1) (Figure 4).

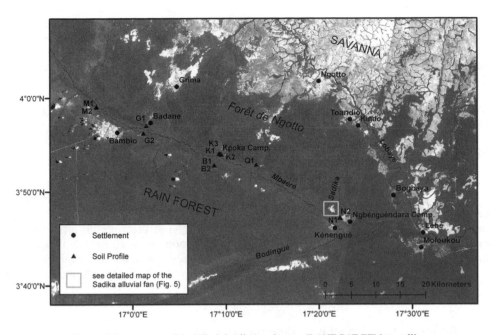

Figure 4. Survey map of the Mbaéré valley study area (LANDSAT ETM+ satellite scenes 181-57, 03.03.2000 and 182-57, 01.04.2002).

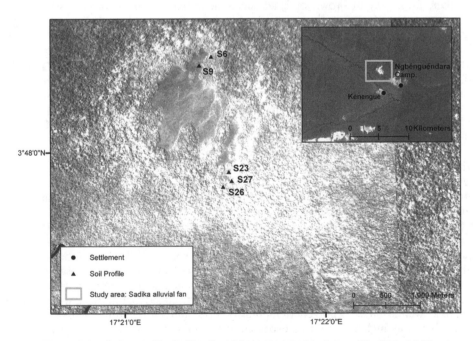

Figure 5. Detailed map of the Sadika alluvial fan coring site (air photos of the Central African Republic 474 and 475, mission AE 1963/64, Institut Géographique National français (IGN), Paris).

The majority of sampling in the Mbaéré valley took place at the marginal zones of the valley. The drilling of profiles N1, N2 was carried out near Ngbénguendara in the sparsely accentuated transition of the flood forest to the savanna that suppressed the primary forest of the plateau in this area by anthropogenic influence. The profiles B1 and B2, taken in the proximity of the Bassamba, intersect sediments of the footslope, under which older sediments of the Mbaéré were assumed as well. The sampling at the abandoned village of Mboum (M1, M2) northwest of Bambio resp. near the forestry road "Route Q" close to Kpoka (Q1) was carried out similarly. For the site selection the small-scale partially abruptly varying geofactors relief, vegetation and soil (moisture) were particularly taken into consideration.

The sampling of the alluvial fan of the Sadika was carried out systematically. It revealed that the distal part of the alluvial fan at the transition to the flood forest of the Mbaéré valley could not be sampled due to the high groundwater table. Even within the central regions of the floodplains west and east of the Sadika in some places smaller drainage channels were conserved until March and intense precipitation provoked an elevation of the groundwater table. Depending on the results of the corings, that primarily provide information on the marginal areas of the alluvial fan distal and near the river, numerous exposures were achieved, that gave more detailed insights into the sediment stratigraphy (S6, S9, S23, S26, S27) (Figure 5).

8.5.2 Results of the conventional laboratory analyses

In contrast to the Mbaéré profiles the sampled coring profiles and exposures on the alluvial fan of the Sadika reveal more heterogeneous colors and distinctive changes from brighter to darker, from grey to brown or black sands at different depths. In contrast to these observations the sampled sediments of the Mbaéré valley exhibit with increasing depth brighter colors. Regarding the grain sizes, all samples are characterised by low skeleton fractions and high sand contents, that amount usually between 85 and 95% in the profiles of the Mbaéré valley and in the profiles located in the area of the alluvial fan between 90 and 100%. The fraction of medium sand predominates. The medium and fine silt contents are usually very low (<1%) within the profiles and accordingly either clay or coarse silt dominate the finer fractions. In addition, the grain size spectrum in all profiles diversifies, but noticeable variations, that could provide evidence of accentuated changes of erosion and sedimentation conditions, are missing.

Enhanced rates of organic matter in low-lying horizons, that represent fossil humus horizons of older soils covered by sediments, appear in case of the samples of the Mbaéré. They are not characterised by a significant change of color. At the surface the rate of organic matter amounts to maximal 6%, but almost clearly below (about 2%). Peaks that appear in more than 100 cm of depth usually don't attain these near-surface values. In contrast the Sadika-samples show increasing carbon contents at different depths that refer to fossil humus horizons. The contents are usual clearly above the near-surface rates on the one hand and correlate with the accordingly darker, partial nearly black color of the sediments in this sample on the other hand. The analysis of the contents of nitrogen shows that they are tendentious low and correspond to the expectations in this temporarily to permanently water-affected milieu. Regarding the C/N ratio it is to be stated that in the alluvions of the Mbaéré as well as on the relatively (temporarily) drier alluvial fan of the Sadika the biotic activity is not as high as would be expected in this climatic environment. Therefore the grade of decomposition of the organic matter is deficiary. The hydromorphic conditions entail, that the microbial processes of decomposition in the soil are inhibited in consequence of an oxygen deficit. The hydromorphic environment also appears to be the reason for the fact that near-surface a not completely decomposed layer of litter is conserved with an adequate sedimentary cover at depths to 100 to 200 cm. This explains the numerous macro remains (roots, pieces of wood, seeds), which were found in the profiles—not evaluated yet—of a field campaign in 2006 at suitable depths in terrace-like locations.

The pH values are in a medium to strongly acid range, which can be interpreted by the petrography and constant eluviation of basic exchangeable cations in the sand-dominated soils under alternating wet climate conditions. The average pH value in the Mbaéré valley amounts to 4,3, the samples of the alluvial fan show an average pH value of 4,5. The soils in the valley of the Mbaéré and particularly within the range of the Sadika usually show fluctuations within the profile between 3,1 and 5,6 (Mbaéré) and between 3,5 and 6,4 (Sadika). The variances of the H^+-concentration within the course of the profiles result in a heterogeneous pattern, which doesn't provide indication for a discussion of the current reactivity of the soils in consequence of processes of pedogenesis, weathering and mass transport.

8.5.3 ^{14}C data and δ^{13}C values

The radiocarbon dated samples from the Mbaéré valley at less than 100 cm of depth represent recent and/or subrecent ages (236 ± 49 BP). By direct comparison older ages do not arise obligatory at greater depths. This refers to the different location of the sites in the middle course section of the Mbaéré between 450 m asl and 425 m asl as well as the varying spatial arrangement in the transverse valley profile. Samples, dated on ^{14}C-ages between 1.450 ± 40 BP and 6.921 ± 71 BP, appear already at a depth of 100 to 120 cm. The sites located closest to the river with supposed high recent morphodynamics cannot be excluded obligatory as potential archives of older material. At terrace-like sites Holocene sediments and charcoal with ages between 1.733 ± 56 and 6.921 ± 71 BP can be found at only 144 cm to 120 cm depth. At sites distal to the river situated in the marginal areas of the valley older archives, buried by slope material, were expected. Data provided shows that despite the recently most moderate fluvial processes no older sediments are preserved at these locations or are beyond the sampled depth.

The datings from the area of the Sadika alluvial fan show that different samples at depths approx. 100 cm represent recent ages, whereas other samples at lower depths contain older material. Datings here range between 951 ± 69 BP (at 37 to 69 cm) and 4.162 ± 78 BP (at 70 to 93 cm), thus they vary stronger than the samples of the Mbaéré. Organic sediment at a depth of 37–69 cm of an exposure was dated between 951 ± 69 BP and is definitely older than a sample originated from the same exposure at a depth of more than 99 cm that is of recent age. This fact clarifies that a relocation of material must have taken place.

The recent sediments of the Mbaéré valley show δ^{13}C values of –29,0‰ and –29,6‰. These values point to conditions indicating forest vegetation (C3) in the humus layer. The δ^{13}C values of the older samples from the Mbaéré valley lie between –24,9‰ and –30,1‰, which suggests forest dominated habitats at the time of humus formation too. Neither the position within the transverse or longitudinal valley profile nor the depth of extraction of the samples contains significant disparities.

The situation on the alluvial fan implies that the material dated as recent correlates regarding its δ^{13}C values with the present vegetation physiognomy at the location. The fossil organic carbon that indicates older humus horizons, analysed on the Sadika alluvial fan provides in contrast δ^{13}C values, which lie between –22,7‰ and –29,8‰. These values consequently point to trees as dominating vegetation in a region that is today mainly dominated by grasses, irrespective of whether it concerns subrecent (348 ± 48 BP) or Holocene (4.162 ± 78 BP) dated layers.

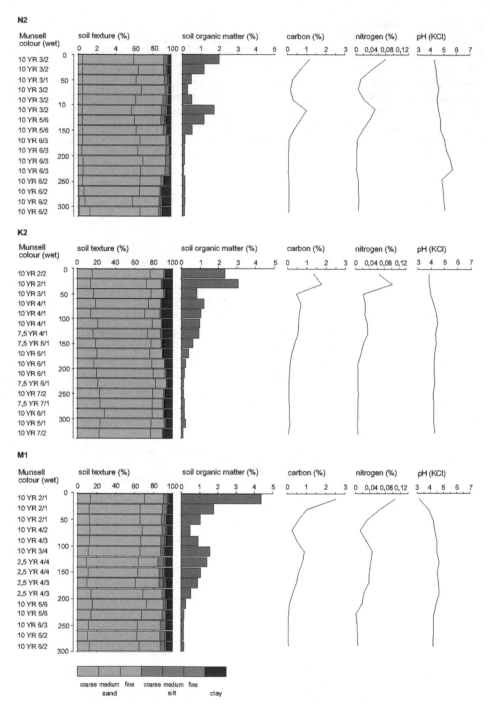

Figure 6. Physical and chemical sediment characteristics of some coring sites within the Mbaéré floodplain.

8.6 INTERPRETATION

The interpretation of the fluvial performance of erosion and sedimentation of tropical rivers and their deposits entails some problems, as the remobilisation and relocation of material, which can redound via radiocarbon datings to a deceptive evaluation of age for the sediment archives. The conditions in the catchments are not uniform as well and can complicate the interpretation, e.g. by anthropogenic influences or tectonic uplift on the discharge performance. Dating gaps emerge frequently, which can be attributed to the fact that at some sites due to occasional local inactivity or material discharge for some main phases of sedimentation no data can be identified. Anyhow, the recovered alluvial sediments, products of the local conditions of discharge and sedimentation provide via the data obtained by them important indication of the development of landscape, history in this ecological system.

The small variations in grain-size within the profiles suggest that on the sandstone plateau coarser weathering products, as they are found in the headwaters of the tributaries Sadika and Kélé (up to walnut-large fragments of sandstone and laterite as well as quartzes from eroded quartz belts), rapidly decrease in size with increasing distance from the spring, will be cut up to small pieces by transport after only a short distance and will be weathered to the characteristic grain size after deposition. Thus explains the absence of coarser sediment layers, that in other investigations in semihumid to semiarid regions often provide indications of modified climatic conditions, weathering processes and discharge situation. Soil chemical analyses suggests, apart from the carbon content, no evidence applicable for the question referring to processes of soil genesis and sedimentation, which is to be seen as result of the parent rock, the coarse pored soil texture and the high material discharge involved. Therefore the petrographic conditions of the plateau are as also evinced by the temporally shifted discharge into the Oubangui system of special relevance for the fluvial processes and the reconstruction of landscape history.

The laboratory data, particularly the datings and the color description of the Sadika samples, indicate intensive relocating processes and processes of soil genesis interrupted repeatedly and attest to small-scale diversifications. This result and the Sadika incised in the alluvial fan as well as localities, where diffuse discharge occurs within the floodplain, suggest that processes of erosion and relocation dominate the current dynamics in this terrain sloping in direction towards the Mbaéré valley. In consideration of the steeper incline compared to the Mbaéré as well as the sparser vegetation it is to be assumed that the discharge particularly in high precipitation years or during intense rain events is carried out across the entire alluvial fan. The deposited and remobilised sediments correspondingly do not contain applicable information of landscape history at a depth of one to two meters. Their accumulation is not climate- but rather position-induced, as shown by inverse datings within a profile. δ^{13}C values referring to forest vegetation are—against the background of a vegetation recently dominated by grassland and savanna—an obvious indication that since 4.162 ± 78 BP changes in vegetation took place. Since for a forest advance no climatic connection can be established, an expansion of the forest within this time frame is to be explained here particularly by the shifting of the Sadika's course and a temporary relatively drier location setting. Singular intense rain events could also have played an important role. The anthropogenic influence, not to be excluded in historical times for this region and today documented by regular slash-and-burn culture of the BaAka pygmies, residentiary in this area until 2005, is conducive to a vegetation alteration and a non-appreciable interference into the fluvial process coherences furthermore.

In contrast the continuity of the color shading, shown by the Mbaéré profiles, as well as the dating results refer to undisturbed and continuous sedimentation conditions, which were interrupted by processes of soil genesis under relatively drier conditions. From the samples dated as mid to recent Holocene ages for the Mbaéré it can be concluded that the preexisting biogeographic conditions, in particular the vegetation conditions repressing the sediment discharge, enhance the preservation of palaeoenvironmental archives in the form of small terraces in this region. The $\delta^{13}C$ values point to a permanent afforestation of the valley during the Holocene. These values legitimate the assumption that within this valley plain neither grass-dominated savanna vegetation nor the flood grasslands, characteristic for valleys under comparable climatic conditions, have developed. This is probably a result of the relative water storage function of the enormous valley fillings that offers a locational advantage for the tree population of the ecological system. Which type of forest was involved at the time of the fossil humus formation, cannot be clarified on the basis of these data. The small-scale changes in vegetation composition, noticeable in this area, are attached to the recent hygric conditions. These changes indicate that the water saturation varies spatially within the aquifer and exerts direct influence on the tree population. It is hence to be assumed that arid phases caused a reduced saturation of the aquifer, compared to today's conditions, and a decrease of the flood forest in favour of the rain forest. In ensuing more humid phases the recolonisation of the valley plain with flood forest must have emanated from regional (gallery forest-like?) moister flood forest cells in proximity of the river. For the closer investigation area there doesn't seem to react a "pulsierendes Vegetationsmosaik" (Runge, 2001:196) of forest and savanna on climatic changes, as in the case of the Mbari (SE of the CAR, Mbomou plateau), but of rain forest and flood forest interchange.

A final temporal evaluation of more arid or more humid conditions is difficult despite the datings referring to the results from the Mbaéré valley taking account of the $\delta^{13}C$ values that cannot be differentiated according to forest lands. Considering the special petrographic and hygric environment it is likely though that the climatic changes to more arid conditions about 5 ka and since 1 ka (Runge, 2002), documented for the Mbari valley, have emerged differently in the landscape. Arising from the discussed research results a multiphase scenario of landscape history (Figure 7) represents an ecological system, that probably shows a temporal disalignment in relation to comparable field studies.

At the end of the Mesozoic the plateau, which drained so far into two basins, the Doba trough (Chad) and the Touboro basin (Cameroon), into the Gulf of Benue (Censier and Lang, 1999), started to tilt. The tilting was intensified during the Tertiary and led to a change in the drainage direction. This was accompanied by the arrangement of the main drainage courses, Mbaéré and Bodingué, and the initial dissection of the sandstone. It is still unexplained when within the Quaternary the deep, probably V-shaped valley-like dissection of the sandstone complex took place. Since it is presumable that changes from drier to more humid conditions already occurred during the Pleistocene, the filling of the large valleys probably began under these conditions. Based on the assumption of a high glacial forest regression between 20 and 15 ka (LGM), as described in several studies of the African tropics, on the plateau of Gadzi-Carnot the equatorial lowland rain forest presumably had to give way, even though a little later, to the savannas expanding to the south to some extent. If a flood forest had already established within the floodplain of the Mbaéré, it has been repressed by the aridification on closely limited insular locations near the river with local hygric favour. The rain forest on the other hand must have colonialized large parts of the valley starting from the valley sides. With a change to more humid climate conditions, as assumed for approx. 12 ka in the SE of the CAR (Runge, 2002), an advance of the rain forest on the plateau towards the savanna and a resettlement of the entire valley plain by the flood forest is presumable. The regionally varying climatic oscillations determined for the Central African area in the course of the Holocene (Gasse, 2000), are due to varying initial conditions in the relevant investigation areas apart from methodically-related differences. These climatic oscillations lead to the presumption that with the repeated transition from more humid to more arid conditions the vegetation succession, as described above, must

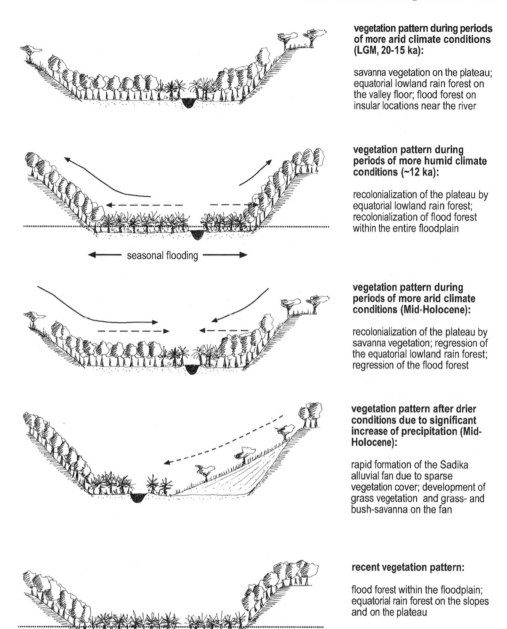

vegetation pattern during periods of more arid climate conditions (LGM, 20-15 ka):

savanna vegetation on the plateau; equatorial lowland rain forest on the valley floor; flood forest on insular locations near the river

vegetation pattern during periods of more humid climate conditions (~12 ka):

recolonialization of the plateau by equatorial lowland rain forest; recolonialization of flood forest within the entire floodplain

seasonal flooding

vegetation pattern during periods of more arid climate conditions (Mid-Holocene):

recolonialization of the plateau by savanna vegetation; regression of the equatorial lowland rain forest; regression of the flood forest

vegetation pattern after drier conditions due to significant increase of precipitation (Mid-Holocene):

rapid formation of the Sadika alluvial fan due to sparse vegetation cover; development of grass vegetation and grass- and bush-savanna on the fan

recent vegetation pattern:

flood forest within the floodplain; equatorial rain forest on the slopes and on the plateau

Figure 7. Scheme of vegetation dynamics in the Gadzi-Carnot region due to climatic oscillations.

have recurred several times here as well in response to altered hydrologic conditions. This points to a permanent and long-ranging competitive situation of the tree species in this region which would explain why the equatorial lowland rain forest in this northern extension stands out by an impoverishment of *"Ayous"* (*Triplochiton scleroxylon*) (de Madron, 2003) and differs in its species composition from the typical primary forest of the Congo basin (Alain Penelon, personal note). The development of the alluvial fan in the lower course of the Sadika cannot have been carried out until the filling of the Mbaéré valley in its today's appearance was essentially terminated.

It can be accepted that a significant increase in precipitation during the mid Holocene after comparatively drier, alternating-wet conditions encouraged the formation of the alluvial fan. The decomposition rate increased by a climatic change and the savanna vegetation still predominating on the plateau will have led—due to excessive high precipitation—to a deep moistening of the slopes destabilized in this way and a rapid development of the palaeo alluvial fan. This happened before the forest vegetation reacted with a succession to the changed parameters.

Under consideration of the "core area" theories and conceptions, discussed since the 1970s, according to which flora and fauna of the tropical rain forest outlasted glacial climatic breaks like cooling down and desiccation in core and refugial areas and again expanded from there after an initially isolated progression of the species, a special importance is attached to the large valley structures within the sandstone plateau at the rain forest savanna boundary as a temporary refugium during more arid conditions in the Holocene. Taking a similar vegetation dynamic as a basis in view of high glacial climatic changes, the valley systems of the sandstone plateau could be refugial areas in terms of Colyn *et al.* (1991) and Maley (1995) due to their hydrological favour. According to them and contrary to other concepts, which postulate closed extensive refugiums of rain forest (Meggers *et al.*, 1973; Hamilton, 1976 and 1982; Littmann, 1987 and 1988; Liedtke, 1990), the high glacial retreat areas are situated gallery forest-like along the large rivers Congo, Oubangui and Kasai inside the basin ("fluvial refuge"). The Mbaéré valley would analogously represent the northernmost known endemism centre of the tropical lowland rain forest in the Congo basin.

ACKNOWLEDGEMENTS

This research was supported by grants of the German Academic Exchange Service (Deutscher Akademischer Austauschdienst, DAAD) and the German Research Foundation (Deutsche Forschungsgemeinschaft, DFG; Ru 555/17-1). Very sincere thanks go to the administration of ECOFAC (Conservation et utilisation rationelle des ECOsystèmes Forestiers d'Afrique Centrale), notably to M. Jérémy Maro and M. Alain Penelon, for supporting fieldworks in 2004 and 2005. In addition we would like to thank the ECOFAC staff at Ngotto, especially M. Denis Passi and M. Landry Nguéta, for their cooperation, assistance and kindness. We are also grateful to our colleagues from the Department of Geography at the University of Bangui, M. Guy-Florent Ankogui-Mpoko, M. Cyriaque Rufin Nguimalet, M. Marcel Koko et M. Marcel Kembé, for their support concerning formalities, information and logistics.

REFERENCES

Alexandre, J., Aloni, K. and de Dapper, M. 1994, Géomorphologie et variations climatiques au Quaternaire en Afrique Centrale, *Revue Internationale de Géographie et d'Écologie Tropicales*, **16**, pp. 167–205.

Alexandre-Pyre, S., 1971, Le Plateau de Biano (Katanga). Géologie et Géomorphologie. *Académie Royale des Sciences Outre-Mer, Classe des Sciences Naturelles et Médicales, N.S.*, **18**, 3, pp. 1–151.

Barrière, P., Nicolas, V. Maro R.K., and Yangoundjara, G., 2000, *Écologie et Structuration des Peuplements de Micro-mammifères Musaraignes et Rongeurs* www.ecofac.org/ Biblio/Download/Micromammiferes_RCA/EcologieMicrommiferes.pdf

Boulvert, Y., 1983, Carte pédologique de la République Centrafricaine à 1:1.000.000. *Orstom, notice explicative*, **100**, pp. 1–126.

Boulvert, Y., 1986, Carte phytogéographique de la République Centrafricaine à 1:1.000.000. *Orstom, notice explicative*, **104**, pp. 1–131.

Boulvert, Y., 1996, Étude géomorphologique de la République Centrafricaine. *Carte à 1:1.000.000 en deux feuilles ouest et est. Orstom, notice explicative*, **110**, pp. 1–258.

Callede, J., Boulvert, Y., and Thiebaux, J.P., 1998, Le Bassin de l'Oubangui. *Éditions Orstom, Collèction Monographies Hydrologiques* (www.mpl.ird.fr/hydrologie/document/ monogras/oubangui/index104.htm).

Censier, C., and Lang, J., 1999, Sedimentary processes in the Carnot formation (Central African Republic) related to the palaeogeographic framework of Central Africa. *Sedimentary Geology*, **127**, pp. 47–64.

Chatelain, C., and Brugière, D., 1999, *Proposition de classement du Parc National Mbaéré-Bodingué et de l'aire d'utilisation durable des écosystèmes de la Mbaéré dans la zone d'intervention de projet ECOFAC-RCA* (*www.ecofac.org/Biblio/Download/Proposition_ Classement.pdf*).

Colyn, M., Gautier-Hion, A. and Verheyen, W., 1991, A re-appraisal of palaeoenvironmental history in central Africa : evidence for a major fluvial refuge in the Zaire basin. *Journal of Biogeography*, **18**, pp. 403–407.

de Dapper, M., 1981, The microrelief of the sandcovered plateau near Kolwezi (Shaba/ Zaïre) II. The microrelief of the crest dilungu. *Revue Internationale de Géographie et d'Écologie Tropicales*, **5**, pp. 1–12.

de Dapper, M., 1985, Quaternary aridity in the tropics as evidenced from geomorphological research using conventional panchromatic aerial photographs. Examples from peninsular Malaysia and Zaïre. *Bulletin de la Sociéte Belge de Géologie*, **3**, pp. 199–207.

Durrieu de Madron, L., 2003, Suivi de la régénération de la forêt en RCA. *Canopée*, **23**, pp. 21–22 (www.ecofac.org/Canopee/N23/N2305_RegenerationForet.pdf).

Degens, E.T. and Hecky, R.E., 1974, Palaeoclimatic Reconstruction of Late Pleistocene and Holocene Based on Biogenic Sediments from the Black Sea and a Tropical African Lake. *Les méthodes quantitatives d'étude des variations du climat au cours du Pleistocène* (Paris : Colloques Internationaux de C.N.R.S, **219**), pp. 13–24.

Elenga, H., Schwartz, D., Vincens, A., Bertaux, J., de Namur, C., Martin, L., Wirrmann D., and Servant, M., 1996, *Diagramme pollinique holocène du lac Kitina (Congo): mise en évidence de changements paléobotaniques et paléoclimatiques dans le massif forestier du Mayombe* (Paris: Comptes rendus de l'Académie des sciences, Ser. IIa), **323**, pp. 403–410.

Fölster, H., 1983, Bodenkunde—Westafrika (4°–8°N, 3°15'–9°30'E). *Afrika-Kartenwerk, Beiheft W4*, (Stuttgart).

Gasse, F., 2000, Hydrological changes in the African tropics since the Last Glacial Maximum. *Quaternary Science Reviews*, **19**, pp. 189–211.

Giresse, P., Maley, J. and Brenac, P. 1994, Late Quaternary palaeoenvironments in the Lake Barombi Mbo (West Cameroon) deduced from pollen and carbon isotopes of organic matter. *Palaeogeography, Palaeoclimatology, Palaeoecology*, **107**, pp. 65–78.

Kadomura, H., 1995, Palaeoecological and Palaeohydrological Changes in the Humid Tropics during the last 20.000 Years, with Reference to Equatorial Africa. In *Global*

Continental Palaeohydrology, edited by Gregory, K.J. *et al.*, (Chichester, New York: Wiley and Sons), pp. 177–202.

Kendall, R.L., 1969, An ecological history of the Lake Victoria basin. *Ecological Monographs*, **39**, pp. 121–176.

Latrubesse, E.M. and Franzinelli E., 2002, The Holocene alluvial plain of the middle Amazon River, Brazil. *Geomorphology*, **44**, pp. 241–257.

Lejoly, 1995, *Utilisation de la méthode des transects en vue de l'étude de la biodiversité dans la zone de conservation de la forêt de Ngotto, République Centrafricaine*, (Bruxelles: ECOFAC).

Les Editions Jeune Afrique (Ed.), 1984, *Atlas de la République Centrafricaine*, (Paris : Les Editions Jeune Afrique), pp. 1–64.

Maley, J. and Brenac, P., 1998, Vegetation dynamics, palaeoenvironments and climatic changes in the forests of western Cameroon during the last 28.000 years BP. *Review of Palaeobotany and Palynology*, **99**, pp. 157–187.

Marret, F., Scourse, J.D., Versteegh, G., Jansen, F.J.H. and Schneider, R., 1998, Integrated marine and terrestrial evidence for abrupt Congo River palaeodischarge fluctuations during the last deglaciation. *Quaternary Research*, **50**, pp. 34–45.

Mehra, O.P. and Jackson, M.L., 1960, Iron oxide removal from soils and clays by dithionite–citrate system buffered with sodium bicarbonate. *Clays and Clay Minerals Proc.*, **7**, (Washington, D.C.: National Academy of Science), pp. 317–377.

Meiwes, K.-J., König, N., Khanna, P.K., Prenzel, J. and Ulrich, B., 1984, Chemische Untersuchungsverfahren für Mineralboden, Auflagehumus und Wurzeln zur Charakterisierung und Bewertung der Versauerung in Waldböden. *Berichte des Forschungszentrums Waldökosysteme/Waldsterben*, **7**, p. 67.

Nanson, G.C. and Price, D.M., 1998, Quaternary Change in the Lake Eyre basin of Australia: an introduction. *Palaeogeography, Palaeoclimatology, Palaeoecology*, **144**, pp. 235–237.

Nguetsop, V.F., Servant-Vildary, S. and Servant, M., 2004, Late Holocene climatic changes in West Africa, a high resolution diatom record from equatorial Cameroon. *Quaternary Science Reviews*, **23**, pp. 591–609.

Nott, J. and Price, D., 1999, Waterfalls, floods and climate change: evidence from tropical Australia. *Earth and Planetary Science Letters*, **171**, pp. 267–276.

Partridge, T.C., Scott, L. and Hamilton, J.E., 1999, Synthetic reconstruction of Southern African environments during the Last Glacial Maximum (21-18 kyr) and the Holocene Altithermal (8-6 kyr). *Quaternary International*, **57/58**, pp. 207–214.

Preuss, J., 1986, Die Klimaentwicklung in den äquatorialen Breiten Afrikas im Jungpleistozän. Versuch eines Überblicks im Zusammenhang mit Geländearbeiten in Zaïre. *Marburger Geographische Schriften*, **100**, pp.132–148.

Roberts, N. and Baker, P., 1993, Landscape stability and biogeomorphic response to past and future climatic shifts in intertropical Africa. In *Landscape Sensitivity*, edited by Thomas, D.S.G. and R.J. Allison, (Chichester, New York: Wiley and Sons), pp. 65–82.

Runge, J., 2001, Landschaftsgenese und Paläoklima in Zentralafrika. Physiogeographische Untersuchungen zur Landschaftsentwicklung und klimagesteuerten quartären Vegetations- und Geomorphodynamik in Kongo/Zaire (Kivu, Kasai, Oberkongo) und der Zentralafrikanischen Republik (Mbomou). In *Relief Boden Paläoklima*, **17**, (Gebrüder Bornträger), pp. 1–294.

Runge, J., 2002, Holocene landscape history and palaeohydrology evidenced by stable carbon isotope ($\delta^{13}C$) analysis of alluvial sediments in the Mbari valley (5°N/23°E), Central African Republic. *Catena*, **48**, pp. 67–87.

Runge, J. and Neumer, M., 2000, Dynamique du paysage entre 1955 et 1990 à la limite forêt-savane dans le nord du Zaïre, par l'étude de photographies aériennes et de données

LANDSAT-TM. In *Dynamique à long terme des écosystèmes forestiers intertropicaux (ECOFIT)*, edited by Servant, M. and S. Servant-Vildary, (Paris : UNESCO, IRD), pp. 311–317.

Soyer, J., 1983, Microrelief de buttes basses sur sols inondés saisonnièrement au Sud-Shaba (Zaïre). *Catena*, **10**, pp. 153–265.

Stager, J.C. and Anfang-Sutter, R., 1999, Preliminary evidence of environmental changes at Lake Bambili (Cameroon, West-Africa) since 24.000 BP. *Journal of Paleolimnology*, **22**, pp. 319–330.

Stevaux, J.C. and dos Santos, M.L., 1998, Palaeohydrological changes in the upper Paraná River, Brazil, during the late quaternary: a facies approach. In *Palaeohydrology and Environmental Change*, edited by Benito, G. *et al.* (Chichester, New York: Wiley and Sons), pp. 273–285.

Steveaux, J.C., 1994, The upper Paraná River (Brazil): Geomorphology, sedimentation and palaeoclimatology. *Quaternary International*, **21**, pp.143–161.

Talbot, M.R., Livingstone, D.A., Palmer, P.G., Maley, J., Melack, J.M., Delibrias G. and Gulliksen, S. 1984, Preliminary results from sediment cores from Lake Bosumtwi, Ghana. *Palaeoecology of Africa*, **16**, pp.173–192.

Thomas, D.S.G. and Shaw, P.A., 1991, "Relict" desert dune systems: interpretation and problems. *Journal of Arid Environments*, **20**, pp.1–14.

Thomas, M.F., 2000, Late Quaternary environmental changes and the alluvial record in humid tropical environments. *Quaternary International*, **72**, pp. 23–36.

Thomas, M.F., 1998, Late Quaternary Landscape Instability in the Humid and Sub-Humid Tropics. In *Palaeohydrology and Environmental Change*, edited by Benito, G. *et al.*, (Chichester, New York: Wiley and Sons), pp. 247–258.

Thorp, M. and Thomas, M., 1992, The timing of alluvial sedimentation and floodplain formation in the lowland humid tropics of Ghana, Sierra Leone and western Kalimantan (Indonesian Borneo). *Geomorphology*, **4**, pp. 409–422.

Tiercellin, J.-J., Mondeguer, A., Gasse, F., Hillaire-Marcel, C. *et al.*, 1988, 25.000 ans d'histoire hydrologique et sédimentaire du lac Tanganyika, Rift east-africain. *Comptes rendus de l'Académie des sciences*, **307**, pp. 1375–1382.

Vincens, A., Schwartz, D., Bertaux, J., Elenga, H. and de Namur, C., 1998, Late Holocene climatic changes in western Equatorial Africa inferred from pollen lake Sinnda, Southern Congo. *Quaternary Research*, **50**, pp. 34–45.

Yongo, O.D., 2003, Contribution aux études floristiques, phytogéographique et phytosociologique de la Forêt de Ngotto (République Centrafricaine). *Acta Botanica Gallica*, **159(1)**, pp.119–124.

Zabel, M., Schneider, R., Wagner, T., Adegbie, A.T., de Vries, U. and Kolonic, S., 2001, Late Quaternary Climate Changes in Central Africa as inferred from Terrigenous Input in the Niger Fan. *Quaternary Research*, **56**, pp. 207–217.

Zeese, R., 1991, Fluviale Geomorphodynamik im Quartär Zentral- und Nordostnigerias. In *Freiburger Geographische Hefte*, **33**, pp.199–208.

Zogning, A., Giresse, P., Maley, J. and Gadel, F., 1997, The Late Holocene palaeoenvironment in the Lake Njupi area, West Cameroon: implications regarding the history of Lake Nyos. *Journal of African Earth Science*, **24/3**, pp. 285–300.

CHAPTER 9

Extension of former tree cover in the today's sudano-sahelian milieu as evidence for late Holocene environmental changes in northern Cameroon

Anselme Wakponou

Department of Geography, University of Ngaoundéré, Ngaoundéré, Cameroon

Bienvenu Dénis Nizesete

Department of History, University of Ngaoundéré, Ngaoundéré, Cameroon

Frédéric Dumay

Laboratory of Zonal Geography for Development, University of Reims Champagne–Ardenne, Reims, France

ABSTRACT: This article focusses on the existence of different types of tree cover in the sudano-sahelian region of Cameroon during the late Holocene. Field work gave evidence for relictual sinks by palaeosoils and vegetation thus proving that the environment used to be slightly wetter than it is today. Sedimentological analysis and dating allowed to propose a palaeoenvironnemental framework for the better comprehension of current evolutions. These results confirm those of palaeogeographical and palynological studies proving the extension of forests and savannas far from their current limits. Reports of explorers of the 19th century and stories mention these; it means that, despite climatic harshness and variations which may explain the degradation of the vegetation, it is most likely that the activities of humans played a major role for this striking environmental change.

9.1 INTRODUCTION

The Cameroonian sudano-sahelian area from the upper Benue river basin (9°N) to the Chad plain (11°N), like all the sahelian margin, is a ecologically highly sensitive milieu according to past and current climatic fluctuations. Sudano-sahelian is an appellation based on phytogeographic considerations. The area has a transitory vegetation where sudanian species (*Daniellia oliveri, Vitellaria paradoxa [Butyrospermum parkii], Cassia sieberiana, Terminalia macroptera, Parkia felicoidea, Prosopis africana* and others) interfere with sahelian species (*Balanites aegyptiaca, Boscia senegalensis, Combretum aculeatum, Calotropis procera, Capparis tomentosa, Capparis corymbosa* and others).

This paper deals successively with the palaeovegetation, the vulnerability of the milieu, the compromising practices for the environment and finally gives some suggestions for a conservatory management of the tree resources for sustainable development.

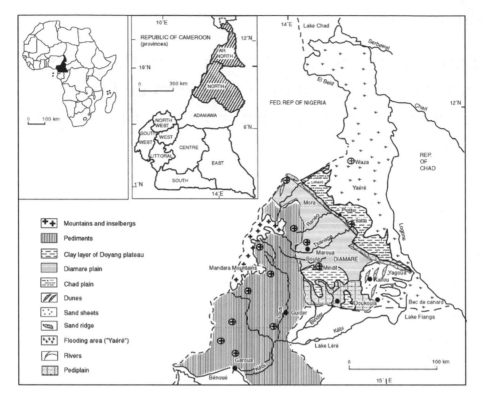

Figure 1. Location map of the study area.

The main objective is to show by extensive field observations, that the area with poor vegetation nowadays experienced a different degree of tree cover during the recent Holocene. Sedimentological analysis and dating gave evidence to propose a palaeoenvironmental reconstitution for a better comprehension of current evolutions. If the activity of humans is important versus the consequences of the natural predispositions (climatic whims?) which contribute to the understanding of the vulnerability of this ecosystem, however, this is the principal transformer of the vegetation's shapes.

9.2 VULNERABILITY OF THE CAMEROONIAN SUDANO-SAHELIEN MILIEU

From the climatic point of view, high annual average temperatures (28 °C), the strong evaporation of more than 2.000 mm/year and the high insolation of 2.750 to 3.000 hours per year are causing up to 7–8 months of a dry season. The high hydrological deficit raised by a weak climatic index (0,21 in Maroua) (Suchel, 1987), worsened by the influence of the Harmattan; a hot and dry wind causes the drying out of xerophytic and thorny vegetation. Bared, desiccated and destructured lithosols and regosols on the slopes, ferruginous soils on pediments and plateaus and the alluvial soils of the plains are hammered by the downpours at the beginning of the 4–5 months rainy season with 400 to 700 mm/year which are mainly concentrated to more than 50% of rainfall in July and August. Thus, the scouring of soils during the rainy season follows the swirls of siliceous dust during the dry season (Wakponou, 2004).

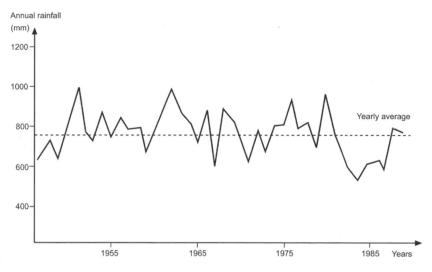

Figure 2. Interannual variability of pluviometry in Maroua. Source: IRA Djarengol (Maroua), after Kaiser quoted by Seiny Boukar (1990).

Generally, years of severe drought e.g. 1968 to 1974 and 1983 to 1985, are more the results of the temporal irregularities and the unequal distribution in space than the total deficits in rainfall. Often the rainy years succeed to the dry years. During 50 years of observation (1945–1990), the Institute of the Agronomic Research's climatic station at Maroua recorded 17 years of rain shortage and 33 years with a surplus (Figure 2).

Without going back to the arid phases related to the Quaternary palaeoclimatic oscillations, droughts of historical times are well-known. Reports of the early explorers in Cameroon at the end of 19th and at the beginning of 20th Century (Petermann, 1854; Barth, 1860; Freydenberg, 1907) quoted by Beauvilain (1986) mentioned them in 1830, 1850–51, 1893, 1908, 1914, 1939–45, 1955, 1972, 1980 and 1984–85. Memories of the food shortage of 1912–1913 that Kotoko (fishing population surrounding the Lake Chad and Logone river) called "*skoum nodoumo*" i.e. "the great hunger" are transmitted from one generation to another. It is the same with the hunger of 1921 that the Choa Arabs (stockbreeder population around the Lake Chad) named "*ankra ahkouk*", i.e. "to disavow his/her brother" (Saïbou, 2001).

9.3 AN ORIGINALLY LUXURIANT VEGETATION

Reports of explorers during the 19th century and oral reports mention the existence of forests in this area which is more or less stripped off trees nowadays. Barth (1865), crossing the far north of Cameroonian plains wrote: "...we then entered a dense forest..."(p.118); "...in this part of the forest the Karage was the most common tree, while besides it there was a considerable variety..." (p. 119). Writing about the vegetation of the Sudan (western and partly Central Africa) Urvoy (1949) wrote that: "... apart from the naked marshes of Chari and Logone, and certain plateau of Baouchi, the small soudanian forest covering hills and plains become denser and darker on rivers..." (p. 9–10).

Research in palynology, palaeogeography and pedology revealed an extension of forests and savannas far from their current limits (5–6°N) (Table 1). Wickens (1982) recognized relics of Guinean vegetation in current sahelian areas and relics of sudanian species up to the Tropic of Cancer. Maley (1981) identified in a cross section at Tjéri in the center of the Chadian basin, pollens of sahelian ecological groups (Mimosacea, Balanitecea), sudanian elements (Sterculiacea, Anacardiacea) as well as Sudano-Guinean elements (Euphorbiaceae, Palmae). Lezine (1989) indicated a major change of vegetation composition during the Holocene around 9.000 BP when the current limits of the wet vegetation (5–6°N) were located approximately 400 to 500 km northwards.

This rising towards the north of wetter vegetation (ombrophilous forests, clear forests and savanna woodlands) corresponds to the increase of at least 300 mm of the annual average of the rains up to 21°N. This is confirmed by high lake levels during the Holocene of Mega Lake Chad around 6.000 years BP (Wakponou, 2004). This climate—significantly wetter than the current one—supports the extension of tropical ferruginous soils between 7.000 and 4.000 years BP (Bocquier, 1973; Millot, 1978) whose scraps remain in the Doyang plateau. These vertisol-like soils are contemporary of fluviatile transport of very fine materials resulting partly from the wind deposits belonging to the arid phase of Kanémian before (20.000–12.000 years BP) (Maley, 1981).

Table 2 highlights palaeoenvironmental fluctuations in the Lake Chad basin during the Pleistocene. Absolute ages, different dating methods and the corresponding vegetation are given. One can then establish a parallelism between arid periods with strong wind activity in the centre of the basin and periods of sediment supply scouring and deposits as

Table 1. Theoretical distribution of the recent vegetation.

Geomorphological unit	Type of soil	Vegetation
Mountains and Inselbergs	Lithosoils and regosolic soils	*Ficus, Lannea, Microcarpa Acacia albida, Parkia biglobosa, Butyrospermum parkii, Tamarindus indica, Ziziphus mauritania, Isoberlinia doka*
Pediments	Fersiallitic soils	*Isoberlinia doka Acacia albida, Boswellia dalzielii, Sterculia setigeria*
Plateaus	Tropical ferruginous soils	*Lannea spp, Anogeissus leicarpus, Balanites aegyptiaca*
Plains	Planosoils, solonetz and vertisoils	*Lannea humulis, Balanites aegyptiaca, Anogeissus leicarpus, Acacia sieberiana, Acacia albida, Acacia senegalensis*
	Hydromorphic soils	*Acacia seyal, Acacia nilotica Echinochloa pyramidalis, Colonna, Stagnina, Hyparrhenia rufa, Oriza longistaminata, Pennisentum ramosum, Vetiveria nigritana*
Dunes and sand ridge	Red sandy soils	*Gueira senegalensis*

consequences of the extension of the semi-arid influences (downpours, floods, scattered vegetation) towards the southern edge of the lake. On the other hand, closer vegetation like forest ecosystems have to be linked to the more humid periods.

It is as true that the Quaternary palaeoclimatic oscillations had an undeniable effect on vegetation patterns. According to Aubreville (1967), the drying of the air due to cold waves during the glacial maximum of the Quaternary, caused "a real cataclysm, [...] destroyed in certain areas with least biological resistance". Lezine (1989) noted an arid period marked by the extension of the sahelian "pseudosteppes" between 15.200 ± 300 and 13.250 ± 200 years BP. However, it is difficult to distinguish in the current situation, the ascribable consequences with the palaeoclimatic variations of those emanating from the influence of humans. The time between the pluvials with luxuriant vegetation and the arid periods with poor vegetation cover are the great moments (?) for vegetation sceneries and landscape evolution. Each vegetation type is indeed related to particular bioclimatic and morpho-edaphic conditions.

The vegetation depends on the nature of the soil which itself depends on lithology, on the pedoclimate and on the relief characteristics. Table 2 gives a precise idea of the distribution of this original vegetation in the far north of Cameroon up to early sixties (Letouzey, 1968).

The areas of the theoretical distribution of this vegetation types are illustrated in figure 3. The degraded savanna with *Lannea acida* and *Microcarpa* developed on young arenaceous skeletal soils of the slopes in granitic, andesitic and syenitic inselbergs. More or less deteriorated savanna with Ficus and *Isoberlinia doka* occurs on lithosoils and regosolic soils. Those sudano-sahelian savannas in high altitude are shrubby with colonizing thorn-bushes (Letouzey, 1968). The piedmonts are covered either by relicts of shrubby savanna or by degraded savanna woodland with *Boswellia dalzielli, Anogeissus leicarpus* and locally with *Acacia albida* dominating or with predominance of *Sterculia setigera* and with thorn-bush as in the buttresses of the extreme-south of the Mandara mountains. All these vegetation patterns occur on planosoils—solonetz and fersiallitic soils. *Acacia seyal* is characteristic in the depressions and other sectors with black cotton soils ("karal"), sandy soils and red tropical ferrugineous soils.

All these woody species were and are continuously cut off by an increasing local population.

9.4 POPULATION GROWTH AND ENVIRONMENTAL DEGRADATION

The far-north of Cameroon experienced since the stone age (advanced Acheulean) the first signs of human occupation (Marliac and Gavaud, 1975). This is proved by the discovery of lithic artefacts older than 50.000 years BP such as the double-side ones incorporated in the duricust in the Doyang plateau. Nowadays, the region is the most densely populated in the country (approx. 3.000.000 inhabitants in 2006) with a density of more than 100 inhabitants/km^2).

The cameroonian sudano-sahelien area has relics of forest vegetations recently limited to 4–5°N of latitude, except the gallery forests which seems similarly be relics of a formerly more extend forest cover; they form finger-like structures along the valley bottoms in the Sudanian savannas. Within this testimony vegetation huge rock blocks and/or rock cavities and especially in the forest reserves of Mayo Louti, Zamay in the Mandara Mountains and the reserve of Gokoro in Mozogo in the pediment of Koza can be found (Figure 4).

In these locations, guinean vegetation is protected from bush fires and clearings thanks to its inaccessibility and the prohibitions. Because of its floristic richness (*Ficus, Lannea acida, Microphone carpa, Acacia albida, Parkia biglobosa, Butyrospermum parkii,*

Table 2. Chronology of the Quaternary palaeoenvironmental evolution of the Lake Chad basin.

Age BP (OBDY Lab-No., radiocarbon dating) TL Age (Thermoluminescence dating) reference and source of proxy data	Palaeoclimatic interpretation	Location & archaeological stages		Vegetation dynamics	Morphodynamic processes
		Chad Basin	Northern Cameroon		
100 ± 160 BP (OBDY 578), (recent?) charcoal in 20 cm depth 1.170 ± 150 BP (OBDY 801), charcoal in 65 cm depth 2.780 ± 90 BP (OBDY 577), charcoal in 170 cm depth 3.330 ± 1.000 TL and 3.550 ± 760 TL, sand on top of dune 4.260 ± 35 BP (OBDY 869), fossil wood in 5 m depth	Rapid alternations of humid and dry periods		Actual, recent (Hervieu, 1969)	Re-expansion of forest	8: recent fluvial dynamics: acceleration of soil erosion, outcrop of the basement complex; continuation of downslope pedimentation; phyto-instability on the slopes by human activities; filling of river beds; beginning wind erosion, deflation
6.000 ± 240 TL (Durand et al., 1981) 8.000 BP (Pullan, 1965) 8.720 ± 420 BP (OBDY 379), nodular limestone 10.000 BP (Grove et al., 1964)	Short episode of tropical humid climate	Nigero-Tchadien	Sub-actual (Hervieu, 1969)	Forest	7: major lacustrine episode: inducing lacustian bar (aeolian fine sand, coarser grains) and filling up of the clay blow-outs; formation of limestones concretions within clayey soils
7.000 BP, Diatomite layer, aeolian sands (north of Mega Lake Chad) (Servant, 1983) Lake regression around 6.300 ± 200 BP 14.500 ± 1.690 and 15.360 ± 2.040 TL, *Pètté*, red sands at the base and 3.330 – 3.350 TL at top of a dune	Dry period subhumid-sudanian: annual rainfall 800 mm per year	Kanemian (Servant, 1983)	Bossoumian (Hervieu, 1969)	forest regression	6: aeolian episode: disappearance of lakes (Servant, 1983); negative sedimentary assessment, setting up longitudinal sandy cords, temporary drying out of the Lake Chad basin; formation of blow-outs; pedogenesis (rubefication) of sands at Kalfù's erg; palaeosoil formation; deposition of palaeo-erg in Cameroon extending to the south and northwest
11.940 ± 750 TL, red dune sand 8.980 ± 1.150 TL and 8.240 ± 1.120 TL, red sand on top of dune10.640 ± 3.930 TL and 8.670 ± 2.140 TL red sands on top of dune	saharo sahelian arid to semi-arid period; annual rainfall below 400 mm per year			formation of an open and discontinuous vegetation	

Age / Period	Climate	(Servant, 1983)	(Hervieu, 1969)	Vegetation	Process
34.460 ± 5.690 BP and 32.950 ± 5.160 BP (TL), palaeosoil at the base of sands dunes in *Pétté*, according to Mathieu (1980), Pleistocene Lake episodes: Megalake Chad at 400–420 m (Pias, 1970); according to Servant (1983) and Servant-Vildary (1978), many lakes within the Lake Chad basin	dry period tropical humid climate	Upper Ghazalian (Servant, 1983)	Peskeborian (Hervieu, 1969)	dense vegetation (forest?)	5: Loess episode: Doyang layer (Mainguet and Wakponou, 2002); fluviale épisode (Marliac and Gavaud (1975); silt-clay layer; kaolinite remobilized by wind. High evapotranspiration of small lakes; discordant sedimentation of silt and clay on the Continental Terminal (CT).
Middle Stone Age advanced Acheuleen (Marliac and Gavaud, 1975), double-side encrusted duricust before 50.000 BP	dry/sub-humid with two contrasting seasons; annual rainfall estimated between 600 and 800 mm per year	Lower Ghazalian (*Servant*, 1983)	Douroumian (Hervieu, 1969)	more open discontinuous vegetation cover	4: Pisolitic duricust formation with coarsed vascuolar structure
Plio-Pleistocene Tertiary, Miocene?	semi-arid	Pre (Ante-) Ghazalian (Servant, 1983)	Pre-Douroumian (Hervieu, 1969)	sparse vegetation cover	3: Pedimentation on inselberg footslopes
Eocene, Oligocene	long arid period			sparse vegetation cover	2: Continental Terminal deposits
Mesozoic	tropical humid climate			dense vegetation cover	1: humid tropical weathering of Precambrian basement

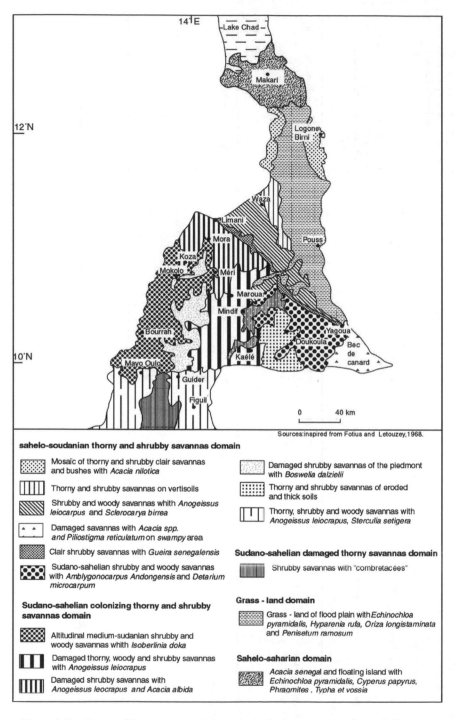

Figure 3. Distribution of the recent natural vegetation in the Cameroonian sudano-sahelian area.

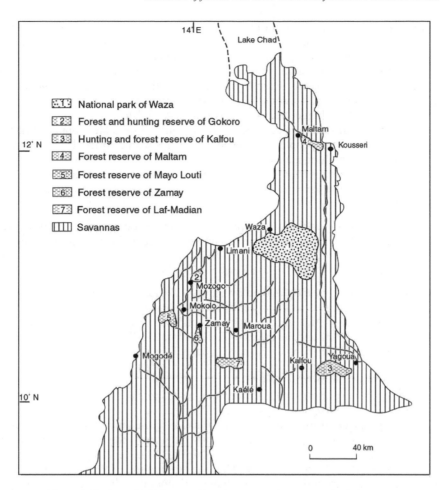

Figure 4. The "so called" forest and hunting preserve in extreme North Cameroon.

tamarindus indica) and its luxuriance due to wetter climatic conditions than in the sectors of Piedmont, this vegetation was qualified by Letouzey (1968) as "primitive vegetation".

For a long time the relatively "slow" vegetation evolution followed palaeoclimatic fluctuations but currently, it is accentuated by the exacerbated use of the land by humans. The transformation of vegetation sceneries are always related to old settlements or to old civilizations.

In a study of palynoflore (Médus *et al.*, 1990) that we carried out on the sub-actual terraces of the rivers Mayo Ranéo and Mangafé in the north of Maroua, the extent of *Poacea, Malvacea, Ipomea, Bibiocus* and the development of the coprophiles (*Chaetomium*) testify human influences. The sporopollinic zones thus reflect an extension of anthropic action on the savannas. *Balanites aegyptiaca* which can even disappear took over from *Acacia albida* and *Acacia seyal*. The extension of *Balanites* around 1.500 years BC indicates deterioration due to the drying out of climate and increasing deforestation by growing human populations supporting *Acacia albida* previously built and then bound to bovine breeding.

Thus the environment is "suffering" from the daily survival struggle of the population. Their compromising acts render the search of new cultivation areas (60% of the population are peasants), the search of new pastures (2nd area of the cattle breeding with 2.000.000 cows and 1st for the small ruminants with 2.500.000 beasts) and in the search of firewood (50.000 t/year: statistics of the Provincial Service of Environment). More than 95% of households use exclusively firewood for the kitchen. This problem of firewood and the safeguarding of harvests against predators like "thieving" birds and monkeys are solved by cutting off trees and shrubs. What follows is the deterioration of the clear forest with disappearance of the bigger trees (*Angeissus leiocarpus*, *Acacia* sp., *Balanites*). In the vegetation sceneries, the halos of disappearance are quite discernible around the hamlets. Firewood scarcity leads the people to use even the species with calorific low value such as *Calotropis procera*, cow dung and stalk of millet.

In this vulnerable ecosystem, a lime factory ("Chaux Roca") established in 1946, furnished for 57 years, uses exclusively firewood (18 m^3 per day) for the cooking of marble, and all this, in spite of:

— "Green Sahel Operation" (trees plantation) launched in the years 1970–1974;
— the fact that the concept of "sustainable development" is the chorus in political speeches;
— the fact that the slogan "fight against poverty" is a leitmotiv.

According to Smith (1976), quoted by Saïbou (2001), the destruction of forests was used long ago as an offensive tactic of war by the great conquerors in the Chad basin at the dawn of the 20th Century. Indeed, barren grounds and grassy savannas constitute a field of predilection for the cavalry and the infantry. For example the "Maï" (King) of Bornu State, Idris Aloma, subjected. Sao (people who have lived in the south of Lake Chad, characterised by baked clay civilization with its apogee between the 9th and the 16th century) by clearing natural vegetation around their settlements.

All the vegetation patterns in the Cameroonian sudano-sahelian area would have been rich in species as is the case nowadays in the forest reserves.

The disappearance of the trees is the initial cause of the environmental deterioration of the dry ecosystems (Mainguet, 2003). The consequences of the regressive dynamics of the vegetation cover on soil surface are its exposure to the meteorological agents and the destructuration which follows from there. The overexploitation of the soils damages them and transforms them into duricrusted, unsuitable soils for agriculture because of very low capacities of water reserve, very defective physical properties and the high risks of erosion. The soils baring exacerbates the evaporation, increase the capillarity ascent of percolating water then generating unfertile crusts.

9.5 SUGGESTIONS FOR A SUSTAINABLE MANAGEMENT OF THE TREE RESOURCE

Of all that precedes, it can be stated that the Cameroonian sudano-sahelian area is predisposed under particularly difficult natural conditions because of the climatic situation, it is also true that the region suffers from inadequate methods of exploitation of the natural resources.

Timid efforts of restoration are proposed by the authorities. Young seedlings especially *Eucalyptus* distributed within the framework of the agroforestery projects do not meet the needs for the peasants who should adopt other "useful" species. The peasants have indeed peculiar knowledge on almost the totality of the trees species—even on grass species—which surrounds them, which is illustrated by table 3.

The cultivation of xerophytic species should be promoted in the parks and not "ghost" trees (e.g. species which cannot support ecologic dry conditions) as those found in "so called" forest reserves created since 1977 within the framework of "Green Sahel Operation". The forest reserve of Kalfou for example is reduced nowadays to scattered trees and can be considered as the "firewood reserve" of the nearby Yagoua town. The situation of other reserves is not even better. Moreover, the example of *Eucalyptus* is edifying. In spite of its use in the making of frame, roof and firewood this tree is dangerous for the sahelian environment. An old *Eucalyptus* absorbs up to 200 litres of water per day, it can pump up underground water up to 15 m of depth. The reddish juice (Kino) that this tree secretes is very rich in tannin and poisonous for insects and other microfaunes with the pedophere. Moreover, *Eucalyptus* is a very dangerous pyromaniac tree in the dry environments. The sparks of bark set ablaze the foliage where the sheets release from volatile oils which propagate fire to the neighbouring trees at the time of the bush fires.

In addition to the parks and forest reserves with the local species, agroforestery with especially leguminous plants such as *Acacia albida* should be extended to all the surfaces of cultures in sudano-sahelian ecosystems. The cutting of trees should be limited to pruning. Moreover, the government should popularize other sources of energy than firewood in the rural dry milieu. If oil, (600 FCFA/litre), electricity (60 FCFA/kWh) and the domestic gas (6.500 FCFA for a bottle of 12 kg) are not available everywhere and are expensive enough for the purse of the peasants, the aeolian wind energy and especially the solar energy should be developed in the sudano-sahelian northern Cameroon which experiences 2.750 to 3.000 hours of average insolation per year and where the mean velocity of the winds is 20 m per second.

9.6 CONCLUSIONS

If the year 2006 had been declared by the United Nations as the "International Year of fight against the desertification", it is that on all the scales (global and local), it would be necessary to wake up the consciences and to create in each one the interest of the management of biodiversity for a sustainable development.

Dating back in history and mostly nowadays, human activities (deforestation, fire) contribute to accelerate the process of desertification of all the sub-Saharan and sahelian area in general and of the Lake Chad basin in particular. In this fragile ecosystem, the environmental deterioration generates conflicts between stockbreeders and farmers on the control of reducing remaining natural resources, migratory movements, problems of water and hydraulic diseases, food insecurity, recurring starvation and poverty.

The Cameroonian sudano-sahelian area experienced different degree of tree cover during the recent Holocene. Nowadays, its relics still exist in inaccessible slopes, between the rock blocks or in the forest reserves. The contracting of this vegetation is due to the influence of humans, even if the fragility of the milieu predisposes it to deterioration.

An efficient management of the natural resources is the key to development which takes into account the environment. The populations of the dry milieu more than those of everywhere else should become aware that vegetation is a potentially renewable natural resource and not inexhaustible. The agricultural monitoring by development companies and other NGO' s (Non Governmental Organizations) should inculcate in the peasants the sense of participative development, by associating them to the problems identification and the search for solutions, all this within the framework of the organizations of the village communities. These populations would have to find new relationships with their ecological environment (Mainguet, 2003). It is in fact a problem of responsibility that of the authorities to put an end to lax management of resources, to stigmatize the environmental problems in these vulnerable milieus and that of the peasants to contribute to the preservation an management of their environment for a sustainable development: i.e. to ensure their existence and survival without compromising that of the future generations.

Table 3. Selection of some trees used in agroforestry according to field investigations.

Scientific names	Local names	Food	Pharmacopeia	Well	Cosmetic	Soil Restoration	Insecticide	Firewood	Construction	Rope	Timber	Shade	Other income
Acacia albida	Ngabdé		X	X		X							
Senegal acacia	Datché	X	X		X	X		X					X
Adansonia digitata	Mbocki	X	X	X						X			
Anogeisus leiocarpus	Kodjoli		X	X				X	X		X		
Anona senegalensis	Laddé	X	X										
Azadirachta indica	Gagné	X	X	X		X	X					X	X
Balanites aegyptiaca	Tanné	X	X	X									X
Bombax constatum	Jééhi	X							X				
Borassusa ethiopum	Mbaassi	X								X	X		X
Boswellia dolzielli	Andakehi			X									
Combretum aculeatum	Mbouski		X					X	X				
Commifora kuntiana	Bannahi			X									

Species	Local name	1	2	3	4	5	6	7	8	9	10	11	12
Ficus gnaphalocarpa	Ibbé	–	–	X	–	–	–	–	–	–	X	X	X
Ficus platyphylla	Doundéhi	–	–	X	–	–	–	–	–	–	–	X	X
Ficus sp.	Tchékéhi	–	–	–	–	–	–	–	–	–	–	X	X
Gueira senegalensis	Geelewki	–	–	X	–	X	X	–	–	–	X	–	X
Hexalabus monopetalus	Boyli	–	–	–	X	X	X	–	–	–	–	–	X
Kaya senegalensis	Ndaléhi	X	–	–	–	X	X	–	–	X	X	X	–
Kigelia africana	Gilaahi	–	–	–	–	–	–	–	–	–	X	–	–
Piliostigma sp.	Barkéhi	–	–	–	X	–	–	–	–	–	–	X	–
Propopis africana	Kohi	–	–	–	–	X	X	–	–	–	–	X	X
Sterculia setigera	Bobori	–	–	–	–	–	–	–	–	–	X	–	X
Stereospermum kunthiana	Golombi	–	–	–	–	–	–	–	–	–	X	–	–
Strychnos spinosa	Narba-tanahi	–	–	–	–	X	X	–	–	–	X	–	X
Terminalia macroptera.	Koulahi	–	–	X	X	–	–	–	–	–	–	–	–
Vitellaria paradoxa	Karité	X	–	–	–	–	X	X	–	–	X	X	X
Vitex doniana	Ngalbidjé	–	–	–	–	X	X	–	–	–	–	X	X
Ximenia americana	Tchabbulé	–	–	–	–	–	X	–	–	–	–	X	X
Ziziphus mauritania	Djaabi	X	–	X	–	X	X	–	–	X	–	X	X

REFERENCES

Aubreville, A., 1967, Savanisation tropicale et glaciations quaternaires. *Adansonia*, Vol. 21, pp. 16–84.

Barth H., 1865, Travels and discoveries in North and Central Africa. Being a journal of expedition undertaken under the auspices of H. B. M'S Government in the years 1849–1855, Centenary Edition. In *Tree Volumes,* Vol. 2, (London: Frank Cass & Co. Ltd), p. 637.

Beauvilain, A., 1986, Les variations du niveau du lac Tchad. In *Revue de Géographie du Cameroun*, Vol. 6, **2**, Yaoundé, pp. 26–34.

Bocquier, G., 1973, Genèse et évolution de deux toposéquences de sols tropicaux au Tchad. *Interprétation biogéodynamique*, (Paris: Mémoire ORSTOM, 62), p. 350.

Durand A and Mathieu P., 1981, Evolution paléogéographique et paléoclimatique du bassin tchadien au Pléïstocène Supérieur. *Revue de géologie dynamique et de géographie physique*, **22**, 4, pp. 97–109.

Grove A. T. and Pullan R. A., 1964, Some aspects of the Pleistocene palaeogeography of the Chad basin. *African Ecology and Human Evolution*, edited by H. C. Howel, F. Bourlière and Cie, London, pp. 230–245.

Hervieu J., 1969, *Le Quaternaire du Nord-Cameroun. Schéma d'évolution géomorphologique et relation avec la pédogénèse.* (Paris: ORSTOM, Série Pédologie, 8, N° 3), p.172.

Letouzey, R., 1968, *Etude phytogéographique du Cameroun*, (Paris: Edition Paul Lechevalier), p. 511.

Lézine, A. M., 1989, Le Sahel; 20.000 ans d'histoire de la végétation. In *Bulletin de la Société géologique de France*, Vol. 8, Tome 5, 1, pp. 35–42.

Mainguet, M., 2003, Les pays secs. Environnement et développement. Carrefours, Ed. Ellipse, Collection dirigée par G. Wackermann, Paris, p. 160.

Mainguet M. and Wakponou A., 2002, Les derniers souffles sahariens à la lisière du Monde Sahélo-Soudanien. *Rapport de Mission 09–30 janv. 2002, Extrême-Nord-Cameroun* + annexes, p. 32.

Maley J., 1981, *Etudes palynologiques dans le bassin du Tchad et paléoclimatologie de l'Afrique nord tropicale de 30.000 ans à l'époque actuelle.* Thèse de Doctorat 3ème cycle, Université de Montpellier, Collection T. et D., **129**, (Paris: ORSTOM), p. 586.

Maley J., 1983, Histoire de la végétation et du climat de l'Afrique nord tropicale au Quaternaire récent. Actes du 10ème Congrès, AETFAT, Bothalia 14, 3 and 4, Prétoria, pp. 377–387.

Marliac, A,. and Gavaud, M., 1975, Premiers éléments d'une séquence paléolithique au Cameroun septentrional. In *Association Sénégalaise pour l'Etude du Quaternaire Africain*, **41**, 1975, pp. 53–66.

Mathieu P., 1980, Données nouvelles sur la sédimentation quaternaire au Sud du lac Tchad. 26 ème *Congrès géologique international* (1er Centenaire), 7-17.07.1980, Vol. 2, (Paris : ORSTOM), pp. 66–72

Millot, G,. 1978, Clay analysis. In R. W. Fairbridge & J. Bourgeois, The encyclopaedia of sedimentology. *Encyclopedia of Earth Sciences,* Vol. 6, pp.152–155.

Médus, J, *et al.*, 1990, *Pollenanalyse de dépôts subactuels au Cameroun septentrional*, (in edit).

Pias, J., 1962, *Les sols du moyen et bas Logone, du bas Chari, des régions riveraines du lac Tchad et du Bahr-El-Gazal*, (Paris: Mémoire ORSTOM, Comité scientifique du Logone et du Tchad), p. 438.

Pias J., 1970, *Les formations sédimentaires tertiaires et quaternaires de la cuvette tchadienne et les sols qui en dérivent*, (Paris: Mémoire ORSTOM, **43**), p. 407.

Pullan R. A., 1965, The recent geomorphological evolution of the south central part of the Chad basin. *Samaru Research Bulletin*, **50**, pp. 115–139.

Saïbou, I., 2001, *Conflits et problèmes de sécurité aux abords sud du lac Tchad. Dimension historique* (XVIème–XXème Siècle). Thèse de Doctorat, Université de Yaoundé I, p. 381.

Seiny Boukar, L., 1990, *Régime et dégradation des sols dans le Nord-Cameroun.* Thèse de Doctorat 3ème Cycle, Université de Yaoundé, p. 226.

Servant M., 1983, *Séquences continentales et variations climatiques : Evolution du bassin du Tchad au Cénozoïque Supérieur* (Paris: ORSTOM, **159**), p. 573.

Servant-Vildary S., 1978, *Etude des diatomées et paléolimnologie du bassin tchadien au Cénozoïque Supérieur.* Thèse Doctorat Sci. Nat., Université de Paris VI, T. et D., (Paris: ORSTOM, **84**, Vol. 2), p. 1346.

Smith, R., 1976, Warfare *and diplomacy in pre-colonial West Africa*, (London: Menthuen and Co. Ltd).

Suchel, J. -.B., 1987, Les climats du Cameroun. Thèse Doctorat d'Etat, Université de Bordeaux III 3 T, +1 Atlas, p. 1186.

Urvoy Y., 1949, *Histoire de l'Empire du Bornou; Mémoires de l'Institut Français d'Afrique Noire*, (Paris: Librairie Larose), p. 166.

Wakponou, A., 2004, *Dynamique géomorphologique des basses terres soudano-sahéliennes dans l'Extrême-Nord-Cameroun.* Thèse de Doctorat NR, Laboratoire de Géographie Zonale pour le Développement, Université de Reims Champagne-Ardenne, p. 229.

Wickens, G. E., 1982, Palaeobotanical speculations and Quaternary environment in the Sudan. In *A land between two Niles*, edited by Williams M. A. J. and Adamson D. A. (Rotterdam: Bakema).

CHAPTER 10

The application of organic carbon and carbonate stratigraphy to the reconstruction of lacustrine palaeoenvironments from Lake Magadi, Kenya

Brahim Damnati

University of Abdelmalek Essadi, Faculté des Sciences et Techniques de Tanger, Department of Earth Sciences, Natural Resources and Risks Observatory (ORRNA), Tangier, Morocco

Michel Icole, Maurice Taieb and David Williamson

CEREGE, Europôle du petit Arbois, Aix-en-Provence Cedex, France

ABSTRACT: The organic and carbonate contents of sediments found in two cores from Lake Magadi, Kenya, are presented within the context of a multidisciplinary study. Organic sediments are analysed for their total organic carbon content and also with reference to the indices of Hydrogen (HI) and Oxygen (OI) from Rock Eval pyrolysis. Carbonate sedimentation is studied through the discrimination of calcite sensu stricto/low Mg-calcite on the one hand and Mg-calcite/ (proto-) dolomite/Mg-carbonate on the other hand.

The results confirm the lake-level fluctuations suggested by earlier work. The transgressive lake-level phases are characterised by an organic sedimentation primarily of terrestrial origin. These high level episodes produced laminated deposits, rich in organic matter with an elevated HI (over 450) and rich in calcite. When the water-balance maintaining these high water levels deteriorated mixed calcium and magnesium carbonates (resembling a protodolomite) and magnesium carbonate form. When the lake regresses further, rhythmic deposition ceases, probably due to mixing of the entire water column. The organic matter content decreases owing to mineralisation and dilution by increased clastic sediment input. Of the several discontinuities in the core, only that at the very summit results in substantial diagenesis which obliterates the original sedimentary characteristics.

The most temporally complete sedimentary sequence found within the cores is the high lake-level phase between 12.500 and 11.000 ^{14}C years BP. The sedimentation of the various forms of carbon allows a short period of hydro-climatic deterioration to be identified. The most humid phase lasted until ca 11.000 ^{14}C years BP, before the initiation of a general lake-level regression.

10.1 INTRODUCTION

Lake fluctuations are one of the most important and widely distributed sources of palaeohydrological, palaeoenvironmental and palaeoclimatic information for continental areas over the Late Quaternary. In this paper we will use organic matter and carbonates in lake sediment from intertropical Africa to provide a sedimentary environments history of the lake fluctuations.

Cores were taken from Lake Magadi, Kenya. The cores have been studied within a multidisciplinary framework allowing a multiproxy approach to palaeoenvironmental reconstruction (Taieb *et al.*, 1991; Roberts *et al.*, 1993). A wide range of indicators have been studied including diatom analysis (Barker, 1990; Roberts *et al.*, 1993), sedimentary magnetism (Williamson, 1991; Williamson *et al.*, 1993), palynology (El Moutaki, 1994), and the structural sedimentology of laminae (Damnati *et al.*, 1992; Damnati and Taieb, 1995). The organic matter and carbonate data presented here complements these earlier studies. The accumulation of organic matter (OM) in lake sediments from the warm regions of the earth generally originate under one of the two following scenarios:

Playa environments: OM is produced in abundance by organisms resistant to salt (algae, cyanobacteria, bacteria), and can become rapidly fossilised and preserved under the anaerobic conditions found beneath evaporite crusts (Bauld, 1980; Reyre, 1984; Warren, 1986; Schreiber, 1988; Busson, 1988).

Stratified meromictic lake environments (Demaison and Moore, 1980; Fan Pu *et al.*, 1980): Anoxic conditions in the hypolimnion and at the sediment-water interface create an extremely reducing bottom environment (Hollander *et al.*, 1992) thus producing conditions very favourable to the preservation of OM. In the absence of bioturbation, such environments are the most favourable for preserving sedimentary structures such as laminations. Many such meromictic lakes were established in the intertropical zone during the Late Pleistocene and Early Holocene (Talbot, 1988).

High lake levels are therefore particularly amenable to the accumulation of OM: productivity is high and preservation under anoxic bottom waters is good. Conversely, low levels of organic matter in sediments may be associated with low productivity or poor preservation. Such conditions are frequently associated with a fall in lake level (Yuretich, 1979; Talbot, 1988).

Lake Magadi and Lake Natron, formerly together creating a large palaeolake, like many of the tropical lakes are known to have varied considerably in level during the Late Pleistocene and Holocene (Baker, 1958; Casanova, 1986; Damnati, 1993). It is therefore possible to use the organic matter pattern as an indicator of palaeoenvironmental changes (Damnati, 2000).

The sedimentary carbonates help reveal the hydrological history of Lake Magadi, which is now almost completely dry with a deep covering of trona ($Na_2CO_3 \cdot NaHCO_3 \cdot 2H_2O$). At the time of the humid episodes during the Late Pleistocene and Early Holocene, the lake had more inflowing freshwater streams than today, was deeper and stratified. Rock weathering associated with the development of soils releases an abundance of chemical elements providing nutrients for phytoplankton. High rainfall that causes a rise in lake level also causes an expanded herb and tree cover that reduces the flux of clastic particles into the lakes. For the alkaline waters of Magadi, chemical evolution with increasing concentration is tied to evaporation and a succession of carbonate precipitates (Eugster and Hardie, 1978): with the increasing salinity the first to precipitate is calcite, bringing about the rapid loss of Ca^{2+}. In consequence the ratio of Mg^{2+}/Ca^{2+} in epilimnetic waters increases to the point that the carbonates forming after the precipitation of calcite are Mg calcites, protodolomites (Müller *et al.*, 1972) and magnesium carbonates (Last and de Dekker, 1990). Therefore, the mineral carbonate sequence at Magadi can register the hydrological condition of the lake from large and deep to its present playa form. However, several other factors can influence the idealised model of carbonate precipitation; these will be discussed below alongside the palaeoenvironmental reconstruction. The carbonate and organic matter data are compared here to that already published by the other specialists involved in this multidisciplinary study. It provides further information on the sedimentary and diagenetic mechanisms, and on the palaeoenvironmental succession.

10.2 THE MAGADI-NATRON BASIN

Lake Magadi (1°52' S, 36°E) is located in Kenya at the lowest point in the Gregory Rift (600 m), along a structurally aligned north-south depression (Figure 1). Lake Magadi is part of the same basin as Lake Natron, lying directly to the south in Tanzania. Lake Natron is larger than Lake Magadi (1.000 km^2 compared to 200 km^2) and the two lakes are separated by a topographic sill at 635 m. The combined basin of the two lakes covers an area of 23.000 km^2 (Figure 1).

It extends to the rift margins and its lithology is comprised of basalt trachyte and phonolite traps on the rim of the valley and Pre-Cambrian basement rocks at the extremities of the catchment. Northwards toward the equator, the basin reaches 3.000 m in altitude and receives more precipitation (over 1.500 mm annually: Vincens and Casanova, 1987). This northern part of the basin is drained by the Ewaso Ngiro River, which today discharges the largest inflow directly into Lake Natron.

In contrast to Lake Natron which receives permanent streams, only hydrothermal springs and some temporary streams which drain the trachyte plain at the foot of the Rift Valley supply Magadi. Lake Magadi is presently arid–hyperarid receiving less than 450 mm/year rainfall whilst potential evaporation is almost 5 times greater than precipitation (Vincens and Casanova, 1987). As a result Magadi is almost entirely covered by a trona body which has been commercially exploited since the beginning of the last century. Open

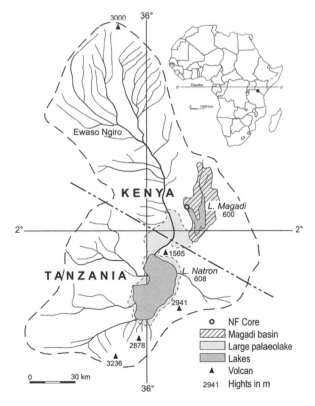

Figure 1. Location map and catchment area of Lakes Natron and Magadi.

water is restricted to lagoons close to springs, an example of which is the Flamingo Nursery (NF) core site (Figure 1). The Lake Natron basin extends over a slightly greater altitudinal range and thus receives more rainfall (>450 mm/year).

Eugster (1970), Surdam and Eugster (1976), Jones *et al.* (1977) and Hillaire-Marcel *et al.* (1986) have studied the chemistry and isotopic geochemistry of freshwater, hot springs and interstitial brines. The hydrochemical evolution of the Magadi brines was interpreted by Eugster during the 1970s: freshwater comparable to that from the Ewaso Ngiro, augmented by precipitation supplies a shallow aquifer. This percolates to supply a much deeper and more saline groundwater reservoir that feeds the hydrothermal springs. Close to the springs the lagoons are relatively dilute but are quickly concentrated by the intense evaporation. The interstitial water retained within the trona crust represents the final stage of this evolution. The chemical composition of these three hydrological units is distinct (Table 1).

Ca^{2+} and Mg^{2+} are found only in the waters descending from the rift in streams like the Ewaso Ngiro River, while the thermal springs are completely impoverished. Temporary local streams carry few alkaline-earth elements given the nature of the trachyte rocks of the rift floor over which they flow. Some traces of Ca^{2+} and Mg^{2+} can be found in interstitial water as a result of the dissolution of calcareous formations exposed at the margin of the palaeolake Natron-Magadi (Figure 1), and from the alteration of ferro-magnesium minerals contained in the ancient lake deposits (Surdam and Eugster, 1976). Table 1 shows also the enrichment of the springs and brines in phosphorus, (10 and 70 ppm of PO_4 respectively). This element is much more abundant in Lake Magadi waters than in many eutrophic lakes of the temperate zone of the world and this abundance in part explains the large primary production found in this evaporitic environment known to feed numerous flamingos. Finally, table 1 demonstrates the elevated level of sulphates in the hydrothermal waters and brines: 150 and 2.100 ppm respectively.

Table 1. Chemical composition of the main types of water in the Natron-Magadi basin. 1) Ewaso Ngiro River, the most important perennial river; 2) Hot springs; 3) Interstitial brines in trona crust of the Lake Magadi. From Surdam and Eugster, (1976). Values in mg/l. TDS = Total dissolved solids. n.d. = not determined.

	Ewaso Ngiro River	Hot springs	Interstitial brines
Na	7,00	12.166,00	119.500,00
K	2,30	212,00	1.470,00
Ca	6,50	0,50	n.d.
Mg	3,70	0,00	n.d.
SiO_2	20,00	91,00	976,00
HCO_3	48,00	14.600,00	3.250,00
CO_3	0,00	3.490,00	94.050,00
SO_4	2,40	159,00	2.100,00
F	0,20	154,00	1.440,00
Cl	4,00	4.740,00	71.750,00
Br	n.d.	114,00	265,00
PO_4	0,03	10,00	70,00
B	n.d.	8,60	88,00
TDS	77,00	29.170,00	294.000,00
pH	7,00	9,20	10,50

Although presently separated by the topographic sill mentioned above, Lakes Natron and Magadi have formed a huge single lake with a surface area of 2.000 km^2 on several occasions during the Quaternary (Figure 1) (Butzer *et al.*, 1972; Casanova, 1986; Icole *et al.*, 1990).

10.3 THE FLAMINGO NURSERY (NF) CORES DESCRIPTION

Flamingo Nursery, situated in the north-western spur of Lake Magadi (Figure 1), has yielded two parallel cores ca. 9 m long. Cores were obtained by a 1 m stationary piston corer aided by chain hoists attached to a scaffolding platform. The synthesis of these two cores shows three lithostratigraphical units resting on sandy material with three probable hiatus at 740 cm, 295 cm, and 6 cm (Figure 2) (Taieb *et al.*, 1991).

The lower unit (870–740 cm) contains laminated sediments on the scale of several mm to a few cm. These laminae are alternately rich in organic matter and sodium silicate (magadiite). The diatoms indicate a general shallow, saline water body reaching intermediate levels toward the top of the zone but remaining highly alkaline throughout (Barker, 1990).

The intermediate unit (740–ca. 295 cm) contains homogenous silts and small pyrite cubes, low in organic matter and carbonate. The high values of magnetic susceptibility suggest a high proportion of clastic material. The diatom flora represents two assemblages, one saline lacustrine community alongside a more dilute water group, possibly from an inflowing stream (Barker *et al.*, 1990).

The upper unit (295–6 cm) is generally laminated, particularly above 273 cm. The transition with the intermediate unit (295–273 cm) is formed from homogenous silts marked by a progressive lowering of magnetic susceptibility. The major change in the diatom assemblages occurs at 273 cm where *Nitzschia* sp. af. *fonticola* becomes dominant and remains so until 130 cm. The cyclical tendency of the sedimentation is established between 280 and 273 cm where the sediment becomes laminated on a millimetre scale with alternate bands of light and dark sediments. Above 130 cm, the diatom population changes to species within the *Nitzschia latens* group, suggesting a return to more saline-alkaline conditions (Barker, 1990; Barker *et al.*, 1990).

The lake level fluctuations indicated by the previous palaeoenvironmental parameters have been placed in a time frame by ^{14}C and U/Th. The latter were made from a model of Uranium lessivage in the open environments typical of soils and sediments (Goetz, 1990). ^{14}C dates have been made on total organic matter from the upper and lower units. Conventional and AMS measurements have produced dates in close accordance with each other for the upper section between 12.500 and 11.000 years BP. Ages confirmed by U/Th measurements on organo-phosphates from a segment between 168–148 cm. The lower unit has been dated between 40.000–48.000 years BP by U/Th analysis of a section 750–770 cm. Unusually high C/N ratios and strongly negative δ^{13}C values suggest contamination of carbon by anoxic bacteria in the lower unit linked with hydrothermal circulation and we consequently prefer to accept the U/Th ages for this part of the sequence (Taieb *et al.*, 1991). The intermediate unit has been dated ca. 24.000–25.000 years BP (440– 470 cm) (Figure 2).

10.3.1 Methods

The cores were bisected lengthways and samples for organic matter and carbonate analysis were taken in 5 cm^3 plastic cubes. Both the organic and inorganic carbon were measured with a CHN Leco 800. This machine measures the quantity of CO_2 given off by the heating of a sample to more than 1.000 °C, on the basis that the CO_2 comes from the combustion

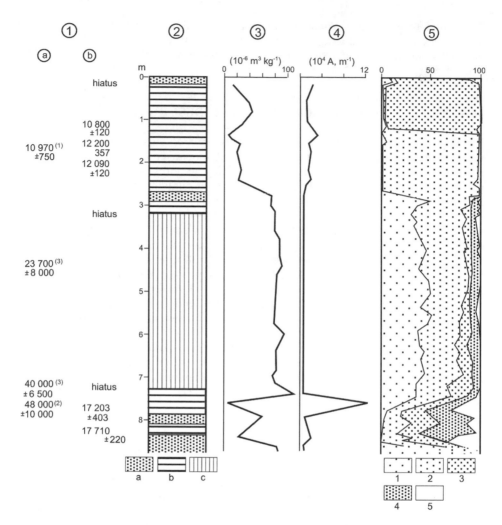

Figure 2. NF core synthesis from Lake Magadi. 1) Chronology (yrs BP); U/Th (a) and ¹⁴C from total organic matter (b). 2) Lithostratigraphy. (a): sands; (b): silty-clay lamina; (c): homogeneous silts; 3) Magnetic susceptibility (X); 4) ARI 1000MT/ARIs = Hematite index; 5) Summary diatom diagrams (after Barker, 1990) with some of the dominant species (distribution of Nitzschia sp. af. fonticola and Nitzschia group latens. Synthetic diagram: (1): freshwater diatoms; (2): planktonic moderately alkaline taxa. (3) planktonic strongly alkaline forms (4): benthic and periphytic strongly alkaline species. (5): others mostly aerophilic.

of organic matter or the decomposition of carbonates (Kelts and Hsü, 1978; Kristensen and Andersen, 1987). The first measurement gives total carbon (C1), the second after heating (at 520 °C) gives the quantity of inorganic carbon, but in the case of Magadi this also includes the inorganic carbon of calcium and magnesium carbonates unstable at 520 °C (Webb and Heysteck, 1957).

Rock Eval pyrolysis (REP) (Allison, 1965) is used here especially in order to determine the nature and the origin of the sedimentary organic products in samples with levels of TOC above 0,5%. Below this level of TOC, the REP data are not significant. The method permits the detection of hydrocarbons (HC) and the CO_2 given off during pyrolysis at 600 °C in an inert atmosphere (Espitalie *et al.*, 1985). These analyses permit two parameters to be calculated about the organic matter: the hydrogen index (HI expressed as mg HC/g. TOC) and the oxygen index (OI in mg CO_2/g. TOC). Plotting HI against OI offers a similar result to elemental analysis of Kerogene and the diagram of Van Krevelen (1950). Essentially, this approach assumes that the organic lake sediments contain a mixture of products originating from higher plants (and indirectly from soil humus), as well as autochthonous algal and bacteriological material. The former give a low HI, the latter one have a much more elevated value and therefore can be used to discriminate the different components of the organic matter (Vandenbrouke *et al.*, 1985). Moreover, as autochthonously produced OM is more liable to mineralisation, abrupt changes in the HI can reveal discontinuities in the sediments (Talbot and Livingstone, 1989).

Carbonate identification is by x-ray diffraction (XRD) on raw, powdered samples, and on minerals. The abundance and composition of calcium and magnesium carbonates has been approximately estimated from the behaviour of ray 104 (XRD) (Goldsmith and Graf, 1958; Périnet, 1974).

In order to better understand the importance of calcium and magnesium to the carbonates, the samples were treated by various acid solutions (after sodium hypochlorite to remove the OM): 1. HCl N/50 28 °C for one hour, 2. as before but at 75 °C, 3. HCl N/5, two hours at 75 °C. Dissolved Ca and Mg were measured by atomic adsorption spectrometry.

These measurements were controlled by XRD which showed that the treatment with weak acids (1.) destroys the calcite sensu stricto, and low magnesium calcite. However, some carbonates with the composition of a protodolomite were also destroyed, notably between 261 and 235 cm. Treatment (2.) removed most of the calcite, the dolomites and eventually magnesium carbonates. Treatment (3.), the most aggressive, destroyed all the carbonates but also some other minerals, in particular some ferromagnesian minerals (frequent release of iron hydroxides).

10.3.2 Results

The distribution of the various forms of carbon in the Flamingo Nursery core (Figure 3 and figure 4) reflects the general stratigraphic outline having three main sections, i.e. a remarkably homogenous silty-clay intermediate unit with very little organic or inorganic carbon dividing two relatively carbon rich sections. Good agreement is shown amongst all forms of carbon except for the top 100 cm in which the level of TOC falls rapidly, leaving only a trace above 65 cm. The opposite trend is found in the carbonates. This suggests post-depositional changes as mentioned above.

The distribution of calcite (cf. Thermo-stable inorganic carbon, figure 3 and figure 4) shows variation in relative abundance in the cores. In the lower unit, calcite is not important except between 760 cm and 740 cm. In the upper unit, the calcite levels increase between 290 and 270 cm; between 270–230 cm levels fall, before increasing again from 230–134 cm. At 130 cm the levels are greatly diminished and stay low until about 100 cm. About 65–70 cm the calcite levels are sharply augmented and maintain a high level until 25 cm.

The distribution of magnesium carbonates (cf. Thermo-unstable inorganic carbon, figure 3) generally follows the pattern established by calcite. There are however a few exceptions; in the segment 865–765 cm in the lower unit magnesium carbonates are abundant whilst calcite is low, and similarly in the sections 270–230 cm and 130–70 cm of the upper unit.

The organic sedimentation characterised by REP reveal clear differences (Figure 5). The upper unit is formed of organic matter largely of the same origin and having a constantly high HI above 450. At the base of this unit between 295 cm and 270 cm the organic matter shows an abrupt change, from a HI of 89 to one of nearly 300 at the base of the laminated

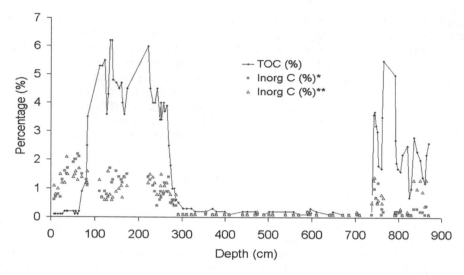

Figure 3. Organic and inorganic Carbon distribution from CHN results after overchecking by chemical analyses. TOC (%) = Total Organic Carbon percentage. Th. St. Inorg. C (%)* = Thermo-stable inorganic carbon percentage (mostly calcite and low Mg calcite, see in text). Th. Unist. Inorg. C. C. (%)** = Thermo-unstable inorganic Carbon percentage (mostly calcium and magnesium carbonates).

section. In the lower unit the HI is more variable oscillating between the high values of the upper unit and close to 200. Note the high value of the HI in the basal sands and the lack of lowering of HI below the likely disconformities (Figure 5).

In summary, the distribution of carbon in the Flamingo Nursery cores allows the distinction of the following zones from the base upwards:

1. basal sands relatively rich in OM with variable and elevated HI (436);
2. a lower unit comprising: a segment 865–765 cm with variable organic matter in both type and quantity, low amounts of calcite and lower magnesium carbonate; the section 760–740 cm with same pattern of OM but carbonates dominated by calcite. The OI is variable oscillating between the high values (300–500) and low values (100–200);
3. an intermediate unit is very homogenous with little OM or carbonate. It has no significant REP data (no significant HI and OI data);
4. an upper unit contains the following sections: from 290–270 cm both inorganic and organic carbon increase progressively as does the HI. High OI is observed ca 290–295 cm drops down and remains relatively stable to the top; from 270–230 cm calcite is low, magnesium carbonates and OM are abundant; from 230–130 cm are all forms of carbon abundant; from 130–70 cm calcite rises again after 100 cm; from 70–6 cm, the section is affected by late diagenesis. OM is only a trace, calcite and magnesium carbonates are abundant.

10.4 DISCUSSION

In order to expand these results to provide palaeoenvironmental interpretations, the mechanisms that have generated the changes in the carbon stratigraphy need to be examined. The organic carbonates indicate that, between 865–756 cm, Lake Magadi was disconnected from its sources of Ca^{2+}. The formation of magnesium carbonates could be

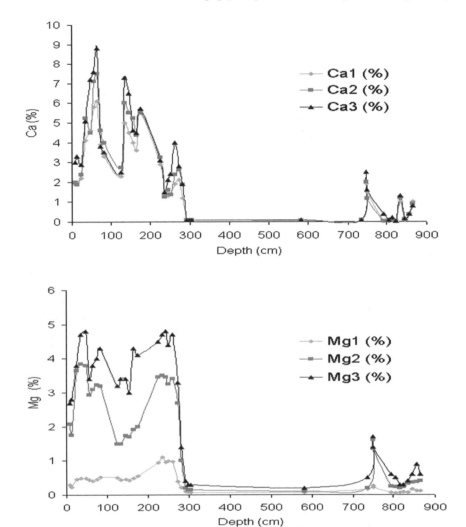

Figure 4. Release of Ca^{2+} and Mg^{2+} by various strength of acid treatment on bulk material (1), (2) and (3), (see in text).

from the reworking of Mg^{2+} from mafic minerals in old lake deposits. On the other hand the calcite-rich segment 755–740 cm marks a higher lake level. The magnetic mineralogy has shown the presence of iron oxides: tracers of the connection with the full drainage basin suggesting the establishment of the connection of Magadi and Natron (Williamson *et al.*, 1993). The diatoms are dominated by the facultative planktonic *Nitzschia* group *latens* to the exclusion of benthic forms, that indicates that although the lake was relatively deep it remained rich in salts. A good correlation exists between the peak in calcite and in *Nitzschia* group *latens* near the top of this unit.

The intermediate unit is homogenous and poor in all forms of carbon, suggesting a lake sufficiently low to have been well mixed by the wind. This precluded the development of sedimentary structures and allowed the mineralisation of the organic matter. The destruction of the organic matter could also have stemmed from the presence of sulphate

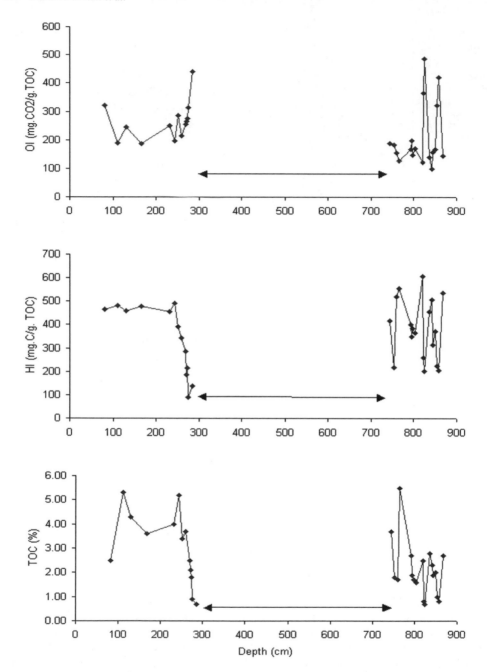

Figure 5. Rock Eval Pyrolysis data. TOC (%) = Total Organic Carbon percentage. HI = Hydrogen Index (mg HC/g. TOC). OI = Oxygen Index (mg CO_2/g. TOC). (NB: In the intermediate unit the TOC is less than 0,5%: in this zone the REP data are not significant).

reducing bacteria which consume OM (Berner, 1980). This would require the presence of anoxic conditions below the sediment-water interface.

The high OI ca 290–295 cm and in some levels at the core bottom is due probably to the oxidation phenomenon during the exposure and desiccation of the sediments.

The upper unit may be divided into a number of sections. At its base there is a section (285–270 cm) characterising transgressive conditions; that is, high in all forms of carbon without stratification becoming firmly established to form laminae. Above this, the laminated sediments indicate the existence of a deepwater lake, a conclusion confirmed by the chronological correspondence with the stromatolite belt 50–58 m above the present lake. The sedimentary structures suggest the lake was stratified and meromictic with an epilimnion fed by fluvial discharge from the higher altitude parts of the catchment. The chemistry of the epilimnion is revealed by the diatom populations which indicate a relatively dilute water body between 270 and 135 cm but a return to more saline-alkaline conditions above 135 cm. The hydrogen index is high for the section 270–70 cm suggesting high primary production and high stratified lake water level as observed in other tropical African lakes (Talbot, 1988; Talbot and Livingstone, 1989).

The chemistry of this large palaeolake is also indicated by the carbonate precipitates. In general the upper unit is rich in calcite, dolomite and magnesium carbonates. The frequent juxtaposition in the samples of calcite sensu stricto or low magnesium, with carbonates much richer in magnesium is problematic within a deep stratified lake. It appears that the chemistry varied sufficiently to allow the precipitation of the full sequence of Ca and Mg carbonates. One explanation is that these reflect short-term (possibly seasonal) deterioration in the hydrological balance during a generally wet episode. The precipitation of dolomite and mixed magnesium carbonates may also explain the high values of ^{13}C in the stromatolites (Hillaire-Marcel *et al.*, 1986). This had previously been thought to reflect the long residence time of the water in the lake, but evaporative concentration may be also invoked as a mechanism. The abundance of calcite sensu stricto fell abruptly above 130 cm and remained low until about 80 cm above which it greatly increased corresponding to the late mineralisation of the organic matter. The drop in calcite at 130 cm is accompanied by an important change in the diatom assemblage (Figure 2), and by a diminution in iron oxides indicating a separation of Magadi from Natron and hence from the high drainage basin (Williamson, 1991; Williamson *et al.*, 1993). Therefore, the fall in calcite may also reflect the relative impoverishment of the lake in base cations. The fall in lake level which caused the separation of the two lakes results from an increase in evaporative concentration caused by a reduction in the precipitation/evaporation ratio.

The palaeoenvironmental reconstruction based on the organic matter and the carbonates is in accord with the other analyses apart from the upper 80 cm where late depletion of the organic matter under a vadose environment is thought to have occurred. For example, and in contrast to other parts of the cores, the rise in secondary calcite above ca. 80 cm is not marked by any significant change in the diatom assemblages (Figure 2). The fall of the phreatic nape indicates a lower water table than is found today, probably as a result of a more arid climate at some time during the Mid-or Late-Holocene. The enhanced aridity is also likely to have caused erosion of the upper part of the lake beds.

The discontinuity at the top of the core (6 cm) is unlike that at 740 cm where the levels of organic carbon stayed high. The truncation occurring in both cases was followed by exposure of the surface sediments. However, no diagenetic phenomena mark the lower deposits and the organic matter is preserved suggesting either the hiatus was brief and/or the water table remained at or above the surface.

10.5 CONCLUSIONS

The sedimentary carbonates and organic carbon in this basin have generally proved to be good indicators of changing environments and lake level. Firstly, stably stratified (meromictic) waters developed even before the lake reached the sill at +35 m marking the connection with Natron. The highest lake levels known are 20 m higher than this sill and these have left evidence of even greater stability in the stratification regime, preserving organic matter in what was by the standards of the East African rift a water body of modest depth. Secondly, the supply of Ca^{2+} and more generally alkaline earths and some magnetic oxides are strictly controlled by the water level. Each time the lake has risen above the sill at 35 m calcite has been able to form in abundance.

Within the context of a multidisciplinary study it has been possible to distinguish the syn-sedimentary phenomena from diagenetic ones. Some unusual sedimentary characteristics have also been found during this study. One of these is the juxtaposition of calcite and calcium and magnesium carbonate, which suggests that even during a relatively wet period the precipitation–evaporation balance must at times have remained negative.

ACKNOWLEDGEMENTS

This work has been funded by INSU programme PIRAT, by LGQ = CNRS (CEREGE, France), and by NERC. The fieldwork has benefited from the support of the Department of Geology of the University of Nairobi, the Magadi Soda Company, and the Cultural Service of the French Embassy in Kenya. R. Lafont and S. Bieda of LGQ (CEREGE, France) and N. Garcia of COM-URA 41 have assisted with some of the analyses. The Rock Eval Pyrolysis was done at IFP-Rueil-Malmaison, with the kind permission of A.Y. Huc. We thank also N. Roberts, P. Barker and D. Williamson for their comments of the manuscript and for English spelling and the Earth Sciences department-ORRNA (FST-Tangier Morocco).

REFERENCES

Allison, L.E., 1965, Organic carbon. In *Methods of soil*, edited by Black, C.A. (Madison: American Society of Agronomy), pp. 1367–1378.

Baker, B.H., 1958, Geology of the Magadi Area. *Geological Survey of Kenya*, **42**, p. 82

Barker, P.A., 1990, *Diatoms as palaeolimnological indicators: a reconstruction of Late Quaternary environments in two East African salt lakes.* (Loughborough University of Technology), p. 267

Barker, P., Gasse, F., Roberts, N. and Taieb, M., 1990, Taphonomy and diagenesis in diatom assemblages: a late pleistocene palaeoecological study from Lake Magadi, Kenya. *Hydrobiologia*, **214**, pp. 267–272.

Bauld, J., 1980, Geobiological role of cyanobacterial mats in sedimentary environments: production and preservation of organic matter. *BMR Journal of Australian Geology and Geophysics*, **6**, pp. 307–317.

Berner, R.A., 1980, *Early diagenesis. A theoretical approach*, (Princeton University Press), p. 237.

Busson., G., 1988, Relations entre les types de dépôts évaporitiques et la présence de couches riches en matière organique (roches-mères potentielles). *Revue de l'Institut Francais du Pétrole*, 2nd, **43**, pp. 181–216.

Butzer, K.W., Isaac, G.L., Richardson, J.L. and Washbourn-Kamau, C., 1972, Radiocarbon dating of East African lake levels. *Science,* **175**, pp. 1069–1076.

Casanova, J., 1986, Les stromatolites continentaux: paléoécologie, paléohydrologie, paléoclimatologie. Application au Rift Gregory, Vol. 2, *Thèse* (Faculté des Sciences de Aix-Marseille II), p. 256.

Damnati, B., Taieb, M. and Williamson, D., 1992, Laminated deposits from Lake Magadi: climatic contrast effect during the maximum wet period between 12.000–10.000 years B.P. *Bulletin de la Société Géologique de France* **163**, 4, pp. 407–414.

Damnati, B., 1993, Sédimentologie et Géochimie de séquences lacustres: Reconstitutions paléoclimatiques. *PhD thesis*, **1**, (Marseille: University of Aix-Marseille II), p. 239.

Damnati, B. and Taieb, M., 1995, Solar and ENSO signatures in laminated deposits from Lake Magadi (Kenya) during the Pleistocene/Holocene transition. *Journal of African Earth Science*, **21**, 3, pp. 373–382.

Damnati, B., 2000, Holocene lake records in the Northern Hemisphere of Africa. *Journal of African Earth Sciences*, **31**, 2, pp. 253–262.

Demaison, G.L. and Moore, G.T., 1980, Anoxic environments and oil source bed genesis. *Organic Geochemistry,* **2**, pp. 9–31.

El Moutaki, S., 1994, Transition glaciaire-interglaciaire et Younger Dryas dans l'Hémisphère Sud (1°–20° Sud): analyse palynologique à haute résolution de sondages marin et continentaux (lac et marécage). *Unpublished thesis*, (Marseille: University of Aix Marseille III), p. 130.

Espitalié, J., Deroo., G. and Marquis, O.F., 1985, La pyrolyse Rock Eval et ses applications. *Revue de l Institut Francais du Petrole*, **40**, pp. 755–784.

Eugster, H.P., 1970, Chemistry and origin of the brines of Lake Magadi, Kenya. *Mineralogical Society of America, Spec. Pap.,* **3**, pp. 215–235.

Eugster, H.P., and Hardie., L.A., 1978, Saline lakes. In *lakes, chemistry, geology, physics*, edited by Lerman, A. (Berlin: Springer Verlag), pp. 237–293.

Fan, P., Luo, B., Huang, R. *et al*, 1980, Formation and migration of continental oil and gas in China. *Scientia Sinica*, **23**, pp. 1286–1295.

Goetz, C., 1990, Traçage isotopique et chronologie des processus d'altération et de sédimentation par l'étude des déséquilibres U et Th. Application aux systèmes lacustres de Magadi (Kenya) et Manyara (Tanzanie). *PhD thesis*, (Marseille: University of Aix-Marseille II), p. 217.

Goldsmith, J.R. and Graf, D.L., 1958, Relation between lattice constants and composition of the Ca-Mg carbonates. *American Mineralogist*, **43**, pp. 84–101.

Hillaire-Marcel, C., Carro, O. and Casanova, J., 1986, ^{14}C and U/Th dating of pleistocene and holocene stromatolites from East african paleolakes. *Quaternary Research*, **25**, pp. 312–329.

Hollander, D.J., McKenzie, J.A. and Haven, H.L., 1992, A 200 years sedimentary record of progressive eutrophication in Lake Greifen (Switzerland). Implications for the origin of organic-carbon rich sediments. *Geology,* **20**, pp. 825–828.

Icole, M., Masse, J.P., Perinet, G. and Taieb, M., 1990, Pleistocene lacustrine stromatolites, composed of calcium carbonate, fluorite and dolomite, from Lake Natron, Tanzania: depositional and diagenetic processes and their paleoenvironmental significance. *Sedimentary Geology*, **69**, pp. 139–155.

Jones, B.F., Eugster, H.P. and Rettig, S.L., 1977, Hydrochemistry of the Lake Magadi Basin, Kenya. *Geochimica et Cosmochimica Acta*, **41**, pp. 53–72.

Kelts, K. and Hsü, K.J., 1978, Freshwater carbonate sedimentation. In *Lakes–Chemistry, Geology and Physics,* edited by Lermann, A., (Berlin: Springer Verlag), pp. 295–323.

Kristensen, E. and Andersen, F.O., 1987, Determination of organic carbon in marine sediments: a comparison of two CHN-analyzer methods. *Journal of Experimental Marine Biology and Ecology*, **109**, pp. 15–23.

Last, W.M. and De Dekker, P., 1990, Modern and holocene carbonate sedimentology of two saline volcanic maar lakes, southern Australia. *Sedimentology*, **37**, pp. 967–981.

Müller, G., Irion, G. and Förstner, U., 1972, Formation and diagenesis of inorganic Ca-Mg carbonates in the lacustrine environment. *Naturwissenschaften*, **59**, pp. 158–164.

Périnet, G., 1974, Contribution à la connaissance minéralogique des formations bauxitiques de Provence. *Thèse* (Faculté des Sciences de Aix-Marseille I), p. 212.

Reyre, D., 1984, Remarques sur l'origine et l'évolution des bassins sédimentaires africains de Busson, Evaporites et hydrocarbures. *Bulletin de la Société Géologique de France*, **26**, pp. 1041–1059.

Roberts, N., Taieb, M., Barker, P., Damnati, B., Icole, M. and Williamson, D., 1993, Timing of Younger Dryas climatic event in East Africa from lake-level changes. *Nature*, **366**, pp. 146–148.

Schreiber, C., 1988, *Relations entre évaporites et accumulations d'hydrocarbures. In Evaporites et hydrocarbures,* edited by Busson G. (Paris: Mémoires du Muséum national d'Histoire naturelle) **55**, pp. 15–18.

Surdam, R.C. and Eugster, H.P., 1976, Mineral reactions in the sedimentary deposits of the Lake Magadi region, Kenya. *Geological Society of America Bulletin*, **87**, pp. 1739–1752.

Taieb, M., Barker, P., Bonnefille, R., Damnati, B., Gasse, F., Goetz, C., Hillaire-Marcel, C., Icole, M.; Massault, M., Roberts, N., Vincens, A., and Williamson, D., 1991, Histoire paléohydrologique du lac Magadi (Kenya) au Pléistocène supérieur. (Paris: Comptes rendus de l'Académie des sciences) **313**, pp. 339–346.

Talbot, M.R., 1988, The origins of lacustrine oil source rocks: evidence from the lakes of tropical Africa. In *Lacustrine Petroleum Source Rocks*, edited by Fleet, A.J., Kelts, K. and Talbot M.R. *Geological Society Special Publication*, **40**, pp. 29–43.

Talbot, M.R. and Livingstone, D.A., 1989, Hydrogen index and carbon isotopes of lacustrine organic matter as lake level indicators. *Palaeogeography, Palaeoclimatology, Palaeoecology*, **70**, pp. 121–137.

Vandenbrouke, M., Pelet, R. and Bebyser, Y., 1985, Geochemistry of Humic substances in Marine sediments. In *Humic substances in soil, sediment, and water: Geochemistry, isolation, and characterization*, edited by Mcknight, D.M., pp. 249–273.

Van Krevelen, D.W., 1950, Graphical statistical method for study of structure and reaction processes of coal. *Fuel*, **29**, pp. 269–284.

Vincens, A. and Casanova, J., 1987, Modern background of Natron-Magadi basin (Tanzania-Kenya): Physiography, climate, hydrology and vegetation. (Strasbourg: Sciences Géologiques Bulletin) **40**, pp. 9–21.

Warren, J.K., 1986, Shallow-water evaporitic environments and their source rock potential. *Journal of Sedimentary Petrology*, **56**, pp. 442–454.

Webb, T.L. and Heystek, H., 1957, The carbonate minerals. In *The differential thermal investigation of clays*, edited by Mackenzie, R.C., (London: Mineralogical Society), pp. 329–363.

Williamson, D., 1991, Propriétés magnétiques de séquences sédimentaires de Méditerranée et d'Afrique intertropicale. Implications environnementales et géomagnétiques pour la période 30–0 ka BP. *Thèse* (Faculté des Sciences de Aix-Marseille II), p. 230.

Williamson, D., Taieb, M., Damnati, B., Icole, M. and Thouveny, N., 1993, Equatorial extension of the Dryas event: evidence from Lake Magadi (Kenya). *Global and Planetary Change*, **7**, pp. 235–242.

Yuretich, R.F., 1979, Modern sediments and sedimentary processes in Lake Rudolf (Lake Turkana), eastern Rift Valley, Kenya. *Sedimentology*, **26**, pp. 313–331.

CHAPTER 11

Forest-savanna dynamics in Ivory Coast

Dethardt Goetze, Klaus Josef Hennenberg and Stefan Porembski

Department of Botany, University of Rostock, Rostock, Germany

Annick Koulibaly

Laboratory of Botany, University of Cocody, Abidjan, Ivory Coast

ABSTRACT: In the transition between the southern Sudanian and northern Guinean zones of West Africa, forest islands are interspersed in extensive savannas. The natural dynamics of this forest–savanna mosaic and how they are altered by human activities were studied in detail in the Comoé National Park (CNP) region, north-eastern Ivory Coast, with respect to past, present and future developments.

Between 1954 and 1996, almost all forest-island contours had remained nearly unchanged in the CNP as well as under traditional subsistence land use. Recently, the extent of deforestation has become larger than the extent of natural reforestation. With multivariate forest–savanna transect data, a combination of a split moving window dissimilarity analysis and a moving window regression analysis, an interlocked sequence of ecotones for grasses, herbs, woody climbers, shrubs and trees within the overall ecotone was revealed. Dominant tree species were arranged in a sequential series from the forest border into the forest interior. A distinct boundary formation was dominated by the tree species *Anogeissus leiocarpus*. In its shade forest tree species regenerated well, like in forest sites. The forest borders were concluded to advance slowly against savanna by sequential succession in which *A. leiocarpus* is an important pioneer. The occurrence of fires only in the savanna seemed to be determined by an elevated amount of grass biomass as fuel. Low grass biomass appeared to result from suppression by competing woody species.

A natural succession from savanna to forest possibly proceeds only very slowly due to the counteracting effects of annual savanna fires and lower climatic humidity in the area compared to the interior Guinean zone. Reduced humidity appears to hamper both linear progression of forest edges and the establishment of new forest initials in the savanna. As the present pattern of Guinean forests has lately been developing under expansion of natural forests, these forests should be considered as habitat islands rather than as habitat fragments. Due to the considerable increase of deforestation, the pre-existing pattern of forest insularization has become overlain by a pattern of forest fragmentation. This will aggravate the ecological and genetic isolation of undisturbed forests like the ones in the CNP.

11.1 INTRODUCTION

In Central and West Africa, the vegetation in the transition zone between equatorial rain forests and Sudanian savannas is prone to dynamic processes related to changes in climate or human land use. Due to the large-scale utilisation of fire in the savanna areas for several millennia (Hopkins, 1992; Wohlfarth-Bottermann, 1994), the boundary between these two vegetation formations has become very distinct, separating areas of very different species composition and ecosystem function. A vegetation mosaic of forest islands, gallery forests and savannas of various types characterizes this transition between the Guineo–Congolian rain forest zone and the drier Sudanian woodland zone (White, 1983). The forest islands and gallery forests consist to varying degrees of deciduous and evergreen tree species, depending on the soil moisture conditions over the year. For the Sudanian and northern Guinean zone in Nigeria, Keay (1959) assumed that moist forest, savanna woodland and transition woodland might have coexisted through interaction of climate and site conditions long before human impact became stronger. Recent palynological studies support this view. In the Sudanian zone in north-eastern Nigeria, for example, savanna vegetation and frequent human-lit fires have occurred throughout the last 11.000 years. It follows that closed semi-deciduous forest has never completely displaced West and Central African savanna vegetation (Salzmann, 2000; Delègue *et al.*, 2001; Salzmann *et al.*, 2002; Ngomanda *et al.*, 2005). In addition, an ecological transition zone between forest and savanna is apparently a long-term natural characteristic of the vegetation, including dynamic processes such as forest succession and savanna encroachment (Gosz, 1991; McCook, 1994).

In West and Central Africa, savannas have often been interpreted as vegetation originally derived from forests, thus colonizing soils that could nourish also forest vegetation (Schnell, 1952; Backéus, 1992; Badejo, 1998; Guinko and Bélem Ouédraogo, 1998). The distribution patterns of forests and savannas seem to be a result of fire occurrence depending on topography and water supply (Spichiger and Pamard, 1973; Menaut and César, 1979, 1982; Fournier *et al.*, 1982). For the existence of most savannas, human-lit fire is a general key factor (Goldammer, 1990; Scholes and Archer, 1997; Jeltsch *et al.*, 2000; Van Langevelde *et al.*, 2003; Mistry and Berardi, 2005). It is in particular responsible for the distinctive physiognomy of humid savannas in West Africa (Swaine *et al.*, 1992; Hochberg *et al.*, 1994; Couteron and Kokou, 1997; Gignoux *et al.*, 1997). In temporarily wet downhill positions and depressions, grasslands with comparatively high productivity and thus high fuel supply can occur. Drier conditions on hilltops, due to faster surface-water runoff and coarser soils, supposedly lead to lower productivity of the herb layer and, consequently, to reduced fire occurrence and intensity, which in turn facilitates tree establishment. In a semi-arid savanna in Central Kenya, the grass layer phytomass declined from lower to upper topographical positions (Augustine, 2003). Accordingly, forest patches are more likely to occur on hilltops than on intermittently wetter slopes with stronger fire influence where savannas dominate.

Furthermore, the West African vegetation mosaic has been profoundly altered by human activities over many centuries (White, 1983). For this reason, forest islands surrounded by savanna are often interpreted as relicts of formerly continuous forests and savannas as degraded sites (see reviews in Hopkins, 1992, Neumann and Müller-Haude, 1999). Fairhead and Leach (1996, 1998) related the existence of forests in several West African regions to human settlements. They provided evidence of new forest establishing within savanna during the last 200 years in the northern Guineo–Congolian zone of Ivory Coast and discussed the colonization of ancient settlement grounds in Central West Ghana by semi-deciduous forests. Agricultural activities may also induce forest establishment. In coastal Gabon, the pioneer tree *Aucoumea klaineana* colonized abandoned cultivation areas (Delègue *et al.*, 2001).

Anthropogenic impacts commonly lead to a decline of natural forests due to habitat fragmentation and isolation, accompanied by changes in their ecological functioning (*cf.* Laurance and Bierregaard, 1997; Primack, 1998). Tropical deforestation has led to an enormous loss of intact natural habitats during the last century (Laurance *et al.*, 1998, 2001; Achard *et al.*, 2002; Fearnside and Laurance, 2003; Curran *et al.*, 2004). As tropical forests are the biologically most diverse ecosystems in the world (*e.g.*, Myers *et al.*, 2000; Brooks *et al.*, 2001; Küper *et al.*, 2004), their species richness serves as an important natural resource and a basic foundation of the livelihood of the local inhabitants. Present biodiversity loss in particular is due to a decline of numerous interior forest species amplified by edge effects in an ecological transition zone between two neighbouring habitats (Laurance and Yensen, 1991; Saunders *et al.*, 1991; Murcia, 1995; Laurance *et al.*, 2002). Edge effects are defined through a change in species composition and ecological conditions at the boundary between adjacent ecosystems due to the interaction of the ecosystems. Loss of interior forest species is furthermore amplified by island ecological processes (MacArthur and Wilson, 1967; Cantrell *et al.*, 2001; Lomolino and Weiser, 2001; Fahrig, 2003) and positive feedbacks among, for example, fragmentation, fire and regional climate (Laurance and Williamson, 2001; Laurance, 2002).

The dynamics of Guineo–Sudanian forest–savanna mosaics have to date scarcely been studied systematically. Some observations on the local scale in the literature primarily focused on forest successional dynamics in the vicinity of the Guineo-Congolian rain forest zone, particularly in the 'Baoulé-V' (Ivory Coast) that includes the Lamto Reserve (Dugerdil, 1970; Spichiger and Pamard, 1973; Avenard *et al.*, 1974; Devineau, 1976; Spichiger and Lassailly, 1981; Menaut and César, 1982; Devineau *et al.*, 1984; Gautier, 1989, 1990). Studies are also available from Central West Ghana (Swaine *et al.*, 1976; Fairhead and Leach, 1998), from South West Nigeria (Clayton, 1958; Hopkins, 1962; Moss and Morgan, 1970, 1977; Adejuwon and Adesina, 1992) and sparsely populated stretches of northern Nigeria (Jones, 1963), from eastern Cameroon (Guillet *et al.*, 2001), as well as from the Mayombe mountains (Schwartz *et al.*, 1996) and a coastal area in Congo (Favier *et al.*, 2004a). Fairhead and Leach (1996) conducted comprehensive research on the dynamics of forest islands in the Guinean savanna of eastern Guinea and drew general conclusions for traditionally used West African landscapes. Generalized descriptions like those made by Adejuwon and Adesina (1992), Hopkins (1992) and Gautier and Spichiger (2004) also mostly refer to the more humid forest–savanna mosaics of the Guinean zone. We will question to what extent these generalizations are applicable to the drier Guineo-Sudanian transition zone.

Most studies on landscape dynamics have so far not considered the Guineo-Sudanian transition zone of West Africa. Instead, one part has focused on the rapidly advancing deforestation and exploitation of the humid dense forests in the Guineo-Congolian zone, that in the meantime have been nearly completely eliminated outside the nature reserves. Wasseige and Defourny (2004) developed a methodical approach using Landsat and SPOT satellite images for large-scale monitoring of selective forest logging in the south-western Central African Republic. Concerning deforestation in Ivory Coast, ecological and economic data were synthesized by Lanly (1969), Arnaud and Sournia (1979), Wohlfarth-Bottermann (1994) and Chatelain *et al.* (1996), as well as by Bertrand (1983), Fairhead and Leach (1998) and Leach and Fairhead (2000). The latter three considered the whole of West Africa. For the Guinean sector of Ivory Coast, Yao *et al.* (2000) described a striking parallelism between a westward shift of isohyetes (due to decreasing annual rainfall), of areas with high rural population density, of centres of cacao and coffee production and of deforestation between the 1960s and 1980s. Chatelain *et al.* (1996) and Anhuf (1997) reported that immigrant populations from other ecological and ethnic regions cleared forest unsustainably in western Ivory Coast. Afikorah-Danquah (1997) suggested that immigrants to western Ghana, provoked by official forest policy, caused degeneration of vegetation and savanna formation. Ghanaian land owners in the same district preserved forest islands by employing sustainable management strategies as a consequence of a different application of forest policy.

Other studies in West Africa have focussed on land-cover dynamics due to climate change, increasing human populations and concomitant land-use changes predominantly in the Sahelian and northern Sudanian zones—areas where forests had already been absent for a long time due to intensive pastoral practices in comparatively arid environments. Thus, many publications deal with the reduction of diverse natural woodland, the spread of agricultural land use and secondary shrub thickets and processes of desertification (*e.g.*, de Wispelaere, 1980; Anhuf *et al.*, 1990; Olsson and Ardö, 1992; Kusserow, 1995; Reenberg *et al.*, 1998; Kusserow and Haenisch, 1999; Lykke *et al.*, 1999; Diouf and Lambin, 2001). Frankenberg and Anhuf (1989) studied areas in the Sahelian, Sudanian and Guinean zone of Senegal where primary and secondary forests had, however, already been eliminated through long-lasting cultivation.

We systematically investigated structure and dynamics of the Guineo-Sudanian forest–savanna mosaic in the Comoé National Park (CNP) region in north-eastern Ivory Coast and complement the results with findings from the Lamto Reserve region in southern central Ivory Coast. In an interdisciplinary approach, remote sensing data, botanic relevés and measurements of biomass and abiotic parameters from field transects were analyzed by directly relating data from the protection area and from the adjacent utilized countryside. Of particular scientific interest is the CNP region with its widespread semi-natural forest–savanna mosaic and, until recently, comparatively sparse population who practised traditional land-use methods (*cf.* Hauhouot, 1982; Wiese, 1988; Anhuf, 1994). Recently, and as with many other tropical regions, land use and exploitation of natural resources have become more intensive due to the growth and immigration of human populations. Moreover, the climate has become drier over past decades (Nicholson, 1989, 2001; Giannini *et al.*, 2003). As yet, only sparse data exist on the vegetation ecology and spatio-temporal dynamics of this type of vegetation, despite the fact that such vegetation is characteristic for the entire Guineo-Sudanian transition zone and for many other of the world's peripheral humid tropical regions. In West Africa, recently unaltered forest–savanna mosaics have been preserved only in a few regions including the southern CNP (*cf.* Anhuf, 1994). We therefore explored forest–savanna dynamics as a means to understand the parameters controlling the natural biodiversity in protected areas and how such biodiversity may be affected by human activities in neighbouring areas. We focused on: (1) general successional trends in vegetation cover and underlying mechanisms on a local and a regional scale, (2) anthropogenic disturbance and fragmentation of predominantly semi-deciduous forests on more local scales and (3) possible future developments. Since direct evidence of successional changes can be provided through long-term studies only, we considered the longest possible time span for which remote sensing data were available, reaching back into the early 1950s.

A main objective of many existing studies as mentioned above, were the analysis and quantification of changes related to well known underlying processes such as the high or increasing human population density. By contrast, our study region had a sparse population and traditional modes of human colonization and land use until recently. Here and under reserve protection, characteristics of landscape dynamics were mostly not readily recognizable from direct field observations.

11.2 THE COMOÉ NATIONAL PARK

Our research was carried out in the southern part of the Comoé National Park (CNP) and adjacent utilized areas to the south and south-west (8°20'–8°50' N, 3°15'–4°30' W, Figure 1). Here, the forest vegetation of the northern Guinean zone is gradually replaced towards the north by savanna vegetation of the Sudanian zone. Annual precipitation nowadays averages 1.050 mm with a large interannual deviation of up to 40% (Fischer *et al.*, 2002), whereas an annual average of 1.150 mm was recorded until the 1960s (Eldin, 1971). Rainfall mainly occurs during one rainy season from March to October. The mean annual temperature is 27°C. Absolute minima during the dry season do not fall below 10°C.

The study area is situated between 200 and 350 m a.s.l. and has a very smooth and level relief, with slope inclinations of *c*. 2%. We identified the soils to be impoverished sandy to loamy Ferralsols above Precambrian granites with a medium cation exchange capacity and a pH ranging from 4 to 6. In general, and as in most areas in north-eastern Ivory Coast, unfertilized soils are poorly to moderately suited for cultivation (*cf.* Wiese, 1988).

The botanical transect studies were executed in the boundaries of seven selected large forest islands in the CNP (Figure 2). The interior of the studied forest islands was dominated by *Diospyros abyssinica* (nomenclature follows Lebrun and Stork,1991–1997) and *Tapura fischeri* in the understorey, *D. mespiliformis*, *Dialium guineense* and *Celtis zenkeri* in the upperstorey and it usually contained scattered *Cola cordifolia, Ceiba pentandra* and *Milicia excelsa* (see also tree inventories in Hovestadt *et al.*, 1999). Many species in the understorey stayed foliated during the dry season, while most species in the upperstorey shed their leaves. Thus, the chosen forest islands were classified as semi-deciduous. They were surrounded by a deciduous forest belt formation of a width of 10 to 55 m. It was characterized by dominant Combretaceae, particularly *Anogeissus leiocarpus*, a lower grass cover than in the neighbouring savanna, and a higher density of shrubs such as *Mallotus oppositifolius* and *Croton membranaceus* than in the neighbouring forest. The adjacent tree savannas were usually characterized by *Crossopteryx febrifuga*, *Daniellia oliveri*, *Detarium microcarpum*, *Lophira lanceolata* and *Terminalia macroptera*. Dominant grass species were *Andropogon gayanus*, *A. schirensis*, *Hyparrhenia subplumosa*, *H. smithiana* and *Panicum phragmitoides*. At forest sites, particularly *Setaria barbata* showed locally high cover values (*cf.* also Poilecot *et al.*, 1991). In general, the combinations of dominant woody species may differ considerably even between adjacent forest islands (*cf.* Mühlenberg *et al.*, 1990), forming mostly semi-deciduous forests and in fewer cases deciduous or evergreen forests, the latter particularly near the Comoé River ('forêt dense sèche', 'forêt claire', 'forêt dense humide'). According to Poilecot *et al.* (1991), they can be classified as an impoverished variant of dense humid semi-deciduous forest of the *Celtis* spp. and *Triplochiton scleroxylon* type (*cf.* Guillaumet and Adjanohoun, 1971).

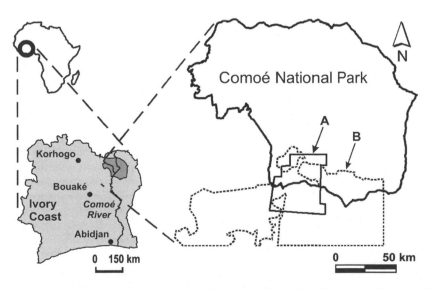

Figure 1. Areas in the Comoé National Park and neighbouring utilized regions for studying dynamics of the forest–savanna pattern (A) and dynamics of forest clearings (B).

The studied forest islands are of very different size, geometry and aggregation (Figure 2). Their sizes vary from some square metres (on ancient termite mounds) to a few square kilometres, with the majority being smaller than 1 ha (median size 0,7 ha, mean size 9,3 ± 40 ha, mean perimeter : area ratio 0,06 m⁻¹, mean distance between the nearest edges of any two forest islands 110 m; data for section A in Figure 1). The forested area in the southern part of the CNP was quantified to approach 11% of the total land surface (8,4% island forests, FGU Kronberg, 1979).

The protected area east of the Comoé River was established as a game reserve/ Classified Forest in 1926 and became national park in 1968 (Poilecot *et al.*, 1991). It has largely been uninhabited (except for a few very small settlements) since at least the 1950s. A direct human impact has been uncontrolled annual savanna burning during the dry season that seldom enters forests. In addition, poaching of large animals has led to a dramatic decrease of wild animals, especially large herbivores and top predators, since the early 1990s (Fischer and Linsenmair, 2001). Large herbivores are known to cause a patchy reduction of grassy biomass in savannas (Adler *et al.*, 2001) that can also lead to a patchy occurrence of fire (Fuls, 1992). Consequently, with declining herbivore numbers, grass biomass distribution and fire occurrence may have become more regular at the studied savanna sites in the CNP.

The forest–savanna mosaic in the CNP can thus be considered a semi-natural system. As this mosaic continues across the southern border of the CNP, the juxtaposition of protected areas and areas utilized by man allows direct comparisons to be made. In the utilized region, forest trees are selectively felled, traditional agriculture of yams and manioc is carried out manually during only one year of cultivation of a field where cashew nowadays is planted simultaneously. Contrary to many other regions in West Africa, there has been no livestock farming due to the presence of the tsetse fly. In 1975 and 1988,

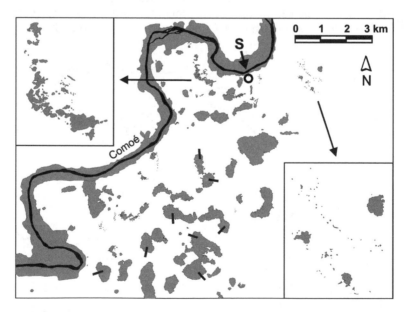

Figure 2. Spatial pattern of forest islands and Comoé gallery forest in a section of the southern Comoé National Park. Perimeters of forest islands were traced with a hand-held GPS. Black bars: Location of the eight forest–savanna transects studied. Inserts: Forest islands in the vicinity of a terrace edge (left) and on ancient termite mounds that mostly are arranged in denser groups (right). The majority of the forest islands are small (area median for this section = 0.01 ha, mean = 2.5 ± 12 ha). S = research station of the University of Würzburg.

human population density amounted to around 10 inhabitants per km² south of the CNP (Wiese, 1988); however, by 1998 it had risen to *c.* 13 inhabitants per km² (Institut National de la Statistique, 1988/2000), although this was still low compared to other regions in tropical Africa.

11.3 FOREST–SAVANNA DYNAMICS UNDER SEMI-NATURAL CONDITIONS

Areas being under protection for a long time and comprising ecosystems in an at least near-natural state are most suited for investigating ecosystem processes because they have been under little human influence. They may serve as reference for investigating the consequences of human land use and ecosystem disturbance. Here, they will be related to nearby areas under traditional land use.

11.3.1 Forest dynamics at the landscape level

Changes in the patterns of forest islands were analysed by comparing the oldest remote sensing data available with the newest data of the same sensor spectral type. The oldest data are panchromatic aerial photographs 1:50.000 from January 1954 (peak of dry season). Lanly (1969) described methodical and technical details of the remote sensing mission. The newest panchromatic data of a comparably high spatial resolution are aerial photographs from November 1996 (very beginning of dry season). A study area of 550 km² in the CNP and 650 km² in the adjacent utilized countryside south of the park (Figure 1, A) was investigated. All forest islands were counted and the size, contour and density of vegetation of each forest was directly compared over time. Because of the very variable image quality, the comparison had to be carried out through visual interpretation, allowing for maximum accuracy. Goetze *et al.* (2006) provide further details.

Within the 1.200 km² study area, a total of 653 forest islands were found. After an interval of 42 years, nearly all of them (648) could still be observed in their original location of 1954 (Table 1). Of these, 95,4% had contours that were, in principle, unaltered and had a stable forest size. At forest borders with evergreen woody species, the details of perimeters had mostly remained identical. In addition, savanna areas with high tree cover mostly stayed unchanged. Only the remaining 4,6% of the forest islands showed an increase or decrease in size; these changes that were the result of human activities were especially evident outside the CNP. However, within this stable forest–savanna pattern, change had taken place primarily due to human impact, and evidently started as early as the 1950s. In some areas outside the CNP, 35,8% of the interior of the forest islands were cut. In these islands, only narrow forest margins of smaller-stature woody species remained (Table 1). They mostly enclosed successional stages towards reforestation and rarely showed signs of intrusion by fire or agriculture that would reverse the forest succession.

An opening-up of 4,8% of the forests located within the CNP in the vicinity of the southern border was apparently caused by selective felling, as revealed by a comparison with CORONA satellite images from 1967. The forest openings could also have been caused by natural disturbances, such as large windthrows. Evidence of such events, however, was not detected on the remote sensing data used.

The studied forest–savanna pattern proved to be remarkably stable, even under extensive land utilization outside the CNP. Although many temporary changes, such as cultivation of savanna and forest boundaries or forest cutting were detected within this pattern at different times, the ecosystems usually returned to the physiognomic starting point a few decades after the disturbance. The observations suggest that forest and savanna systems are both resilient

Table 1. Spatio-temporal development of 653 forest islands in savanna within 550 km² of the southern Comoé National Park and 650 km² of an adjacent region under human use from 1954 to 1996 (Goetze *et al.*, 2006).

	Number of islands studied	Newly established (%)	Increase in size (%)	Equal in size (%)	Loss in size (%)	Vanished (%)	Forest interior cleared 1954 (%)	Forest interior cleared 1996 (%)	Agriculture in forest area (%)
Inside CNP	379 = 100%	0,5	0,5	97,9	0,8	0,3	0,8	4,8	0,0
Outside CNP	274 = 100%	0,0	2,9	92,0	4,4	0,7	2,6	31,0	0,4

to change (*cf.* Gigon and Grimm, 1997; Kratochwil and Schwabe, 2001), including that resulting from naturally occurring disturbances and other fluctuations, for example episodic mass herbivory (*cf.* Poilecot *et al.*, 1991). These results contribute to the characterization of savanna resilience and support its apparent stability against degradation and desertification in the Sudan and Sahel zones (Walker and Noy-Meir, 1982).

The frequency of manual cultivation on each plot of land outside the CNP was obviously low enough to allow regeneration of the earlier vegetation during the long fallow periods (*cf.* Mitja and Puig, 1993). Moss and Morgan (1970, 1977) made the same observation in south-western Nigeria over a 10-year period when population density and land cultivation were even increasing. In contrast, Spichiger and Lassailly (1981) working in Central Ivory Coast observed that following a considerable reduction of the fallow period, the woody vegetation was not able to regenerate when compared to regeneration levels observed during previously less intense cultivation phases. Mitja and Puig (1993) from western Ivory Coast described a repression of tree regeneration and a decrease in species richness. This was primarily due to mechanized cultivation practices that additionally encouraged *Trinervitermes* termites to colonize, exerting deleterious effects on soil conditions. On leached soils in northern Zambia, the woody species that vividly regenerated directly after abandonment were largely replaced by grasses from the second fallow year on before woody species slowly regained dominance during the subsequent 20 years of fallow (Stromgaard, 1986).

11.3.2 Forest border dynamics at the local level

In order to elucidate in detail processes and possible causes of forest border dynamics it is essential to relate dynamics at the landscape level to detailed field data from representative sampling points within the study area. For this purpose we collected data on vegetation composition, structure, biomass and environmental parameters such as soil depth and fire occurrence along eight continuous transects perpendicularly crossing the forest borders between the savanna and the forest interior (Figure 2; for details see Hennenberg *et al.*,

2005a). The transects were located in the intact mosaic of forest islands and savanna in the CNP and were chosen such that on both sides of the transects the forest border was as straight as possible to provide standardized conditions. Data were recorded in 2001 and 2002.

At tropical forest borders the spatial structure of woody plant communities is a strong indicator of possible community dynamics at the border ecotone. Species zonation and age distribution may provide information on ongoing community dynamics, particularly when soil and humidity variations in space do not correspond with the observed community patterns. Furthermore, plant communities as a whole are an integral indicator for many past and ongoing ecological factors and processes that influenced or are influencing a site. Changes in plant species composition along ecotones should reflect associated gradients (Fagan *et al.*, 2003).

Borders and ecotones at forest–savanna boundaries

Community variation along ecological gradients is described by multivariate plant species data that are to be analyzed with multivariate numerical procedures. To date, Hennenberg *et al.* (2005a) in the CNP study area have conducted the only such ecotone study from tropical forests. The methodical concept was based on existing multivariate methods developed for detecting borders along ecological gradients (Cornelius and Reynolds, 1991; Jacquez *et al.*, 2000; Csillag and Kabos, 2002). The currently widely-used moving window approach has a great potential also in the analysis of floristic data across transitional areas (Kent *et al.*, 1997). For one type of analysis—the split moving window dissimilarity analysis (SMWDA)—Cornelius and Reynolds (1991) provided a statistic that tests the significance of a detected border.

The SMWDA was applied to the CNP data to detect dissimilarities along forest–savanna transects (Cornelius and Reynolds, 1991). From a window split in two halves, the dissimilarity of species composition was calculated by Euclidean distance for each window-midpoint position (*cf.* Hennenberg *et al.*, 2005a). In this SMWDA, window sizes from 2 to 20 plots were used. Only the dissimilarity profiles of the mean Z scores are presented here that were computed as the mean of the z-transformed Euclidean distance of the different window sizes. For each dataset, the standard deviation was computed above the overall expected mean by a Monte Carlo procedure with 1.000 replicates (Cornelius and Reynolds, 1991). The one-tailed 95% confidence interval was used to detect significant peaks of the mean Z score indicating borders along the transects.

In a second step, the width of the ecotones that are always associated with ecosystem borders has to be determined. For this purpose, Walker *et al.* (2003) used moving window regression analysis (MWRA) along a coastal vegetation gradient in New Zealand. For the CNP transects, ecotone width was determined correspondingly with MWRA (Hennenberg *et al.*, 2005a). Based on mean Z scores, the slope of the linear least-square regression with a window size of $n = 5$ was calculated at each midpoint position of the SMWDA. The slope at a window midpoint position gives the degree of change of mean Z scores. The maximum value above 0 and the minimum value below 0 of the MWRA between a detected border and the next change of sign (or hitting 0) of the slope on the left and right side of a detected border are interpreted as ecotone borders between two non-ecotone zones.

The dissimilarity profile of a representative transect, computed on the basis of the cover of all species in 64 relevés, revealed one clear and significant peak of dissimilarity (Figure 3a). The location of the borderline between savanna and forest belt visually observed in the field (0–m indication) agreed well with the one detected by SMWDA using the cover of all species. MWRA yielded an ecotone width of 125 m, i.e. 50 m into the savanna and 75 m into the forest (Figure 3a). Thus, the depth-of-edge influence (DEI) determined from the cover of all species was 75 m into the forest.

The dissimilarity profiles presented in figure 3b and c were computed from the cover values of grasses and herbs, respectively. Both profiles show a single border, but for the grass dataset, both the position of the maximum of dissimilarity and the location of the ecotone are indicated further in the savanna than for the herb dataset. For herbs, ecotone detection was somewhat fuzzy with a second peak at 5 m on the savanna that could not be separated as a border on its own (Figure 3c). The observed pattern was a result of a continuous species turnover, particularly involving the abundant grasses *Andropogon gayanus* and *Panicum phragmitoides* (savanna), *Setaria barbata* (forest belt/closed forest) and *Sporobolus pyramidalis* (forest belt), as well as the characteristic herbs *Mitracarpus villosus* (savanna), *Justicia insularis* (savanna/forest belt), *Blepharis maderaspatensis* and *Geophila repens* (closed forest). The dissimilarity profile computed for the cover of woody climbers (Figure 3d) showed a significant border at 25 m which was in good accordance with the border between closed forest and forest belt visible in the field. The associated ecotone of 30 m was rather narrow. This distinct pattern was caused by a regular appearance of woody climbers such as *Campylostemon warneckeanum, Motandra guineensis, Salacia baumannii* and *Saba comorensis* in forest plots and their absence in savanna plots.

The cover of trees and shrubs revealed an ecotone of 100 m width (window midpoint position from 20 m into the savanna to 80 m into the forest, figure 3e), wherein several significant peaks were detected. In addition, a border with a very narrow ecotone (15 m) occurred at 115 m. The observed change in tree and shrub species composition between savanna and forest belt was associated with an increase in stem density and tree and shrub cover towards the forest interior.

The border detection by tree and shrub cover could be clarified by examining the pattern of the size classes of trees and shrubs (Figure 3f–h). The dissimilarity profile computed for trees and shrubs of 1–10 cm DBH revealed two significant borders (Figure 3f). The first border at 25 m and its associated ecotone were congruent with the detected border and ecotone for woody climbers and reflected the location of the border between closed forest and forest belt that was visually identified in the field. Between the two detected ecotones with a width of 25 m and 15 m, a non-ecotone zone of 15 m occurred (Figure 3f). It was characterized by a high density of shrubs, mainly *Croton membranaceus* and *Mallotus oppositifolius*. The DEI could be marked at 75 m, which was the same as detected for the cover of all species (Figure 3a).

The dissimilarity profile computed for tree and shrub species of 10–20 cm DBH showed three significant borders (Figure 3g). Near the stratified border between savanna and forest belt (window midpoint position of –5 m), a peak of the mean Z score occurred that was just above the significance level (ecotone width of 20 m, figure 3g). A second significant border was detected at the window midpoint position of 50 m (ecotone from 30–55 m). Between these two borders, the tree species *Anogeissus leiocarpus* was very abundant. At the window midpoint position of 75 m a third significant peak of dissimilarity occurred. Between the second and third border *Diospyros abyssinica* was dominant. Further to the forest interior, *Tapura fischeri* became the most abundant species besides *D. abyssinica*. The ecotone associated with the third border had a width of 60 m and reached up to 120 m into the forest interior (Figure 3g).

SMWDA for tree species larger than 20 cm DBH revealed two significant borders. Between them, *Anogeissus leiocarpus* was very abundant (Figure 3h). Detected ecotones at the first and second border had widths of 20 m and 25 m, respectively. For this tree size classes, the DEI was 60 m.

The median of the DEI values detected by MWRA was 55 m with a maximum of 100 m and a minimum of 35 m. The DEI values determined for tropical semi-deciduous forest islands with a closed character at the forest-savanna border are in a similar order to other values mentioned for temperate and tropical forests (*cf.* Baker and Dillon, 2000; Laurance *et al.*, 2002).

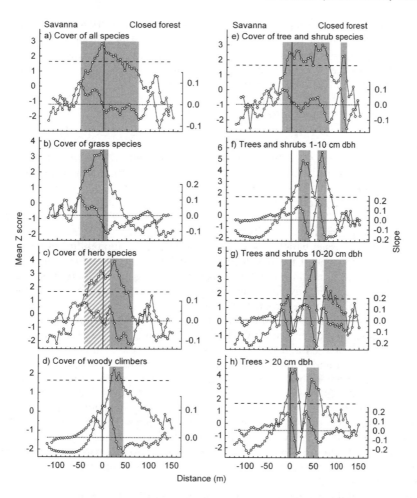

Figure 3a–h. Pooled mean Z score (circles) computed by SMWDA (window sizes from 2 to 20 were included) and slopes calculated by MWRA (diamonds) along one transect for cover of (**a**) all species (133 species); (**b**) grasses (22 species); (**c**) herbs (45 species); (**d**) woody climbers (14 species) and (**e**) trees and shrubs (43 species) and for counted tree and shrub individuals of different size classes: **f** –18 species, **g** –11 species; **h** –12 species. The upper horizontal line marks the one-tailed 95% confidence interval that is used as a significance level of the SMWDA to detect borders. The lower horizontal line reflects the zero-baseline of slopes of the MWRA. The borderline between savanna and forest belt visible in the field is placed at a distance of 0 m (vertical line). Ecotones are marked grey (fuzzy detected ecotones: dashed grey) (after Hennenberg *et al.*, 2005a).

It follows that combining SMWDA and MWRA provides a coherent set of analyses for reliably detecting borders, more objectively characterizing the width of associated ecotones, and allowing better comparisons between studies (Hennenberg *et al.*, 2005a). The spatial structure of the forest–savanna border proved to be determined by an interlocked sequence of ecotones for different plant life forms and tree ages, occurring along an overall ecological gradient with a continuous species turnover. This spatial pattern indicates that forests may slowly encroach onto adjacent savanna by sequential succession.

Fire, biomass and soil depth at forest–savanna borders

Fire plays an important role in facilitating the co-occurrence of forests and savannas typical for many parts of West and Central Africa (White, 1983; Salzmann, 2000; Salzmann *et al.*,

2002). Experimental fire exclusion yielded closed forests after 30–60 years (Ramsey and Rose Innes, 1963; Devineau *et al.*, 1984; Swaine *et al.*, 1992; Louppe *et al.*, 1995). In tropical forests edge-related fires are reported to penetrate into forests for a few metres (Kellman and Meave, 1997; Biddulph and Kellman, 1998) up to several kilometres (Cochrane and Laurance, 2002). Occurrence and intensity of surface fires in savannas are strongly depending on the availability of plant material as fuel, with the amount of grass biomass playing the most important role (Stocks *et al.*, 1996; Stott, 2000).

In the CNP, surface fires occurred only in the savanna with a high amount of grass biomass, but not in closed forests even though a relevant amount of leaf litter was available as fuel (Hennenberg *et al.*, 2006). This highlights the importance of grass biomass for the occurrence of surface fires in the forest–savanna system. Along with litter, grasses represent the most important combustible material in most tropical ecosystems, especially as grasses play a vital role for ignition (Stott, 2000). Fires usually start and spread in fine fuels due to their low ignition temperature as a result of a large surface-to-volume ratio (Anderson, 1982). Leaf litter with its more compact structure may need a higher ignition temperature than grasses which is also likely to reduce the ignitability of the vegetation if the grassy component of the fuel load drops below a certain value. This is a possible explanation of why fires did not enter deeper into forests where there was a re-increase of grass biomass, and of why fire-sensitive species can establish more easily towards the forest interior. A higher grass production in wetter years at forest boundaries and interiors may favour a deeper penetration of fire into the forest (*cf.* Van Langevelde *et al.*, 2003).

Along the eight transects in the CNP, above-ground grass biomass amounted to some 900 gm^{-2} at savanna sites and to less than 400 gm^{-2} at forest sites, leaf litter averaged 150 gm^{-2} at savanna and 600 gm^{-2} at forest sites (Hennenberg *et al.*, 2006). Phytomass values obtained in other correspondingly structured savannas in Ivory Coast are similar, amounting to 400–1.2000 gm^{-2} for the herb layer (see Menaut and César, 1979; César, 1981; Villecourt *et al.*, 1979, 1980; Fournier, 1991), partly depending on the topographic position with the highest values occurring at inland valleys (Fournier *et al.*, 1982). In the Lamto Reserve (Central Ivory Coast), Fournier (1991) measured a maximal biomass of the herb layer of 1.400 gm^{-2} in a wooded savanna at the lowest part of a slope. For semi-deciduous forests (Lamto Reserve) the weight of annual leaf litter fall varied from 510 to 810 gm^{-2} (Devineau, 1976).

Grass growth in savannas and, thus, the amount of grass biomass is influenced by complex interactions of biotic and abiotic parameters (Higgins *et al.*, 2000; Scholes and Archer, 1997). Scholes and Archer (1997) pointed out that in general an increase in woody plant cover or density results in a strong decline of grass production. This is mainly a consequence of competition between trees and grasses for resources (such as light, soil water and nutrients). Also along the forest–savanna transects studied in the CNP, a strong negative relation between the vegetation cover above 1 m and grass biomass was detected, which was also clearly visible in the field where tall savanna grasses changed to low forest grasses (Figure 3b). It may be interpreted as suppression of grasses by woody species (Mordelet and Menaut, 1995). In contrast, grass biomass showed no clear relation to soil depth that, in addition, did not vary systematically between forest and savanna. This may indicate that in the studied system the interaction between grasses and trees is of greater importance than are soil conditions.

Grass biomass production is known to be related to annual rainfall. Breman and Dewit (1983) confirmed this relation for pasture sites along a rainfall gradient from the Sahel to the Sudanian zone. In their study, however, grass growth was more strongly limited by nutrient availability than by annual rainfall. Yet, annual variation in local grass-biomass production without fertilization mainly depended on annual rainfall conditions (César, 1981; Sturm, 1993; Van de Vijver, 1999). This topic should be investigated in more detail.

The role of *Anogeissus leiocarpus* (Combretaceae) *in forest–savanna dynamics*

In the forest section of the transects studied in the CNP, a clear spatial sequence of the dominant tree species *Anogeissus leiocarpus*, *Combretum nigricans*, *Diospyros mespiliformis*, *D. abyssinica* and *Dialium guineense* was encountered from the forest border towards the forest interior (Hennenberg *et al.*, 2005b). To some extent, the diameter classes for the latter four species were arranged in a spatial sequence, with young individuals (smaller 1 cm DBH) being more frequent towards the savanna. The distribution of this smallest diameter class indicates the location of regeneration sites of tree species. Correlation analysis of tree density values of this diameter class and studied environmental parameters yielded high correlation coefficients, making it possible to separate two species groups. Firstly there is *Detarium microcarpum*, *Crossopteryx febrifuga* and *Vitellaria paradoxa* that regenerated in a savanna environment, where fuel load is high and shading by upper tree layers is low, and secondly *Diospyros mespiliformis*, *D. abyssinica* and *Dialium guineense* that regenerated in a forest environment where fuel load is low and shading by upper tree layers is high. *Anogeissus leiocarpus* had a high potential to regenerate at savanna sites with an intermediate fuel load (Figure 4).

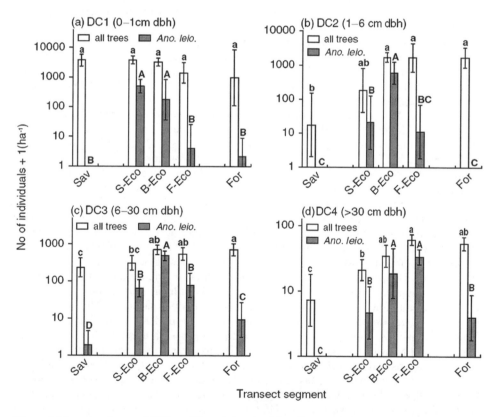

Figure 4. Mean abundance and 95% confidence intervals of four diameter classes (DC1-DC4) of *Anogeissus leiocarpus* and of all trees (including *A. leiocarpus*) along eight forest-savanna transects (transect segments: Sav = savanna, S-Eco = savanna ecotone, B-Eco = boundary ecotone, F-Eco = forest ecotone, For = forest) in the Comoé National Park. Different letters indicate statistically significant differences between transect segments (multiple range test (α = 0.05, TUKEY, SPSS 11.0) subsequent to a two-way ANOVA). No differences between transects and no interaction effects occurred (Hennenberg *et al.*, 2005b).

Its regeneration sites were also correlated with comparatively shallow soils like the ones of *Detarium microcarpum*. *A. leiocarpus* was almost completely absent from the forest interior and the open savanna (Figure 4). The diameter class composition of *A. leiocarpus* indicated a gradual change. Individuals smaller than 1 cm DBH were most abundant in the savanna close to the forest border (Figure 4a), trees of 1 to 30 cm DBH at the forest border (Figure 4b and c), and the largest trees in the forest belt (Figure 4d), most likely reflecting a former position of this borderline.

These detailed distribution data reveal a sequence of size classes in space of *A. leiocarpus* and of other dominant forest-tree species and their regeneration in the understorey of *A. leiocarpus* stands. This can be interpreted as a sign of sequential succession of the forest border into the savanna with *A. leiocarpus* facilitating the establishment of the other forest species. In general, succession is characterized by a sequential turnover of species in time (Huston and Smith, 1987; Sheil *et al.*, 2000). The studied type of forest border appears to encroach onto the adjacent savanna in the CNP. This agrees with interpretations by Neumann and Müller-Haude (1999). However, the high number of juveniles of *A. leiocarpus* at the forest–savanna border could also be interpreted as fire-suppressed juveniles (see Higgins *et al.*, 2000), supporting an interpretation of the forest border as being more stable.

The contrary assumption that savanna formations in the CNP encroach onto forests by sequential succession is to be rejected for two reasons. Firstly, no similar sequential patterns could be found for tree species dominating the sequence of savanna plots and, secondly, the fuel load encountered in the forest belts was clearly similar to that in forest situations and not to that in savanna situations, making an intrusion of fire generally improbable. Thus, forest decline should occur by sudden stochastic events like, for example, drought periods with increased tree mortality (Swaine *et al.*, 1992; Condit *et al.*, 1995; Hovestadt *et al.*, 1999; Gascon *et al.*, 2000).

The correlation analysis for the density of young *A. leiocarpus* individuals and selected environmental parameters (fuel load, shading of the upper tree layer, soil depth) in the CNP showed that juveniles of *A. leiocarpus* occur at sites with an intermediate amount of inflammable material and rather shallow soils. These conditions can typically be found at borders of the studied CNP forest islands. In addition, Sobey (1978) points out that regeneration of *A. leiocarpus* seems to be favoured at sites with reduced fire intensity or frequency (north-western Ghana, see also Hall and Swaine, 1981). In West Nigeria, Letouzey (1969) observed a high recruitment rate of *A. leiocarpus* in savannas adjacent to forest-island borders dominated by *A. leiocarpus*. Soil conditions at *A. leiocarpus* sites were more beneficial than in the open savanna, however this might be a result of positive effects of *A. leiocarpus* on soil conditions (Sobey, 1978). In southern Burkina Faso, *A. leiocarpus* was frequent at locations with shallow, poor soils, unfavourable for agriculture (Neumann and Müller-Haude, 1999). Even a 'humid' climate with an annual rainfall of above 900–1.000 mm may allow for fires of high intensities in open savannas. This might be the reason why *A. leiocarpus* is missing at such sites, while under a drier climate with fire of lower intensity *A. leiocarpus* regenerates well in open savannas (*e.g.*, at 500–700 mm y^{-1} in Burkina Faso; Couteron and Kokou, 1997). However, in the Sahel region low water availability seems to restrict *A. leiocarpus* to gallery forests (Müller and Wittig, 2002).

Factors maintaining savannas are complex and may be unique for each savanna (Walter, 1979; Scholes and Archer, 1997; Higgins *et al.*, 2000; Van Langevelde *et al.*, 2003). Looking for general patterns, Jeltsch *et al.* (2000) have identified two groups of 'buffering mechanisms' that reduce the probability of a shift either from savanna to forest (for example, buffered by fire, elephants or seed predators) or from savanna to plain grassland (for example, buffered by micro-sites favouring tree establishment or grazers) and thus lead to a grass–tree coexistence. Fire as a prominent buffering mechanism (Jeltsch *et al.*, 2000) is of paramount importance for the distinctive physiognomy of humid savannas in West Africa (Swaine *et al.*,

1992; Hochberg *et al.*, 1994; Couteron and Kokou, 1997; Gignoux *et al.*, 1997). Reduced fire intensity and frequency that facilitate tree establishment can occur on termite mounds of the genus *Macrotermes* (Harris, 1971; Bloesch, 2002), at abandoned village sites (Lawson *et al.*, 1968; Sobey, 1978) and in the shade of trees (Mordelet, 1993; Mordelet and Menaut, 1995; San Jose and Montes, 1997). In addition, stochastic variation in fire frequency and the date of burning are of importance (Louppe *et al.*, 1995; Bloesch, 2002).

Among the tree species encountered in the CNP, *A. leiocarpus* has the highest potential to break through the buffering mechanism of fire by regenerating in savanna sites, being moderately fire resistant, and killing out grass effectively (Irvine, 1961). *Combretum nigricans*, that showed a regeneration peak directly at the borderline of the CNP forests, did not regenerate in savanna plots and was not correlated with fuel load. Savanna trees such as *Crossopteryx febrifuga* and *Detarium microcarpum* may also suppress grasses, but their abundance, however, is usually too low to initiate a succession from savanna to forest. On the other hand, forest tree species clearly reduced grass biomass in the forest interior, but regenerated at shady sites with a low fuel load including sites in the understorey of *A. leiocarpus*, and not in the savanna.

Anogeissus leiocarpus (Combretaceae)

The genus *Anogeissus* comprises eight species in subtropical Africa and Asia (Mabberley, 1997). *Anogeissus leiocarpus* colonizes the Guinean and Sudanian domain from West through East Africa (Wickens, 1976). White (1983) describes *A. leiocarpus* as a typical element of woodlands and savannas of the Sudanian regional centre of endemism. From the northern Guinean zone up to the Sahelian zone, the species can be found in savannas, dry forests and gallery forests (Couteron and Kokou, 1997; Hahn-Hadjali, 1998; Müller and Wittig, 2002; Neumann and Müller-Haude, 1999).

Anogeissus leiocarpus is a deciduous tree that has been reported to be highly abundant at forest borders like at the studied semi-deciduous forest islands in the CNP and at gallery forests (MacKay, 1936; Clayton, 1958; Jones, 1963; Letouzey, 1969; Hall and Swaine, 1981; Poilecot *et al.*, 1991; Porembski, 2001). Nansen *et al.* (2001) and Neumann and Müller-Haude (1999) point out that *A. leiocarpus* may play an important role in forest succession. It can grow up to a height of 15–18(–30) m (Arbonnier, 2002) and exhibits a usually sparse crown (Neumann and Müller-Haude, 1999). Fruits of *A. leiocarpus* contain about 40 wind-dispersed seeds of 10 mg each (Hovestadt *et al.*, 1999). Seeds ripen during the dry season and germinate mainly at the beginning of the rainy season. A seed bank is absent (Thies, 1995).

In total, these results suggest that *A. leiocarpus* may act as an important pioneer species, which plays a significant role in a succession from savanna to forest and, thus, in the dynamics of the studied forest–savanna mosaic in the CNP. However, not all forest borders are fringed by *A. leiocarpus*. Tree inventories (Hovestadt *et al.*, 1999) show that *A. leiocarpus* can also be absent from small forest islands. This underlines that the successional pathway (Gibson, 1996) proposed here is only one possibility in forest succession at forest–savanna borders in tropical Africa.

11.3.3 Discussion

The general colonization pattern of *Anogeissus leiocarpus* and further dominant woody species indicates a stability of the forest borderlines or their slow advance into established savanna in the CNP. Also the latter option is in agreement with the results from the time series analysis with aerial photographs from the corresponding CNP area, revealing stability of the forest–savanna pattern over a 42-year period. Due to an error margin of around 10 m in the image interpretation caused by phenological differences, varying lighting conditions and varying technical quality among the photographs, an advance of some forest borders by only several metres during the 42 years tentatively appears to be possible. Hence, interpreting all CNP data synoptically, a slow encroachment of forest on savanna by very few decametres per century is indicated. Working in the more humid mountainous Congo, Schwartz *et al.* (1996) also determined low advancement rates of forest pioneer species of 20–50 m per century into edaphically non-contrasting savanna enclosures. In a large-scale remote sensing study of an 8.600-km² section of northern Zaïre, Runge and Neumer (2000) detected for the period from 1955 to 1990 only little dynamics of either direction between forest and savanna but also no general trend of change. By contrast, based on a comparison with a historical vegetation map, Hopkins (1962) estimated an annual forest advance of more than 3 m into a protected Guinean savanna in south-western Nigeria. Miège (1966) observed an encroachment of *c.* 2,5 m year⁻¹ into a moist laguna savanna at the Atlantic coast in Ivory Coast and Delègue *et al.* (2001) quantified annual forest progression into coastal savanna of Gabon to be nearly 1 m. In all three cases, however, climate conditions were more humid than in the CNP, and this presumably promoted forest establishment.

On the other hand, other authors have documented considerable progress in forest fragmentation during the past decades in West Africa, due to anthropogenic deforestation and/or intensive land use, often combined with population growth (Wohlfarth-Bottermann, 1994; Chatelain *et al.*, 1996; Anhuf, 1997). In the CNP region, these processes did not take place until the early 1990s. Consequently, the forests in the CNP region should be considered as habitat islands rather than fragments, concerning their more recent spatial development and their surrounding habitats (corresponding to definitions by Laurance and Bierregaard, 1997; Schaefer, 2003). The hypothesis that fragmentation of previously larger or continuous forest habitats might have caused the present forest pattern during the past decades can be rejected. Although the origin of the present forest pattern is not known definitely, we conclude that it can be called semi-natural. In general, this is in accordance with the findings of Wohlfarth-Bottermann (1994), who worked on ancient vegetation cover and land occupation in Ivory Coast. His investigation of the National Archives, Abidjan and the Overseas Archives, Aix-en-Provence, revealed that in the Guineo-Sudanian transition zone of north-eastern Ivory Coast, a similarly structured forest–savanna mosaic also existed in the nineteenth century under a relatively high human colonization intensity. In contrast, in the north-west of the country woody vegetation may have been more continuous. At this pre-colonial time, human population density in the dry-forest/savanna zone was higher than in the humid rain forest zone, with the urban centres of Kong, Bouna, Bondoukou and Odienné being of overwhelming importance. These population centres presumably had no major impact on our study area as they are located at least 50 km away. The general situation began to reverse in the 1880s with dramatic population losses in the north and south due to war (see also Fairhead and Leach, 1998), and the beginning of the plantation and timber industry in the south. Whilst this accelerated the twentieth century degradation of the southern rain forests, the forest island pattern in the subsequently sparsely populated north-east was maintained under the continued annual burning of the entire savanna (Wohlfarth-Bottermann, 1994).

In addition, for many Guinean forest–savanna mosaics there is some evidence that they have not changed greatly under continued use (Adjanohoun, 1964, for the 'Baoulé-V') and climate oscillations (Schöngart *et al.*, 2006) during the last centuries. This implies that, as with north-eastern Ivory Coast, they also cannot be considered—as is sometimes suggested—as relics of more closed forest cover in the nineteenth or even eighteenth century (*cf.* Aubréville, 1950; Ekanza, 1981; Wohlfarth-Bottermann, 1994; Fairhead and Leach, 1996; Bassett and Boutrais, 2000; Leach and Fairhead, 2000; Bassett *et al.*, 2003; Bassett and Zuéli, 2003).

Anhuf (1994) undertook a very rough comparison of ground cover directly east of the areas studied in the CNP by means of aerial photographs from 1954 and 1972. In general, our results from outside the CNP match his findings. General trends of savanna vegetation becoming more dense, as in Anhuf's (1994) study area inside the CNP, could be due to the varying intensity of grazing by savanna herbivores like, for example, in northern Ivory Coast (Bassett and Boutrais, 2000; Bassett *et al.*, 2003; Bassett and Zuéli, 2003). Here woody cover in the savanna increased over some 30 years due to increasing livestock farming. Intensified grazing reduced the amount of combustible biomass and savannas were burned earlier during the dry season to provide the cattle with resprouting grasses as soon as possible. Both processes reduced fire intensity, which allowed woody species to establish more easily. Thus, also in the CNP, regional shifting of grazing patterns of herds of wild herbivores may possibly have induced similar fluctuations before poaching increased dramatically. A much more humid forest–savanna mosaic in the Kissidougou prefecture in eastern Guinea has also remained largely unchanged since pre-colonial times. It shows a quite stable landscape pattern in some parts and an increase in forest surface area since 1952 in others, at least partly because of human population increase (Fairhead and Leach, 1996). Reforestation on cleared forest and abandoned settlement areas, as well as forest expansion into savanna, was observed there corresponding to our findings in the CNP region. However, the villagers have promoted forest islands as they traditionally establish their villages at the centres of the forests, so to profit from ecological and economic benefits of the surrounding forests and to pursue social and cultural activities within them. Monnier (1981) presented an aerial view of such a forest island, which is also from the same climate zone in western Ivory Coast. Jones (1963) reported on a formerly analogous settlement pattern within forest islands in the Abuja and Jema districts of northern Nigeria, where land use differs from Ivory Coast. Blanc-Pamard (1979) pointed out that the majority of villages in the southern 'Baoulé-V' in Ivory Coast are at least attached to forest islands. Whilst such forest patterns can clearly be a testimony of the settlement history over the last centuries, they contrast in this respect with north-eastern Ivory Coast, where the remote sensing data do not show a comparable relationship between recent settlements and forests; however, for the more distant past this relationship cannot be ruled out.

Fairhead and Leach (1998) provided evidence of new forest establishing within savanna during the last 200 years in the northern Guineo-Congolian zone of Ivory Coast, and discussed the colonization of ancient settlement grounds in Central West Ghana by semi-deciduous forests. In general, this process also appears to happen on ancient settlement sites within the CNP region and here it is correspondingly initiated by *Anogeissus leiocarpus*. However, such a process can only account for a small fraction of the large number of forest islands in the area (*cf.* Figure 2). In addition, a notable source of forest islands appears to be the colonization of ancient termite mounds by certain fire-tender forest species (*cf.* Menaut and César, 1982; Mühlenberg *et al.*, 1990).

The examples of short-term encroachment of forest onto savanna as discussed above, as well as findings from other studies, all represent the Guineo-Congolian zone that is more humid than the CNP region (also Guillaumet, 1967; Gautier, 1989, 1990; Fairhead and Leach, 1996). This suggests that savanna–forest succession progresses faster there than near the climatic limit of the forest in the north (see also Favier *et al.*, 2004b). Also the forest–savanna transects we surveyed inside and outside the more humid Lamto Reserve, southern central Ivory Coast, revealed a distinct zonation of woody species with a regeneration peak on the savanna side of the forest boundaries. Moreover, some savanna tree species were frequently encountered with old individuals only in the outer parts of the forest interior zones, indicating a more rapid forest encroachment on savanna. The savanna inside the reserve even included a notable number of Guinean species, indicating a high potential of colonization by forest species in the open savanna, which is most likely due to the recently declining frequency of fire-use in the reserve. The initialization of small forest clumps by forest species in humid savanna was discussed by Favier *et al.* (2004a) and Ginoux *et al.* (2006) and, apart from termite mounds, was not observed in the drier CNP.

Furthermore, subsistence farming may promote subsequent natural reforestation of the fallows in the more humid Guinean savannas (Adjanohoun, 1964; Spichiger and Pamard, 1973; Blanc-Pamard, 1979; Spichiger and Lassailly, 1981; Hopkins, 1992). In the Lamto region, we recorded a higher proportion of Guineo-Congolian species in agricultural areas than inside the reserve, whilst in the drier CNP region this was reversed, with a higher percentage of Sudanian species on fallows compared to the undisturbed interior of the national park.

11.4 FOREST DYNAMICS DUE TO LOGGING

In many regions of West Africa, logging of forest trees began in the 1950s (Arnaud and Sournia, 1979). Also on the photographs reproducing the CNP region in 1954, a few small cutting areas are visible. During the subsequent decades, wood cutting generally increased. In the region south of the CNP many forests and successional thickets had remained within the old forest contours until more recently. The areas of deforestation and reforestation are variously distributed across the region. For detecting and quantifying related forest dynamics with automated procedures, the oldest and most recent high-resolution satellite images available for the region were analyzed. These were a Landsat TM scene from December 1988 and an ETM+ scene from December 2002 (dry season). With supervised classifications, all forested areas and areas undergoing reforestation after cutting were differentiated at both dates within an area of 5.250 km². From that, the entire area was also determined that could potentially be covered with island forest if no cutting had occurred. Change detection procedures identified areas of deforestation and reforestation in 2002 compared to 1988. The changes in forest cover and the potentially forested areas were quantified pixel-wise. This was done separately in two selected areas inside and outside the CNP, covering 854 km² and 4.320 km², respectively (Figure 1, B). Goetze *et al.* (2006) provide further details.

In addition, deforestation of gallery forests was quantified in the general study region along 75 km of the Comoé River south of the CNP. This was done by quantifying the formerly continuous areas of gallery forest on CORONA satellite scenes from 1967 and relating them to the remaining forest areas encountered on the Landsat ETM+ image from 2002 (Goetze *et al.*, 2006).

Deforestation and reforestation were not noteworthy in the investigated southern part of the CNP within the 14-year time span under consideration (Table 2). Instead, outside the CNP both processes took place throughout the investigation area, with deforestation mostly dominant over reforestation, averaging 40% against 14% of the area potentially covered by forests. Although reforestation occurred to a similar degree on all study areas outside the CNP, deforestation varied between 20% and 60% of the potential forest area.

Of the 40 km² of gallery forests that were nearly continuously developed along both banks of the 75-km section of the Comoé River in 1967, 24,7 km² (62%) had been cleared by 2002, often over the full width of the gallery. In contrast to most forest islands, these areas mostly remained un-wooded. Consequently, river sections that are missing gallery forests on both banks have developed for stretches of up to 13 km.

Table 2. De- and reforestation processes in forest islands between 1988 and 2002 in two areas in the Comoé National Park (CNP) and south of it (*cf.* Figure 1, B) (Goetze *et al.*, 2006).

Research area	Size (km²)	Reforestation		Deforestation		Potentially forested	
		Area (km²)	% Potential forest cover	Area (km²)	% Potential forest cover	Area (km²)	% Study area
Inside CNP	854	42,7	2	82,4	3	260	30
Outside CNP	4.320	230,3	14	671,3	40	1.698	39

By 2002, considerable and progressive deforestation had occurred within the whole study area outside the park. The areas of deforestation and reforestation from 1988 to 2002 were already equal to 54% of the entire area potentially covered by island forests and the area of natural reforestation amounted to only about one-third of the area of deforestation. Therefore, very few (if any) undisturbed forests have survived outside the CNP since the cutting of forests started in the 1950s (Arnaud and Sournia, 1979). Moreover, forest clearance accelerated during the 1990s in some regions. This is associated with an incipient growth of the human population of *c.* 30% between 1988 and 1998 and the introduction of cashew and new cultivation practices due to immigration.

The recent deforestation processes outside the CNP indicate that island forests have become increasingly fragmented. The existing pattern of insularization has been superimposed by a pattern of fragmentation, as the forests within the unaltered island contours increasingly consist of various disturbed and early-successional stages. This will aggravate the ecological and genetic isolation of undisturbed forests in the CNP, which are among the last remaining natural forests of the entire Guineo-Sudanian transition zone. As the previously less disturbed Comoé gallery forests (with many floristic elements of the Guineo-Congolian rain forests) have become both highly fragmented outside the CNP and separated from the southern rain forests, their functioning as a phytogeographical migration corridor and extrazonal biodiversity pool of rain forest species (Porembski, 2001; Natta and Porembski, 2003) has also become increasingly jeopardized.

11.5 GENERAL CONCLUSIONS

In consequence of increased forest fragmentation, legal conservation in protection areas and, in particular, continued traditional protection by local populations in land use areas (*cf.* Koulibaly *et al.*, 2006) have become an issue of paramount importance in order to conserve the last intact island and gallery forests, being important recruitment pools of Guineo-Sudanian biodiversity. On the other hand, in order to keep the forest–savanna vegetation pattern, the anthropogenic burning of the savannas should continue, with early dry-season fires, as has been the practice for thousands of years in West Africa (Hopkins, 1992; Wohlfarth-Bottermann, 1994). This allows for their continued existence through counteracting the natural spread of forests (*cf.* Favier and Dubois, 2004; Favier *et al.*, 2004b).

The results obtained serve as a reference point for landscape dynamics in regions with low human population density and traditional subsistence land use without livestock farming in the Guineo-Sudanian transition zone of West Africa. It became evident that forest patterns and forest dynamics differ regionally, necessitating a general distinction to be made between climatically drier and more humid zones, as well as recognizing strong regional distinctions when synthesizing findings on historical, ecological, demographical and socio-economic development (for the latter point *cf.* Leach and Fairhead, 2000).

11.6 SUMMARY

In the transition between the southern Sudanian and northern Guinean zones of West Africa, numerous islands of predominantly semi-deciduous forests are interspersed in extensive savannas. Little is known about the natural dynamics of this widespread forest–savanna mosaic and how they are altered by human activities. This was investigated with respect to past, present and future forest–savanna dynamics, which presumably influence northern Guinean biodiversity.

Forest–savanna patterns were studied in detail in the Comoé National Park (CNP) region (north-eastern Ivory Coast). A comparison of aerial photographs from 1954 and 1996 revealed a remarkable stability over this 42-year period, even in regions under traditional subsistence land use outside the CNP. The contours towards the surrounding savanna and, thus, the size of 95% of the forest islands remained nearly unchanged. Outside the CNP, anthropogenic deforestation and subsequent natural reforestation within the contours of the existing forests occurred to a noteworthy extent. A change detection procedure with satellite scenes from 1988 and 2002 revealed a much greater extent of deforestation than of reforestation during this more recent period (40% vs. 14%). By 2002, 62% of the original gallery forest along a 75-km section of the Comoé River was cleared.

Along eight forest–savanna transects at semi-deciduous forest islands in the CNP study area, plant species composition, the spatial structure of tree sizes and heights, aboveground phytomass, the environmental parameters as fire occurrence, shading by upper tree layers and soil depth were recorded. Split moving window dissimilarity analysis and moving window regression analysis were combined to detect statistical significance of borders in multivariate vegetation data along continuous transects, to determine the width of associated ecotones, and, thus, the penetration depth of edge effects towards the forest interior. The penetration averaged 55 m towards the forest interior. Ecotone detection with all species present revealed an interlocked sequence of ecotones for grasses, herbs, woody climbers, shrubs and trees, with each of these ecotones being narrower than the overall ecotone. For dominant tree species, a sequential series was observed from the forest border into the forest interior. For large trees a distinct boundary formation was detected, dominated by the semi-fire resistant tree species *Anogeissus leiocarpus*. At the forest border, *A. leiocarpus* was the most abundant tree with juveniles reaching highest density values at the outer periphery of the forests. Forest tree species regenerated well at forest sites, but also in the shade of *A. leiocarpus* stands.

Successional processes at the forest-savanna border may explain the patterns found. The borders of the studied forest islands are concluded to advance slowly against savanna by sequential succession. This is not contradicting the before-mentioned stability of forest borders observed by means of aerial photographs because the advancement rate appears to be as slow as to lie within the spatial error margin of the photographs compared. In this successional process *A. leiocarpus* is most likely to act as an important pioneer due to its effective regeneration at savanna–forest boundaries under moderate fire impact and on rather shallow soils. Once established, it appears to modify site conditions, especially intensity of fire, by shading out savanna grasses. The occurrence of early dry-season fires seems to be determined mainly by the amount of grass biomass as fuel, because fires occurred only in the savanna while forest sites remained unaffected. Low grass biomass appears to be primarily the result of suppression by competing woody species (i.e. shading) and not of shallow soil.

Thus, a natural succession from savanna to forest possibly proceeds only very slowly due to the counteracting effects of annual savanna fires and the lower climatic humidity of the area compared to the interior Guinean zone. Reduced humidity appears to hamper both linear progression of forest edges and the establishment of new forest initials in the savanna. As also proved by newer studies on Holocene vegetation dynamics, the present pattern of Guinean forests has lately been developing under expansion of natural forests. Consequently, these forests should be considered as habitat islands rather than as habitat fragments. As deforestation has increased considerably, the pre-existing pattern of forest insularization has become overlain by a pattern of forest fragmentation. This will aggravate the ecological and genetic isolation of undisturbed forests like the ones in the CNP, which are among the last remaining natural forests of the entire Guineo-Sudanian transition zone.

ACKNOWLEDGEMENTS

The German Federal Ministry of Education and Research, BMBF, funded the research in the CNP through its interdisciplinary BIOTA Africa research programme (project ID: 01 LC 0017/01 LC 0409). The Ivorian Ministère d' Eaux et Forêts kindly gave the permission to conduct research in the CNP. We express our gratitude to our Ivorian partners at the Universities of Cocody and Abobo-Adjamé (Abidjan), namely Dossahoua Traoré, Laurent Aké Assi, Edouard N'Guessan and Martine Tahoux Touao (Centre of Ecological Research, Abidjan) for their helpful co-operation. We are indebted to K. Eduard Linsenmair and Frauke Fischer (Würzburg), and employees of the 'Projet Biodiversité' (Kakpin) for providing shelter and logistics to the BIOTA scientists at their research station in the CNP. We are especially grateful to Bianca Hörsch (Rome) for giving GIS support, to Philippe Kersting (Mainz) for providing soil data and to our field assistant Lucien K. Kouamé (Kakpin). Frauke Fischer (Würzburg) and Koffi Kouadio (Kakpin) contributed GPS tracks to the forest contour map. Louis Scott (Bloemfontein) provided valuable comments on the manuscript. We warmly thank the organisers of the workshop 'Dynamics of forest ecosystems in Central Africa during the Holocene' of the German Research Foundation (DFG) in Yaoundé 2006, Jürgen Runge and Thorsten Herold (Frankfurt), for inviting us to contribute to this volume.

REFERENCES

Achard, F., Eva, H.D., Stibig, H.J., Mayaux, P., Gallego, J., Richards, T. and Malingreau, J.P., 2002, Determination of deforestation rates of the world's humid tropical forests. *Science*, **297**, pp. 999–1002.

Adejuwon, J.O. and Adesina, F.A., 1992, The nature and the dynamics of the forest-savanna boundary in south-western Nigeria. In *Nature and dynamics of forest-savanna boundaries*, edited by Furley, P.A., Proctor, J. and Ratter, J.A., (London a.o.: Chapman & Hall), pp. 331–351.

Adjanohoun, E., 1964, Végétation des savanes et rochers découverts en Côte d'Ivoire centrale. *Mémoires ORSTOM*, **7**, (Paris: ORSTOM).

Adler, P.B., Raff, D.A. and Lauenroth, W.K., 2001, The effect of grazing on the spatial heterogeneity of vegetation. *Oecologia*, **128**, pp. 465–479.

Afikorah-Danquah, S., 1997, Local resource management in the forest-savanna transition zone: the case of Wenchi district, Ghana. *IDS Bulletin*, **28**, pp. 36–46.

Anhuf, D., 1994, Zeitlicher Vegetations- und Klimawandel in Côte d'Ivoire. Veränderung der Vegetationsbedeckung in Côte d'Ivoire. *Erdwissenschaftliche Forschung*, **30**, edited by Lauer, W., (Stuttgart: Steiner), pp. 1–299.

Anhuf, D., 1997, Satellitengestützte Vegetationsklassifizierung zur Analyse von Vegetationsveränderungen im Bereich der Côte d'Ivoire. *Mannheimer Geographische Arbeiten*, **45**.

Anhuf, D., Grunert, J. and Koch, E., 1990, Veränderungen der realen Bodenbedeckung im Sahel der Republik Niger (Regionen Tahoua und Niamey) zwischen 1955 und 1975. *Erdkunde*, **44**, pp. 195–209.

Arbonnier, M., 2002, *Arbres, arbustes et lianes des zones sèches d'Afrique de l'Ouest*, 2nd., (Paris: CIRAD, MNHN).

Arnaud, J.-C. and Sournia, G., 1979, Les forêts de Côte-d'Ivoire: une richesse en voie de disparition. *Cahiers d'Outre-Mer*, **31**, pp. 281–301.

Aubréville, A., 1950, *Flore forestière soudano-guinéenne*, (Paris: Société d'Éditions Géographiques, Maritimes et Coloniales).

Augustine, D.J., 2003, Spatial heterogeneity in the herbaceous layer of a semi-arid savanna ecosystem. *Plant Ecology*, **167**, pp. 319–332.

Avenard, J.-M., Bonvallot, J., Latham, M., Renard-Dugerdil, M. and Richard, J., 1974, Aspects du contact forêt-savane dans le Centre et l'Ouest de la Côte d'Ivoire. *Travaux et documents de l'ORSTOM*, **35**, (Paris: ORSTOM).

Backéus, I., 1992, Distribution and vegetation dynamics of humid savannas in Africa and Asia. *Journal of Vegetation Science* **3**, pp. 345–356.

Badejo, M.A., 1998, Agroecological restoration of savanna ecosystems. *Ecological Engineering*, **10**, pp. 209–219.

Baker, W.L. and Dillon, G.K., 2000, Plant and vegetation responses to edges in the southern Rocky Mountains. In *Forest fragmentation in the southern Rocky Mountains*, edited by Knight, R.L., Smith, F.W., Buskirk, S.W., Romme, W. H. and Baker, W. L., (Boulder: University Press of Colorado), pp. 221–245.

Bassett, T.J. and Boutrais, J., 2000, Cattle and trees in the West African savanna. In *Contesting forestry in West Africa*, edited by Cline-Cole, R. and Madge, C., (Aldershot a.o.: Ashgate), pp. 242–263.

Bassett, T.J. and Zuéli, K.B., 2003, The Ivorian savanna: global narratives and local knowledge of environmental change. In *Political ecology: an integrative approach to geography and environment-development studies*, edited by Zimmerer, K. S. and Bassett, T.J., (New York, London: Guilford), pp. 115–136.

Bassett, T.J., Zuéli, K.B. and Ouattara, T., 2003, Fire in the savanna: environmental change and land reform in northern Côte d'Ivoire. In *African savannas: global narratives and local knowledge of environmental change*, edited by Bassett, T. J. and Crummey, D., (Oxford/Portsmouth: Currey/Heinemann), pp. 53–71.

Bertrand, A., 1983, La déforestation en zone de forêt dense en Côte d'Ivoire. *Bois et Forêts des Tropiques*, **202**, pp. 3–17.

Biddulph, J. and Kellman, M., 1998, Fuels and fire at savanna gallery forest boundaries in southeastern Venezuela. *Journal of Tropical Ecology*, **14**, pp. 445–461.

Blanc-Pamard, C., 1979, Un jeu écologique différentiel: les communautés rurales du contact forêt-savane au fond du 'V Baoulé' (Côte d'Ivoire). *Travaux et Documents de l'ORSTOM*, **107**.

Bloesch, U., 2002, *The dynamics of thicket clumps in the Kagera savanna landscape, East Africa*, (Aachen: Shaker).

Breman, H. and Dewit, C.T., 1983, Rangeland productivity and exploitation in the Sahel. *Science*, **221**, pp. 1341–1347.

Brooks, T., Balmford, A., Burgess, N., Fjeldsa, J., Hansen, L. A., Moore, J., Rahbek, C. and Williams, P., 2001, Toward a blueprint for conservation in Africa. *Bioscience*, **51**, pp. 613–624.

Cantrell, R.S., Cosner, C. and Fagan, W.F., 2001, How predator incursions affect critical patch size: the role of the functional response. *American Naturalist*, **158**, pp. 368–375.

César, J., 1981, Cycles of the biomass and regrowths after cutting in savanna (Ivory-Coast). *Revue d'Elevage et de Médecine Vétérinaire des Pays Tropicaux*, **34**, pp. 73–81.

Chatelain, C., Gautier, L. and Spichiger, R., 1996, Deforestation in southern Côte d'Ivoire: a high-resolution remote sensing approach. In *The Biodiversity of African Plants–Proceedings XIVth AETFAT Congress*, edited by Van der Maesen, L.J.G., Van der Burgt, X. M. and Van Medenbach de Rooy, J.M., (Dordrecht a.o.: Kluwer), pp. 259–266.

Clayton, W.D., 1958, Secondary vegetation and the transition to savanna near Ibadan, Nigeria. *Journal of Ecology*, **46**, pp. 217–238.

Cochrane, M.A. and Laurance, W.F., 2002, Fire as a large-scale edge effect in Amazonian forests. *Journal of Tropical Ecology*, **18**, pp. 311–325.

Condit, R., Hubbell, S.P. and Foster, R.B., 1995, Mortality rates of 205 neotropical tree and shrub species and the impact of a severe drought. *Ecological Monographs*, **65**, pp. 419–439.

Cornelius, J.M. and Reynolds, J.F., 1991, On determining the statistical significance of discontinuities within ordered ecological data. *Ecology*, **72**, pp. 2057–2070.

Couteron, P. and Kokou, K., 1997, Woody vegetation spatial patterns in a semi-arid savanna of Burkina Faso, West Africa. *Plant Ecology*, **132**, pp. 211–227.

Csillag, F. and Kabos, S., 2002, Wavelets, boundaries, and the spatial analysis of landscape pattern. *Ecoscience*, **9**, pp. 177–190.

Curran, L.M., Trigg, S.N., McDonald, A.K., Astiani, D., Hardiono, Y. M., Siregar, P., Caniago, I. and Kasischke, E., 2004, Lowland forest loss in protected areas of Indonesian Borneo. *Science*, **303**, pp. 1000–1003.

Delègue, M.-A., Fuhr, M., Schwartz, D., Mariotti, A. and Nasi, R., 2001, Recent origin of a large part of the forest cover in the Gabon coastal area based on stable carbon isotope data. *Oecologia*, **129**, pp. 106–113.

Devineau, J.-L., 1976, Principales caractéristiques physionomiques et floristiques des formations forestières de Lamto (moyenne Côte d'Ivoire). *Annales de l'Université d'Abidjan, série E*, **9**, pp. 274–303.

Devineau, J.-L., Lecordier, C. and Vuattoux, R., 1984, Evolution de la diversité spécifique du peuplement ligneux dans une succession préforestière de colonisation d'une savane protégée des feux (Lamto, Côte d'Ivoire). *Candollea*, **39**, pp. 103–134.

Diouf, A. and Lambin, E.F., 2001, Monitoring land-cover changes in semi-arid regions: remote sensing data and field observations in the Ferlo, Senegal. *Journal of Arid Environments*, **48**, pp. 129–148.

Dugerdil, M., 1970, Recherches sur le contact forêt-savane en Côte-d'Ivoire. 1. Quelques aspects de la végétation et de son évolution en savane préforestière. *Candollea*, **25**, pp. 11–19.

Ekanza, S.-P., 1981, Le Moronou à l'époque de l'administrateur Marchand: aspects physiques et économiques. *Annales de l'Université d'Abidjan, série I*, **9**, pp. 53–70.

Eldin, M., 1971, Le climat. In *Le milieu naturel de la Côte d'Ivoire. Mémoires ORSTOM*, **50**, (Paris: ORSTOM), pp. 73–108.

Fagan, W.F., Fortin, M.J. and Soykan, C., 2003, Integrating edge detection and dynamic modeling in quantitative analyses of ecological boundaries. *Bioscience*, **53**, pp. 730–738.

Fahrig, L., 2003, Effects of habitat fragmentation on biodiversity. *Annual Review of Ecology Evolution and Systematics*, **34**, pp. 487–515.

Fairhead, J. and Leach, M., 1996, *Misreading the African landscape: society and ecology in a forest-savanna mosaic*, (Cambridge a.o.: Cambridge University Press).

Fairhead, J. and Leach, M., 1998, *Reframing deforestation: global analysis and local realities: studies in West Africa*, (London, New York: Routledge).

Favier, C. and Dubois, M.A., 2004, Reconstructing forest savanna dynamics in Africa using a cellular automata model, FORSAT. In *Cellular Automata: 6th International Conference on Cellular Automata for Research and Industry, ACRI 2004, Amsterdam, The Netherlands, October 25–28, 2004. Proceedings*, edited by Sloot, P. M. A., Chopard, B. and Hoekstra, A. G., (Berlin/Heidelberg: Springer), pp. 484–491.

Favier, C., Namur, C. de and Dubois, M.A., 2004a, Forest progression modes in littoral Congo, Central Atlantic Africa. *Journal of Biogeography*, **31**, pp. 1445–1461.

Favier, C., Chave, J., Fabing, A., Schwartz, D. and Dubois, M. A., 2004b, Modelling forest-savanna mosaic dynamics in man-influenced environments: effects of fire, climate and soil heterogeneity. *Ecological Modelling*, **171**, pp. 85–102.

Fearnside, P.M. and Laurance, W.F., 2003, Comment on 'Determination of deforestation rates of the world's humid tropical forests'. *Science*, **299**, p. 1015a.

FGU-Kronberg, 1979, Comoé-Nationalpark-Teil 1: Bestandsaufnahme der ökologischen und biologischen Verhältnisse. In *Gegenwärtiger Status der Comoé- und Taï-Nationalparks sowie des Azagny-Reservats und Vorschläge zu deren Erhaltung und Entwicklung zur Förderung des Tourismus*, **2**.

Fischer, F. and Linsenmair, K.E., 2001, Decreases in ungulate population densities: examples from the Comoé National Park, Ivory Coast. *Biological Conservation*, **101**, pp. 131–135.

Fischer, F., Gross, M. and Linsenmair, K.E., 2002, Updated list of the larger mammals of the Comoé National Park, Ivory Coast. *Mammalia*, **66**, pp. 83–92.

Fournier, A., 1991, *Phénologie, croissance et production végétales dans quelques savanes d'Afrique de l'Ouest. Variation selon un gradient climatique*, (Paris: ORSTOM), 312 pp.

Fournier, A., Hoffmann, O. and Devineau, J.-L., 1982, Variations de la phytomasse herbacée le long d'une toposéquence en zone soudano-guinéenne, Ouango-Fitini (Côte d'Ivoire). *Bulletin de l'Institut Fondamental d'Afrique Noire, série A*, **44**, pp. 71–77.

Frankenberg, P. and Anhuf, D., 1989, Zeitlicher Vegetations- und Klimawandel im westlichen Senegal. *Erdwissenschaftliche Forschung*, **24**, (Stuttgart: Steiner).

Fuls, E.R., 1992, Ecosystem modification created by patch-overgrazing in semiarid grassland. *Journal of Arid Environments*, **23**, pp. 59–69.

Gascon, C., Williamson, G.B. and Da Fonseca, G.A.B., 2000, Receding forest edges and vanishing reserves. *Science*, **288**, pp. 1356–1358.

Gautier, L., 1989, Contact forêt-savane en Côte d'Ivoire centrale: évolution de la surface forestière de la réserve de Lamto (sud du V-Baoulé). *Bulletin de la Société Botanique de France, Actualitiés botaniques*, **136**, pp. 85–92.

Gautier, L., 1990, Contact forêt-savane en Côte-d'Ivoire centrale: évolution du recouvrement ligneux des savanes de la Reserve de Lamto (sud du V-Baoulé). *Candollea*, **45**, pp. 627–641.

Gautier, L. and Spichiger, R., 2004, The forest-savanna transition in West Africa. In *Biodiversity of West African forests: an ecological atlas of woody plant species*, edited by Poorter, L., Bongers, F., Kouamé, F.N. and Hawthorne, W.D., (Wallingford, Cambridge: CABI), pp. 33–40.

Giannini, A., Saravanan, R. and Chang, P., 2003, Oceanic forcing of Sahel rainfall on interannual to interdecadal time scales. *Science*, **302**, pp. 1027–1030.

Gibson, D.J., 1996, Textbook misconceptions: the climax concept of succession. *American Biology Teacher*, **58**, pp. 135–140.

Gignoux, J., Clobert, J. and Menaut, J.-C., 1997, Alternative fire resistance strategies in savanna trees. *Oecologia*, **110**, pp. 576–583.

Gignoux, J., Barot, S., Menaut, J.-C. and Vuattoux, R., 2006, Structure, long-term dynamics, and demography of the tree community. In *Lamto: structure, functioning, and dynamics of a savanna ecosystem*, edited by Abbadie, L., Gignoux, J., Le Roux, X. and Lepage, M., (New York: Springer), pp. 335–364.

Gigon, A. and Grimm, V., 1997, Stabilitätskonzepte in der Ökologie: Typologie und Checkliste für die Anwendung. In *Handbuch der Umweltwissenschaften: Grundlagen und Anwendungen der Ökosystemforschung*, edited by Fränzle, O., Müller, F. and Schröder, W., (Landsberg: Ecomed), pp. 1–19.

Goetze, D., Hörsch, B. and Porembski, S., 2006, Dynamics of forest-savanna mosaics in north-eastern Ivory Coast from 1954 to 2002. *Journal of Biogeography*, **33**, pp. 653–664.

Goldammer, J.G. (ed.), 1990, Fire in the tropical biota: ecosystem processes and global changes. *Ecological Studies*, **84**, (Berlin: Springer).

Gosz, J.R., 1991, Fundamental ecological characteristics of landscape boundaries. In *Ecotones: the role of landscape boundaries in the management and restoration of changing environments*, edited by Holland, M.M., Risser, P.G. and Naiman, R.J., (New York: Chapman & Hall), pp. 8–30.

Guillaumet, J.-L., 1967, Recherches sur la végétation et la flore de la région du Bas-Cavally (Côte d'Ivoire), *Mémoires ORSTOM*, **20**, (Paris: ORSTOM).

Guillaumet, J.-L. and Adjanohoun, E., 1971, La végétation de la Côte d'Ivoire. *Le Milieu Naturel de la Côte d'Ivoire. Mémoires ORSTOM*, **50**, (Paris: ORSTOM), pp. 157–263.

Guillet, B., Achoundong, G., Youta Happi, J., Beyala, V.K.K., Bonvallot, J., Riéra, B., Mariotti, A. and Schwartz, D., 2001, Agreement between floristic and soil organic carbon isotope ($^{13}C/^{12}C$, ^{14}C) indicators of forest invasion of savannas during the last century in Cameroon. *Journal of Tropical Ecology*, **17**, pp. 809–832.

Guinko, S. and Bélem Ouédraogo, M., 1998, La flore du Burkina Faso. *AAU Reports*, **39**, pp. 43–65.

Hahn-Hadjali, K., 1998, Les groupements végétaux des savanes du sud-est du Burkina Faso (Afrique de l'Ouest). *Etudes sur la Flore et la Végétation du Burkina Faso et des Pays Avoisinants*, **3**, pp. 3–79.

Hall, J.B. and Swaine, M.D., 1981, *Distribution and ecology of vascular plants in a tropical rain forest: forest vegetation in Ghana*, (The Hague, Boston, London: Junk).

Harris, W.V., 1971, *Termites: their recognition and control*, 2nd edn., (London: Longman, Green and Co.).

Hauhouot, A.D., 1982, Problématique du développement dans le pays Lobi (Côte d'Ivoire). *Cahiers d'Outre-Mer*, **35**, pp. 307–334.

Hennenberg, K.J., Goetze, D., Kouamé, L.K., Orthmann, B. and Porembski, S., 2005a, Border and ecotone detection by vegetation composition along forest-savanna transects in Ivory Coast. *Journal of Vegetation Science*, **16**, pp. 301–310.

Hennenberg, K.J., Goetze, D., Minden, V., Traoré, D. and Porembski, S., 2005b, Size-class distribution of *Anogeissus leiocarpus* (Combretaceae) along forest-savanna ecotones in northern Ivory Coast. *Journal of Tropical Ecology*, **21**, pp. 273–281.

Hennenberg, K.J., Fischer, F., Kouadio, K., Goetze, D., Orthmann, B., Linsenmair, K.E., Jeltsch, F. and Porembski, S., 2006, Phytomass and fire occurrence along forest-savanna transects in the Comoé National Park, Ivory Coast. *Journal of Tropical Ecology*, **22**, pp. 303–311.

Higgins, S.I., Bond, W.J. and Trollope, W.S.W., 2000, Fire, resprouting and variability: a recipe for grass-tree coexistence in savanna. *Journal of Ecology*, **88**, pp. 213–229.

Hochberg, M.E., Menaut, J.C. and Gignoux, J., 1994, The influences of tree biology and fire in the spatial structure of the West African savannah. *Journal of Ecology*, **82**, pp. 217–226.

Hopkins, B., 1962, Vegetation of the Olokemeji Forest Reserve, Nigeria. 1. General features of the reserve and the research sites. *Journal of Ecology*, **50**, pp. 559–598.

Hopkins, B., 1992, Ecological processes at the forest-savanna boundaries. In *Nature and dynamics of forest-savanna boundaries*, edited by Furley, P.A., Proctor, J. and Ratter, J.A. (London a.o.: Chapman & Hall), pp. 21–33.

Hovestadt, T., Yao, P. and Linsenmair, K.E., 1999, Seed dispersal mechanisms and the vegetation of forest islands in a West African forest-savanna mosaic (Comoé National Park, Ivory Coast). *Plant Ecology*, **144**, pp. 1–25.

Huston, M. and Smith, T., 1987, Plant succession: life-history and competition. *American Naturalist*, **130**, pp. 168–198.

Institut National de la Statistique, 1988/2000, Données socio-démographiques et économiques des localités. Tome 1: Résultats définitifs par localité: Région du Zanzan. *Recensement général de la population et de l'habitation 1988/2000*, **3**, (Abidjan).

Irvine, F.R., 1961, *Woody plants of Ghana*, (London: Oxford University Press).

Jacquez, G.M., Maruca, S. and Fortin, M.-J., 2000, From fields to objects: a review of geographic boundary analysis. *Journal of Geographical Systems*, **2**, pp. 221–241.

Jeltsch, F., Weber, G.E. and Grimm, V., 2000, Ecological buffering mechanisms in savannas: a unifying theory of long-term tree-grass coexistence. *Plant Ecology*, **161**, pp. 161–171.

Jones, E.W., 1963, The forest outliers in the Guinea zone of northern Nigeria. *Journal of Ecology*, **51**, pp. 415–434.

Keay, R.W.J., 1959, Derived savanna: derived from what? *Bulletin de l'Institut Fondamental d' Afrique Noire, série A*, **21**, pp. 427–438.

Kellman, M. and Meave, J., 1997, Fire in the tropical gallery forests of Belize. *Journal of Biogeography*, **24**, pp. 23–34.

Kent, M., Gill, W.J., Weaver, R.E. and Armitage, R. P., 1997, Landscape and plant community boundaries in biogeography. *Progress in Physical Geography*, **21**, pp. 315–353.

Koulibaly, A., Goetze, D., Traoré, D. and Porembski, S., 2006: Protected versus exploited savanna: characteristics of the Sudanian vegetation in Ivory Coast. *Candollea*, **61**, pp. 425–452.

Kratochwil, A. and Schwabe, A., 2001, *Ökologie der Lebensgemeinschaften: Biozönologie*, (Stuttgart: Ulmer).

Küper, W., Sommer, J.H., Lovett, J.C., Mutke, J., Linder, H.P., Beentje, H. J., Van Rompaey, R.S.A.R., Chatelain, C., Sosef, M. and Barthlott, W., 2004, Africa's hotspots of biodiversity redefined. Annals of the Missouri Botanical Garden, 91, pp. 525–535.

Kusserow, H., 1995, Einsatz von Fernerkundungsdaten zur Vegetationsklassifizierung im Südsahel Malis: Ein multitemporaler Vergleich zur Erfassung der Dynamik von Trockengehölzen. *Wissenschaftliche Schriftenreihe Umweltmonitoring*, **1**, (Berlin: Köster).

Kusserow, H. and Haenisch, H., 1999, Monitoring the dynamics of 'tiger bush' (brousse tigrée) in the West African Sahel (Niger) by a combination of Landsat MSS and TM, SPOT, aerial and kite photographs. *Photogrammetrie–Fernerkundung–Geoinformation*, **2**, pp. 77–94.

Lanly, J.P., 1969, Régression de la forêt dense en Côte d'Ivoire. *Bois et Forêts des Tropiques*, **127**, pp. 45–59.

Laurance, W.F., 2002, Hyperdynamism in fragmented habitats. *Journal of Vegetation Science*, **13**, pp. 595–602.

Laurance, W. F. and Bierregaard Jr., R.O. (eds), 1997, *Tropical forest remnants: ecology, management, and conservation of fragmented communities*, (Chicago: University of Chicago Press).

Laurance, W.F. and Williamson, G.B., 2001, Positive feedbacks among forest fragmentation, drought, and climate change in the Amazon. *Conservation Biology*, **15**, pp. 1529–1535.

Laurance, W.F. and Yensen, E., 1991, Predicting the impacts of edge effects in fragmented habitats. *Biological Conservation*, **55**, pp. 77–92.

Laurance, W.F., Ferreira, L.V., Rankin-de Merona, J.M. and Laurance, S.G., 1998, Rain forest fragmentation and the dynamics of Amazonian tree communities. *Ecology*, **79**, pp. 2032–2040.

Laurance, W.F., Lovejoy, T.E., Vasconcelos, H.L., Bruna, E.M., Didham, R.K., Stouffer, P.C., Gascon, C., Bierregaard, R.O., Laurance, S.G. and Sampaio, E., 2002, Ecosystem decay of Amazonian forest fragments: a 22-year investigation. *Conservation Biology*, **16**, pp. 605–618.

Lawson, G.W., Jeník, J. and Armstrong-Mensah, K.O., 1968, A study of a vegetation catena in Guinea savanna at Mole game reserve (Ghana). *Journal of Ecology*, **56**, pp. 505–522.

Leach, M. and Fairhead, J., 2000, Challenging neo-Malthusian deforestation analyses in West Africa's dynamic forest landscapes. *Population and Development Review*, **26**, pp. 17–43.

Lebrun, J.P. and Stork, A.L., 1991–1997, *Enumération des plantes à fleurs d'Afrique tropicale*, Vol. I–IV, (Geneva: Conservatoire et Jardin Botaniques de Genève).

Letouzey, R., 1969, Observations phytogéographiques concernant le plateau africain de l'Adamaoua. *Adansonia série 2*, **14**, pp. 321–337.

Lomolino, M.V. and Weiser, M.D., 2001, Towards a more general species-area relationship: diversity on all islands, great and small. *Journal of Biogeography*, **28**, pp. 431–445.

Louppe, D., Ouattara, N. and Coulibaly, A., 1995, Effet des feux de brousse sur la végétation. *Bois et Forêts des Tropiques*, **245**, pp. 59–74.

Lykke, A.M., Fog, B. and Madsen, J.E., 1999, Woody vegetation changes in the Sahel of Burkina Faso assessed by means of local knowledge, aerial photos, and botanical investigations. *Geografisk Tidsskrift, Danish Journal of Geography*, **2**, pp. 57–68.

Mabberley, D.J., 1997, *The plant book: a portable dictionary of the vascular plants*, 2nd edn., (Cambridge: Cambridge University Press).

MacArthur R.H. and Wilson, E.O., 1967, *The theory of island biogeography*, (Princeton (NJ): Princeton University Press).

Mackay, J.H., 1936, Problems of ecology in Nigeria. *Empire Forestry Journal* **15**, pp. 190–200.

McCook, L.J., 1994 Understanding ecological community succession: causal-models and theories, a review. *Vegetatio*, **110**, pp. 115–147.

Menaut, J.C. and César, J., 1979, Structure and primary productivity of Lamto savannas, Ivory Coast. *Ecology*, **60**, pp. 1197–1210.

Menaut, J.-C. and César, J., 1982, The structure and dynamics of a West African savanna. In *Ecology of tropical savannas*, edited by Huntley, B. J. and Walker, B. H., (Berlin: Springer), pp. 80–100.

Miège, J., 1966, Observations sur les fluctuations des limites savanes-forêts en basse Côte d'Ivoire. *Annales de la Faculté des Sciences Dakar*, **19**, pp. 149–166.

Mistry, J. and Berardi, A., 2005, Assessing fire potential in a Brazilian savanna nature reserve. *Biotropica*, **37**, pp. 439–451.

Mitja, D. and Puig, H., 1993, Essartage, culture itinérante et reconstitution de la végétation dans les jachères en savane humide de Côte d'Ivoire (Booro-Borotou, Touba). In *La jachère en Afrique de l'Ouest*, edited by Floret, C. and Serpantié, G., (Paris: ORSTOM), pp. 377–392.

Monnier, Y., 1981, *La poussière et la cendre: paysages, dynamique des formations végétales et stratégies des sociétés en Afrique de l'Ouest*, (Paris: Agence de Coopération Culturelle et Technique).

Mordelet, P., 1993, Influence of tree shading on carbon assimilation of grass leaves in Lamto savanna, Côte d'Ivoire. *Acta Oecologica*, **14**, pp. 119–127.

Mordelet, P. and Menaut, J.C., 1995, Influence of trees on above-ground production dynamics of grasses in a humid savanna. *Journal of Vegetation Science*, **6**, pp. 223–228.

Moss, R.P. and Morgan, W.B., 1970, Soils, plants and farmers in West Africa. *Human ecology in the tropics*, edited by Garlick, J.P. and Keay, R.W.J., (Oxford, London: Pergamon), pp. 1–31.

Moss, R.P. and Morgan, W.B., 1977, Soils, plants and farmers in West Africa. *Human ecology in the tropics*, 2nd edn., edited by Garlick, J.P. and Keay, R.W.J., (London: Taylor & Francis), pp. 27–77.

Mühlenberg, M., Galat-Luong, A., Poilecot, P., Steinhauer-Burkart, B. and Kühn, I., 1990, L'importance des îlots forestiers de savane humide pour la conservation de la faune de forêt dense en Côte d'Ivoire. *Revue d'Ecologie—la Terre et la Vie*, **45**, pp. 197–214.

Müller, J. and Wittig, R., 2002, L'état actuel du peuplement ligneux et la perception de sa dynamique par la population dans le Sahel burkinabé—présenté à l'exemple de Tintaboora et de Kollangal Alyaakum. *Etudes sur la Flore et la Végétation du Burkina Faso et des Pays Avoisinants*, **6**, pp. 19–30.

Murcia, C., 1995, Edge effects in fragmented forests: Implications for conservation. *Trends in Ecology and Evolution*, **10**, pp. 58–62.

Myers, N., Mittermeier, R.A., Mittermeier, C.G., Da Fonseca, G. A. B. and Kent, J., 2000, Biodiversity hotspots for conservation priorities. *Nature*, **403**, pp. 853–858.

Nansen, C., Tchabi, A. and Meikle, W.G., 2001, Successional sequence of forest types in a disturbed dry forest reserve in southern Benin, West Africa. *Journal of Tropical Ecology* **17**, pp. 525–539.

Natta, A.K. and Porembski, S., 2003, Ouémé and Comoé: forest-savanna border relationships in two riparian ecosystems in West Africa. *Botanische Jahrbücher für Systematik, Pflanzengeschichte und Pflanzengeographie*, **124**, pp. 383–396.

Neumann, K. and Müller-Haude, P., 1999, Forêts sèches au sud-ouest du Burkina Faso: végétation—sols—action de l'homme. *Phytocoenologia*, **29**, pp. 53–85.

Ngomanda, A., Chepstow-Lusty, A., Makaya, M., Schevin, P., Maley, J., Fontugne, M., Oslisly, R., Rabenkogo, N. and Jolly, D., 2005, Vegetation changes during the past 1300 years in western equatorial Africa: a high-resolution pollen record from Lake Kamalete, Lope Reserve, Central Gabon. *Holocene*, **15**, pp. 1021–1031.

Nicholson, S.E., 1989, Long-term changes in African rainfall. *Weather*, **44**, pp. 46–56.

Nicholson, S.E., 2001, Climatic and environmental change in Africa during the last two centuries. *Climate Research*, **17**, pp. 123–144.

Olsson, L. and Ardö, J., 1992, Deforestation in African dry lands: assessment of changes in woody vegetation in semi arid Sudan. In *Proceedings of the Central Symposium of the 'International Space Year' Conference, Munich, Germany, 30 March–4 April 1992* (Noordwijk: ESA SP-341, July 1992), pp. 85–89.

Porembski, S., 2001, Phytodiversity and structure of the Comoé river gallery forest (NE Ivory Coast). In *Life forms and dynamics in tropical forests*, edited by Gottsberger, G. and Liede, S., (Berlin, Stuttgart: Cramer), pp. 1–10.

Poilecot, P., Bonfou, K., Dosso, H., Lauginie, F., N'Dri, K., Nicole, M. and Sangare, Y., 1991, *Un écosystème de savane soudanienne: le Parc National de la Comoé (Côte d'Ivoire)*, (Paris: UNESCO).

Primack, R.B., 1998, *Essentials of conservation biology*, 2nd edn., (Sunderland: Sinauer).

Ramsey, J.M. and Rose Innes, R., 1963, Some quantitative observations on the effects of fire on the Guinea savanna vegetation of northern Ghana over a period of eleven years. *African Soils*, **8**, pp. 41–86.

Reenberg, A., Nielsen, T.L. and Rasmussen, K., 1998, Field expansion and reallocation in the Sahel: land use pattern dynamics in a fluctuating biophysical and socio-economic environment. *Global Environmental Change-Human and Policy Dimensions*, **8**, pp. 309–327.

Runge, J. and Neumer, M., 2000, Dynamique du paysage entre 1955 et 1990 à la limite forêt/savane dans le nord du Zaïre, par l'étude de photographies aériennes et de données

LANDSAT-TM. In *Dynamique à long terme des écosystèmes forestiers intertropicaux*, edited by Servant, M. and Servant-Vildary, S., (Paris: UNESCO), pp. 311–317.

Salzmann, U., 2000, Are modern savannas degraded forests? A Holocene pollen record from the Sudanian vegetation zone of NE Nigeria. *Vegetation History and Archaeobotany*, **9**, pp. 1–15.

Salzmann, U., Hoelzmann, P. and Morczinek, I., 2002, Late Quaternary climate and vegetation of the Sudanian zone of northeast Nigeria. *Quaternary Research*, **58**, pp. 73–83.

San Jose, J.J. and Montes, R.A., 1997, Fire effect on the coexistence of trees and grasses in savannas and the resulting outcome on organic matter budget. *Interciencia*, **22**, pp. 289–298.

Saunders, D.A., Hobbs, R.J. and Margules, C.R., 1991, Biological consequences of ecosystem fragmentation: a review. *Conservation Biology*, **5**, pp. 18–32.

Schaefer, M., 2003, *Wörterbuch der Ökologie*, 4th edn., (Heidelberg, Berlin: Spektrum).

Schnell, R., 1952, Contribution à une étude phytosociologique et phytogéographique de l'Afrique occidentale: les groupements et les unités géobotaniques de la région guinnéenne. *Mémoires de l'Institut Fondamental d'Afrique Noire*, **18**, pp. 45–234.

Scholes, R.J. and Archer, S.R., 1997, Tree-grass interactions in savannas. *Annual Review of Ecology and Systematics*, **28**, pp. 517–544.

Schöngart, J., Orthmann, B., Hennenberg, K. J., Porembski, S. and Worbes, M., 2006, Climate–growth relationships of tropical tree species in West Africa and their potential for climate reconstruction. *Global Change Biology*, **12**, pp. 1139–1150.

Schwartz, D., de Foresta, H., Mariotti, A., Balesdent, J., Massimba, J. P. and Girardin, C., 1996, Present dynamics of the savanna-forest boundary in the Congolese Mayombe: a pedological, botanical and isotopic (^{13}C and ^{14}C) study. *Oecologia*, **106**, pp. 516–524.

Sheil, D., Jennings, S. and Savill, P., 2000, Long-term permanent plot observations of vegetation dynamics in Budongo, a Ugandan rain forest. *Journal of Tropical Ecology*, **16**, pp. 765–800.

Sobey, D.G., 1978, *Anogeissus* groves on abandoned village sites in the Mole National Park, Ghana. *Biotropica*, **10**, pp. 87–99.

Spichiger, R. and Lassailly, V., 1981, Recherche sur le contact forêt–savane en Côte d'Ivoire: note sur l'évolution de la végétation dans la région de Béoumi (Côte d'Ivoire centrale). *Candollea*, **36**, 145–153.

Spichiger, R. and Pamard, C., 1973, Recherches sur le contact forêt–savane en Côte d'Ivoire: Etude du recrû forestier sur des parcelles cultivées en lisière d'un îlot forestier dans le sud du pays Baoulé. *Candollea*, **28**, 21–37.

Stocks, B.J., Van Wilgen, B. W., Trollope, W.S.W., McRae, D. J., Mason, J. A., Weirich, F. and Potgieter, A.L.F., 1996, Fuels and fire behavior dynamics on large-scale savanna fires in Kruger National Park, South Africa. *Journal of Geophysical Research–Atmospheres*, **101**, pp. 23541–23550.

Stott, P., 2000, Combustion in tropical biomass fires: a critical review. *Progress in Physical Geography*, **24**, pp. 355–377.

Stromgaard, P., 1986, Early secondary succession on abandoned shifting cultivator's plots in the Miombo of South Central Africa. *Biotropica*, **18**, pp. 97–106.

Sturm, H.J., 1993, Auswirkungen von Feuer- und Beweidungsausschluß auf die Produktion der Krautschicht einer Strauchsavanne. *Verhandlungen der Gesellschaft für Ökologie*, **22**, pp. 323–328.

Swaine, M.D., Hall, J.B. and Lock, J.M., 1976, The forest–savanna boundary in west-central Ghana. *Ghana Journal of Science*, **16**, pp. 35–52.

Swaine, M.D., Hawthorne, W.D. and Orgle, T.K., 1992, The effects of fire exclusion on savanna vegetation at Kpong, Ghana. *Biotropica*, **24**, pp. 166–172.

Thies, E., 1995, Principaux ligneux (agro-)forestiers de la Guinée—zone de transition. *Schriftenreihe der GTZ*, **253**, (Rossdorf: GTZ).

Van De Vijver, C.A.D.M., 1999, Fire and life in Tarangire. Effects of burning and herbivory on an East African savanna system. Dissertation, Wageningen University.

Van Langevelde, F., Van De Vijver, C.A.D.M., Kumar, L., Van De Koppel, J., De Ridder, N., Van Andel, J., Skidmore, A.K., Hearne, J.W., Stroosnijder, L., Bond, W.J., Prins, H. H.T. and Rietkerk, M., 2003, Effects of fire and herbivory on the stability of savanna ecosystems. *Ecology*, **84**, pp. 337–350.

Villecourt, P., Schmidt, W. and César, J., 1979, Research on chemical composition (N, P, K) of the grass savanna of Lamto, Ivory Coast. *Revue d'Ecologie et de Biologie du Sol*, **16**, pp. 9–15.

Villecourt, P., Schmidt, W. and César, J., 1980, Losses of ecosystem during bush fire (tropical savanna of Lamto, Ivory Coast). *Revue d'Ecologie et de Biologie du Sol*, **17**, pp. 7–12.

Walker, B.H. and Noy-Meir, I., 1982, Aspects of the stability and resilience of savanna ecosystems. *Ecology of tropical savannas*, edited by Huntley, B.J. and Walker, B.H., (Berlin a.o.: Springer), pp. 556–590.

Walker, S., Wilson, J.B., Steel, J.B., Rapson, G.L., Smith, B., King, W. M. and Cottam, Y. H., 2003, Properties of ecotones: evidence from five ecotones objectively determined from a coastal vegetation gradient. *Journal of Vegetation Science*, **14**, pp. 579–590.

Walter, H., 1979, *Vegetation und Klimazonen*, (Stuttgart: Ulmer).

Wasseige, C. de and Defourny, P., 2004, Remote sensing of selective logging impact for tropical forest management. *Forest Ecology and Management*, **188**, pp. 161–173.

White, F., 1983, The vegetation of Africa: a descriptive memoir to accompany the UNESCO/AETFAT/UNSO vegetation map of Africa. *Natural Resources Research*, **20**, (Paris: UNESCO).

Wickens, G.E., 1976, The flora of Jebel Marra (Sudan Republic) and its geographical affinities. *Kew bulletin/Additional series*, **5**, (London: Her Majesty's Stationery Office).

Wiese, B., 1988, Elfenbeinküste: Erfolge und Probleme eines Entwicklungslandes in den westafrikanischen Tropen. *Wissenschaftliche Länderkunden*, **29**, edited by Storkebaum, W., (Darmstadt: Wissenschaftliche Buchgesellschaft).

Wispelaere, G. de, 1980, Les photographies aériennes témoins de la dégradation du couvert ligneux dans un géosystème sahélien sénégalais: influence de la proximité d'un forage. *Cahiers ORSTOM, série Sciences Humaines*, **17**, pp. 155–166.

Wohlfarth-Bottermann, M., 1994, Anthropogene Veränderungen der Vegetationsbedeckung in Côte d'Ivoire seit der Kolonialisierung. *Veränderungen der Vegetationsbedeckung in Côte d'Ivoire. Erdwissenschaftliche Forschung*, **30**, edited by Lauer, W., (Stuttgart: Steiner), pp. 301–480.

Yao, T.B., Servat, E. and Paturel, J.-E., 2000, Evolution du couvert forestier ivoirien sur la période 1950–1990, en relation avec la variabilité du climat et les activités anthropiques. In *Dynamique à long terme des écosystèmes forestiers intertropicaux*, edited by Servant, M. and Servant-Vildary, S., (Paris: UNESCO), pp. 57–62.

CHAPTER 12

The impact of land use on species distribution changes in North Benin

Konstantin König
Institute for Physical Geography, Johann Wolfgang Goethe University, Frankfurt am Main, Germany

Jürgen Runge
Centre for Interdisciplinary Research on Africa (CIRA/ZIAF), Johann Wolfgang Goethe University, Frankfurt am Main, Germany

Marco Schmidt and Karen Hahn-Hadjali
Institute for Ecology and Geobotany, Johann Wolfgang Goethe University, Frankfurt am Main, Germany

Pierre Agbani and Didier Agonyissa
Faculty of Agronomic Sciences, Laboratory of Applied Ecology, University of Abomey Calavi, Cotonou, Benin

Annika Wieckhorst
Institute for Ethnology and Africa Studies, Johannes Gutenberg University, Mainz, Germany

ABSTRACT: In this paper we intend to examine the potential distribution of tree species in the sudanian woodland savannas of North Benin with remote sensing data. In a spatial modelling approach we combine botanical field data of mature savanna tree individuals with environmental data layers to assess changes in their spatial distribution. Therefore georeferenced occurrence points of tree species with high socio-economic relevance for the local population were collected from 1999 to 2002. To account for land-use changes at local scale we used derivates of historical (1986) and recent (2000) LANDSAT satellite images as environmental data layers to assess changes in the potential distribution of tree species. For both time steps we acquired scenes of the wet and the dry season to include different phenological aspects in the modelling process. Distribution maps of single species were evaluated with independent test data and selected by statistical analysis. Recent and historical distributions of tree species were modelled with high accuracy. Changes of the potential distribution of tree species were analysed in regard to land use changes and can be used to assist in management and conservation measures of tree species.

12.1 INTRODUCTION

To assess the impact of land-use/cover changes on the ability of biological systems to support human needs, fine resolution, spatially explicit data on distribution of species are required. Field investigations alone can only provide information of species occurrence on localized sampling points. In response to the demand for detailed information of species distribution on landscape scale environmental niche modelling approaches have been developed to assist in conservation and management measures of biodiversity (Elith *et al.*, 2006). Environmental niche models (ENMs) predict the potential distribution ranges of species based on documented occurrences (field data, museum specimens) and spatially continues environmental variables. For continental and nationwide modelling of species distributions applying climatic data layer proved to be useful (e.g. Schmidt *et al.*, 2005). Unfortunately these data are only available at coarse spatial scales (1 km²) and have no predictive power at finer spatial scales. At these scales remote sensing data have been used to assess environmental parameters and thus might be an interesting complement for ENMs, because they provide the only means to deliver information on land cover (Tappan *et al.*, 2000; Mayaux *et al.*, 2004) and its changes (Mayaux *et al.*, 2005), soil (Houssa *et al.*, 1996) and vegetation properties. Despite its potential, studies which integrate satellite images as environmental data layers in ENMs are relatively rare.

The objective of this paper is twofold: to model the potential distribution of West African savanna tree species with LANDSAT satellite data at two points in time and to analyse the effects of land use changes on the distribution of these species.

12.2 METHODS

12.2.1 Study area

The study site in North Benin (West Africa) is part of the southern Sudanian Zone. Climatically, the northern limit is defined by the 800 mm/year isohyet and the southern limit by the 1.100 mm/year isohyet. The active vegetation period is between 5 and 7 months from May to October.

Characteristic vegetation types are woodland savannas and dry forests. Anthropogenic impact in the form of pastoralism and agriculture has been shaping the vegetation for several thousand years. Shifting cultivation results in a small scale mosaic of fields, fallows of different ages and edaphic savannas. Due to the current population growth and immigration, the pressure on natural resources is steadily increasing.

12.2.2 Classification of land use

LANDSAT satellite TM and ETM+ data, from November1986 and 1999, were used to classify agriculture in the study area. In November, the beginning of the dry season, fields are already harvested and can be clearly distinguished from surrounding savannas.

Digital raw data values were transformed to radiances by using the COST atmospheric correction model (Chavez, 1996) and calibration coefficients of Markham and Barker (1986). Geometric accuracy of the recent ETM+ image was improved to 20 m with 23 ground control points, sampled on street crossings and dams of water reservoirs. Historical images were co–registered to the recent scene with a relative accuracy of 30 m.

During two field campaigns in October 2000/01 intensive ground truthing was conducted in the study area. Therefore we selected 183 sites by a stratified random sampling procedure and determined land use and structural vegetation parameters. The data set was separated randomly in two parts. The first data set (95 training sites) was used for supervised classification of recent land use, the second (88 test sites) for testing of classification accuracy.

Spectral signatures of land use classes (fields and young fallows) were extracted and implemented in an unsupervised classification of the historical satellite scene from 1986. The resulting 15 spectral classes of the unsupervised classification were interpreted with aerial photos from 1986. Accuracy assessment of land use classes was conducted by visual comparison of land use classes between the classified map and aerial photos at 400 randomly distributed test sites.

All analysis were accomplished with the software ERDAS/IMAGINE version 8.6 and ArcView 3.2 with the Spatial Analyst extension.

12.2.3 Species data

Our data set comprises 150 botanical sample sites georeferenced with GPS-coordinates (phytosociological relevees from Didier Agonyssa unpublished, specimen data from Pierre Agbani and Annika Wiekhorst, unpublished).

In this study we modelled the distribution of 20 tree species, which have been selected because of their high utilization value for the local population. Only occurrence points (N = 1.126) of mature tree individuals have been utilized for modelling, because these individuals have already grown at the time of the first set of satellite images.

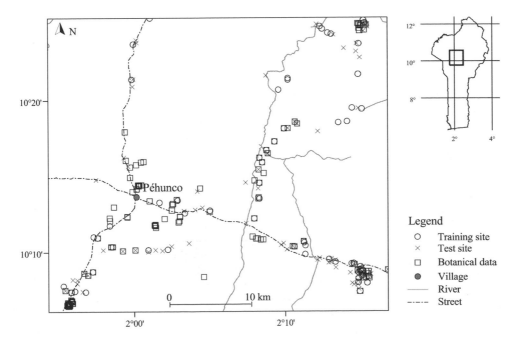

Figure 1. Map of the study area in North Benin.

12.2.4 Modelling approach

For modelling we used the Genetic Algorithm of Rule-set Production (GARP: Stockwell, 1999). Environmental data layers were derived from the prescribed recent and historic LANDSAT satellite data and from dry season images (February 1989/2000). For modelling we include the first three components of a tasselled cap transformation of the rainy season images and radiances from band 3 to 6 of dry season images.

We equally divided the species occurrences (at least 30 occurrence points/species) in two statistically independent data sets: one for modelling and one for testing. For each species we created 20 incidence maps representing absence (0) and presence (1) to account for the variability of possible model results. Due to different phenological conditions historic and recent environmental data sets might not be directly comparable. Therefore we decided not simply to project the recent niche definitions onto the old dataset, but to define the historic niche with an independent model run (historic environmental layer and occurrence points of mature tree individuals).

Consequently, a careful evaluation of the resulting distribution maps was conducted, following the proposed methodology of Anderson *et al.* (2003) using omission (error of underprediction) and commission error (error of overprediction) values. Models with an omission error higher than 20% were rejected. We ranked the remaining models according to their omission and commission values and superimposed the best five models of each species to create a composite map showing the number of optimal models in each pixel. A threshold of at least three models predicting presence was determined as appropriate for gaining a suitable balance between omission and commission errors. Incidence maps of species were added to generate a map of species diversity. Processing of maps written in ASCII format was conducted with VBA scripts implemented in Access 2000 software.

Table 1. Contingence matrix of land use. a) 1986, b) 1999.

a)

Reference (aerial photos)	Classified land use (fields and young fallows)	
	Present	Absent
Present	78	20
Absent	13	289

b)

Reference (ground truth data)	Classified land use (fields and young fallows)	
	Present	Absent
Present	19	2
Absent	2	65

12.2.5 Spatial analysis

To analyse effects of land use on the distribution of tree species, we determined the proportion of land use (fields and young fallows) and the mean species number within hexagonal grid cells (N = 240, size = 100 ha) randomly distributed over the whole study area. Hexagons were preferred over quadratic grid cells to minimize edge effects. To avoid effects of spatial autocorrelation minimum distance between neighbouring hexagons was set to 900 m.

12.3 RESULTS

12.3.1 Analysis of land use change

In 1986 agricultural fields and young fallows were classified with an overall classification accuracy, defined as the percentage of correctly classified sites of 92%, in 1999 with 95% (see table 1). In 1986 10% of the study area (310 km²) are classified as fields and young fallows (age <5 a). In 1999, the agricultural area has almost doubled to 570 km² (19% of the study area). For the time span from 1986 to 1999 this means an average annual land cover conversion rate of 0,6%. This increase is mainly caused through the extension and intensification (increase of field density) of traditional agricultural areas near population rich villages. Today, only regions classified as locally protected areas, like the forest of Alibori (IUCN status IV) in the northeast and branches of the Oueme catchment in the southeast of the study area are not so much affected by agriculture (see figure 2).

12.3.2 Maps of species distribution

In total, 400 recent and 400 historic distribution maps of 20 tree species were obtained. The average value of producer accuracy of all models reached 77% (corresponding to 23% omission error). After the evaluation and selection procedure of best performing models, average omission error was reduced below 15%. Distribution models of two tree species have been rejected, because they did not pass the evaluation criteria (at least 5 models with an omission error below 20%).

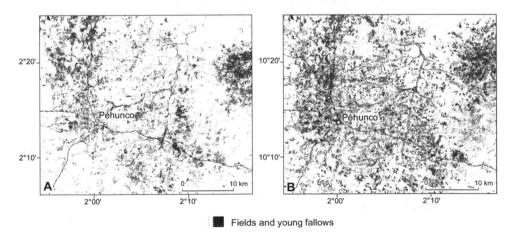

Figure 2. Classification of land use; A = 1986, B = 1999.

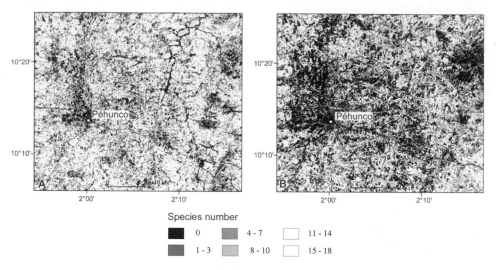

Species number

■ 0	■ 4 - 7	□ 11 - 14
■ 1 - 3	▨ 8 - 10	□ 15 - 18

Figure 3. Distribution of tree diversity; A = 1986, B = 1999.

The map composite of the modeled distribution of 18 savanna tree species is shown in figure 3. As confirmed by the modelled distribution maps, most species can potentially occur in a wide range of different savanna types allover the study area. Negative impact of land use on modelled species diversity of the selected savanna tree species is shown in figure 4. In 1986 only agricultural areas around villages and forests along river systems are excluded from the modelled distribution. In 1999 the areas of potentially suitable habitats have been severely reduced due to the extension of cropland.

12.4 DISCUSSION

In this study we use LANDSAT satellite images and field data to model the recent and historic distribution of savanna tree species in the North Sudanian Zone of Benin. We demonstrate the ability of widely available satellite data to model species distributions with high accuracy and at fine spatial resolutions. However, relationships between causal factors determining species distributions and the signal of satellite images remain unclear, thus extrapolation of findings from one area to another must be performed with caution. Despite these constrains satellite images should be considered as valuable complement in environmental niche modelling, because these data provide the only means to assess land cover changes with a sufficient temporal and spatial resolution and can reflect a considerable proportion of the environmental envelopes of species.

In a recent meta-analysis Geist and Lambin (2004) review the proximate causes of land degradation mentioned in 132 case studies. In Africa 48% of the considered studies suggest intensification of agriculture and extension of cropland production as one the main causes of degradation. Thus land cover/land use change is seen as one of the main causes for the decline of species richness in many tropical countries. Our study supports these findings. In the study area land cover conversion has strongly accelerated during the last 30 years. This results in the conversion of former rangelands like dry forests, woodlands, and savannas to arable land and settlements. Main causes for this are changes in agricultural policies (increase of cash crop farming) and an increase in population due to population growth and migration.

a)   b)

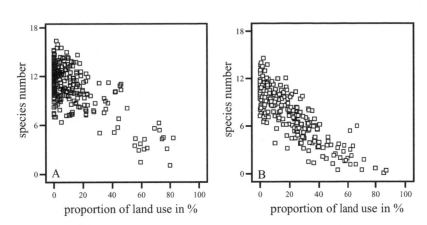

Figure 4. Relation between species diversity and land use. a) 1986, b) 1999.

Effects on land use/land cover change on species diversity have rarely been tested within spatial explicite analysis. We believe that through the implementation of high resolution satellite images within ENMs, effects of land use changes on species diversity can be assessed directly and might give useful information for management measures of biodiversity.

ACKNOWLEDGEMENTS

The investigations are part of the BIOTA research network on biodiversity, sustainable use and conservation in Africa, funded by the German Federal Ministry of Research and Education (BMBF).

REFERENCES

Anderson, R.P., Lew, D. and Peterson, A.T., 2003, Evaluating predictive models of species' distributions. Criteria for selecting optimal models. *Ecological Modelling,* **162(3)**, pp. 211–232.

Chavez, P.S., 1996, Image-based atmospheric corrections-revisited and revised. *Photogrammetric Engineering and Remote Sensing,* **62**, pp. 1025–1036.

Elith, J., Graham, H., Anderson, P., Dudik, M., Ferrier, S., Guisan, A., Hijmans, J., Huettmann, F., Leathwick, R., Lehmann, A., Li, J., Lohmann, G., Loiselle, A., Manion, G., Moritz, C., Nakamura, M., Nakazawa, Y., Overton, C.M., Townsend Peterson, A., Phillips, J., Richardson, K., Scachetti–Pereira, R., Schapire, E., Soberon, J., Williams, S., Wisz, S., Zimmermann, E., 2006, Novel methods improve prediction of species' distributions from occurrence data. *Ecography,* **29**, pp. 129–151.

Geist, H.J. and Lambin, E.F., 2004, Dynamic causal patterns of desertification. *Bioscience,* **54**, pp. 817–829.

Houssa, R.; Pion, J.C. and Yesou, H., 1996, Effects of granulometric and mineralogical composition on spectral reflectance of soils in a Sahelian area. *Photogrammetric Engineering and Remote Sensing,* **51(6)**, pp .284–298.

Markham, B.L. and Barker, J.L., 1986, Landsat MSS and TM post-calibration dynamic ranges, exoatmospheric reflectances and at-satellite temperatures. *EOSAT Technical Notes*, pp. 3–8.

Mayaux, P., Holmgren, P., Achard, F., Eva, H., Stibig, H-J. and Branthomme, A., 2005, Tropical forest cover change in the 1990's and options for future monitoring. *Philosophical Transactions of the Royal Society of London. Series B, Biological Science,* **360** (**1454**), pp. 373–384.

Mayaux, P., Bartholomé, E., Fritz, S. and Belward, A., 2004, A new land–cover map of Africa for the year 2000. *Journal of Biogeography,* **31**, pp. 861–877.

Schmidt, M., Kreft, H., Thiombiano, A. and Zizka, G., 2005, Herbarium collections and field data-based plant diversity maps for Burkina Faso. *Diversity and Distribution,* **11**, pp. 509–516.

Stockwell, D.R.B., 1999, Genetic algorithms II. In *Machine Learning Methods for Ecological Applications*, edited by Fielding, A.H., (Boston: Kluwer Academic Publishers), pp. 123–144.

Tappan, G.G., Hadj, A. and Wood, E.C., 2000, Use of Argon, Corona, and Landsat imagery to assess 30 years of land resource changes in West–Central Senegal. *Photogrammetric Engineering and Remote Sensing,* **66**(**6**), pp. 727–735.

CHAPTER 13

Potentials of NigeriaSat-1 for Sustainable Forest Monitoring in Africa: A Case Study from Nigeria

Ayobami T. Salami

Space Applications and Environmental Science Laboratory (SPAEL), Institute of Ecology and Environmental Studies, Obafemi Awolowo University, Ile-Ife, Nigeria

ABSTRACT: The developed countries had to rely on satellite images and Geographical Information Systems for sustainable forest monitoring. But in Africa, funding constraint has often been the reason for not making use of these methods. With the advent of Disaster Monitoring Constellation (DMC) satellites, in which Nigeria (in the West Africa Sub-region) and Algeria (in the North Africa Sub-region) are involved, the problem of data access for these countries or regions seems to have been addressed. Hence, this paper examines the potentials of NigeriaSat-1 (which was launched in September 2003) for regular forest assessment in Sub-Saharan Africa, with a particular reference to Nigeria. NigeriaSat-1 has a spatial resolution similar to LANDSAT ETM+ and spectral resolution similar to SPOT XS. This paper demonstrates the potentials of these characteristics with the unique advantage of low cost, easy access and high temporal resolution for sustainable forest monitoring in Africa.

13.1 INTRODUCTION

There are approximately 3.400 million ha of forest globally, representing about 30% of the world's land area (WRI, 2003). Pressures on forests to provide economic resources are increasing rapidly. The rate of tropical forest destruction is not known with any accuracy, but is estimated by the Food and Agriculture Organisation (FAO) as around 15,4 million ha per year. Its destruction has many serious long-term environmental implications and hence there is need for proper monitoring of this invaluable resource also known as the global ecological lung.

The growing decline in the areal coverage of forest has been a source of concern to various governments in Sub-Saharan Africa and many of these countries have accepted, at least in principle, the need to address the problem. For instance, the National Forestry Programme (NFP) and Forestry Component of the Environmental Management Programme (EMP) in Nigeria, as well as the Tanzania Forestry Action Plan (TFAP), identified the need to update land use and vegetation maps, monitor deforested lands and undertake national forest resources survey. But most of these proposals have stopped at the stage of need identification, as there are no concrete projects in the field to achieve these laudable objectives. Experience from most of the African countries show that lack of reliable data base and systematic monitoring system has been a major problem inhibiting effective management of the forest estate.

In Nigeria, the eco-climatic zones range from the very humid freshwater mangrove swamps in the south to the semi-arid sahelian zone in the north. These varied zones support a variety of vegetation, among which the most extensive vegetation zones are savannas mainly in the north and forests mainly in the south (Akinbami *et al.,* 2003). Apart from the traditional role of the Nigerian luxuriant forest in regulating the microclimate and environmental value in terms of biodiversity, they represent an important source of domestic energy.

The southern rainforest, the source of the country's timber resources, covers only about 2% of the total land area today. Even this forest cover is being depleted at an estimated rate of 1,36% per year (Salami, 2006). This implies deforestation on a large scale and has brought with it significant negative impacts on the flora and structure of Nigeria's forest. Unpublished surveys in the arid northern zones of the country show that farm tree densities have declined from about 15 to 3 per ha since the 1950s. It is estimated that about 120.000 acres of arable land in the north is lost to desertification yearly. Desert encroachment is advancing at an estimated rate of 0,6 km per year in the northern Nigeria while uncontrolled logging and tree felling is the order of the day in the south; and hence, about 350.000 hectares of Nigeria's forest cover in the south is lost annually (Salami, 2004). This leads to loss of biological diversity. It has been shown that the number of threatened and endangered species is increasing due to unbridled exploitation of the forest resources. FAO (1990) concluded that if the current rate of deforestation is maintained, the remaining forest area in Nigeria may disappear by the year 2020. The value of forest cover loss in Nigeria (based on Total Cost Accounting procedure i.e. TCA) is now estimated at $750 million annually. The World Bank also estimated the value (in terms of avoided costs) of forest cover, which protects and regulates soil, water, wildlife, biodiversity, and carbon fixation, at over $5 billion annually.

Nigerian government has embarked a number of measures and projects to stem the trend although; it is doubtful whether significant progress has been made in this regard. For instance, the Protected Area Programme (PAP) identifies areas of conservation interest, but only four out of the 36 states in the country have been so far inventoried. The Nigerian Biodiversity Strategy and Action Plan (NBSAP) is an attempt to bridge the gaps identified in the conservation efforts. Part of the needs identified include inventory and identification of threatened and endangered species. The National Forestry Programme (NFP) and the forestry component of the Environmental Management Project (EMP) also identified the need to update land use and vegetation maps of the country, monitor deforested lands and undertake national forest resources survey.

The major problem with forestry management in Nigeria, as in most African countries, is the lack of geo-spatial database required for regular monitoring. This is largely due to the cost of the imagery required for such monitoring, except where external funding for such exercise is secured. But even if and where external funding was available for such assessment, monitoring beyond the project face would be a major problem. For sustainable monitoring therefore, regular access to image data sources is a prerequisite. It is against this background that this paper examines the importance of remote sensing techniques and assesses the usefulness of the newly launched NigeriaSat-1 in providing a reliable base for regular forest monitoring in Nigeria. The paper attempts to validate the performance of NigeriaSat-1 for forest monitoring by comparing its classification accuracy with that obtained from LANDSAT ETM+, with which it has similar spectral resolution.

13.2 REMOTE SENSING DATA FOR FORESTRY MONITORING IN NIGERIA

Forest monitoring or change detection has benefited immensely from advances in remote sensing techniques. Remote sensing for change detection is a technique for determining and assessing the differences in a variety of land cover characteristics over time (Jha and Unni, 1994). Detecting, describing and understanding such changes are of considerable interest, not only to ecologists or environmental conservationists, but also to resource managers (Salami, 1999; Salami, 2000). Change detection studies are usually focused on identification and classification of the biotic and abiotic components of the spectral and temporal changes that are occurring within the ecosystem (Monat *et al.*, 1993; Salami *et al.*, 1999). Christenson *et al.* (1998) stated that recording land cover dynamics is probably one of the most important applications of remote sensing data.

In Nigeria, the Federal Government initiated the Nigerian Radar (NIRAD) project during 1976–1978. The project was based in the Federal Department of Forestry, and employed Side-looking Airborne Radar (SLAR) to obtain radar imagery covering the whole country, for the purpose of obtaining a land use/vegetation map of Nigeria. A similar initiative was coordinated by Forest Monitoring, Evaluation and Coordination Unit (FORMECU) of the Federal Ministry of Environment between 1993 and 1995. This initiative led to the production of land use maps at a scale of 1:250.000 but the effort has not been sustained and this has resulted in data gaps in the country.

Despite the emergence of satellite products for environmental monitoring since the 1970s, it has hitherto been very difficult to monitor the Nigerian forest estate using this means. This was due to a number of reasons including the problem of accessing the much necessary image data, since the country lacked its own satellite or ground receiving station. The cost of obtaining the satellite products was also highly prohibitive for researchers. Consequently, the often-quoted rates of deforestation for Nigeria were largely based on mere estimates or surrogate data rather than empirical studies. Most of the vegetation maps produced by international organizations such as FAO or International Steering Committee for Global Mapping (ISCGM), are nothing more than broad generalizations which are not usually in tandem with local realities and are therefore of little use to local authorities for planning purposes. Some empirical studies (see figures 1–3 for example) suggest that both the global and national estimates being quoted for Nigeria by FAO are under estimates (Salami, 1999; Salami *et al.*, 1999). But since NigeriaSat-1 was launched in September 2003 and the products are now available, it should now be much easier to obtain up-to-date information on most environmental parameters including the forests.

13.3 NIGERIASAT-1 AND DISASTER MONITORING CONSTELLATION (DMC) SATELLITES

Since the development of satellite technology, the available sensors have been increasing both in number and sophistication. The development of DMC heralds a novel north-south mutualistic (rather than parasitic) co-operation in space applications for environmental monitoring. The objective is to provide a daily global coverage with a constellation of 5 low-cost micro satellites at a medium spatial resolution of 30–40 m. This collaboration involves Algeria, China, Nigeria, Thailand, Turkey, United Kingdom and Vietnam and has led to the launching of two satellites owned by African countries. These are ALSAT-1 launched by Algeria on November 28, 2002 and NigeriaSat-1 launched by Nigeria on September 27, 2003. This development has far reaching implications for forest monitoring in Africa. At least theoretically, this is capable of ameliorating the problems such as prohibitive costs of image data source and lateness in arrival of such data for near real-time monitoring in northern and West Africa. For instance, the launching of NigeriaSat-1 was immediately followed by the establishment of a Ground Receiving Station in Abuja, Nigeria.

The NigeriaSat-1 with a spatial resolution of 32 m and 3 bands with a swath width of 600 km is already being utilized for environmental monitoring. For instance, when

Figure 1a. SPOT XS (1986) False Colour Composite showing deforestation within Shasha Forest Reserve in Nigeria.

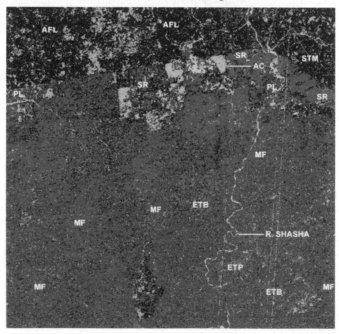

Figure 1b. Classified SPOT XS (1986) showing deforestation within Shasha Forest Reserve in Nigeria (source: Salami *et al.*, 1999); MF: Mixed Forest; ETP: Exotic Tree Species Plantation; PL: Ploughed Land; AFL: Agroforest Land; STM: Settlement/Open Space; SR: Secondary Regrowth.

Hurricane Katrina wrecked havoc on the Gulf Coast of the USA on August 29, 2005, NigeriaSat-1 was the first satellite that captured the disaster in near real-time on September 2, 2005. This underscores the need to employ the capability of this satellite for ecological assessment in general and forest monitoring in particular within the African context. NigeriaSat-1 has great potentials especially for forest monitoring because its multi-spectral imagers are similar to SPOT HRV and bands 2, 3 and 4 of LANDSAT TM with an added advantage of better temporal resolution than either SPOT XS or LANDSAT TM. Table 1 provides a summary of some technical characteristics of NigeriaSat-1 in comparison with some other earth observation (EO) satellites. This shows that NigeriaSat-1 has about 3 times the swath width of LANDSAT ETM+ and ten times that of SPOT as well as improved spatial resolution. This emphasises its importance for inventory as well as rapid and timely assessment, monitoring and environmental management.

13.3.1 Examples from Nigeria

NigeriaSat-1 has been used for a number of pilot studies on forest monitoring in Nigeria. For example, Salami and Balogun (2005) showed that comparable results were obtained from the vegetation mapping of a part of southern Nigeria using the NigeriaSat-1 and LANDSAT-ETM+ of the same area. The quality of the False Color Composites of the two images as shown by figures 2a and 2b is good enough for ground truthing. The major settlements and hydrological networks are clearly discernible.

NigeriaSat-1 was also utilized to assess the current status of Yankari National Park in northern Nigeria and the output was compared to the results obtained from the assessment based on LANDSAT-ETM+. The results are as shown by figures 3 and 4. It was observed that dry woodland was increasing while the evergreen woodland and coverage by water bodies were declining.

Salami and Balogun (2006) made a preliminary assessment of the NigeriaSat-1 image of December 2003 in comparison with ASTER image of January 12, 2002 and of the LANDSAT ETM+ image of January 8, 2001 for ecosystem monitoring in the mangrove belt of Nigeria. The images used were of the same season to avoid the effect of seasonal variation. The images were first geo-referenced to the same coordinate system using the topographical maps of the area (Saadi and Abolfazi, 2003). The images were geo-referenced to Universal Transverse Mercator (WGS84- Zone 32N) and a common window covering the same geographical coordinates was then extracted from each of the images.

Supervised classification was carried out based on the Maximum Likelihood algorithm. For this purpose, a set of 32 field observation points were used for training sample selection. This produced the major land cover classes in the area. Accuracy

Table 1. Comparison of characteristics of NigeriaSat-1 with other EO Satellites.

Satellite (Instrument)	LANDSAT ETM+	SPOT (HRV)	NigeriaSat-1 (DMC Imager)
Swath	185 km	60 km	600 km
Typical Revisit	16 days	26 days	3–5 days
Spatial Resolution	30 m	20 m	32 m
Spectral Resolution	0,45–12,5 µm	0,5–0,89 µm	0,52–0,90 µm
Pixels	6.000	3.000	19.000

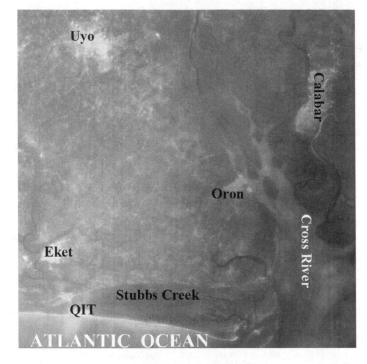

Figure 2a. NigeriaSat-1 (2003).

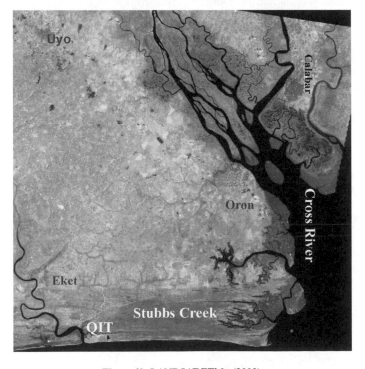

Figure 2b. LANDSAT-ETM+ (2003).

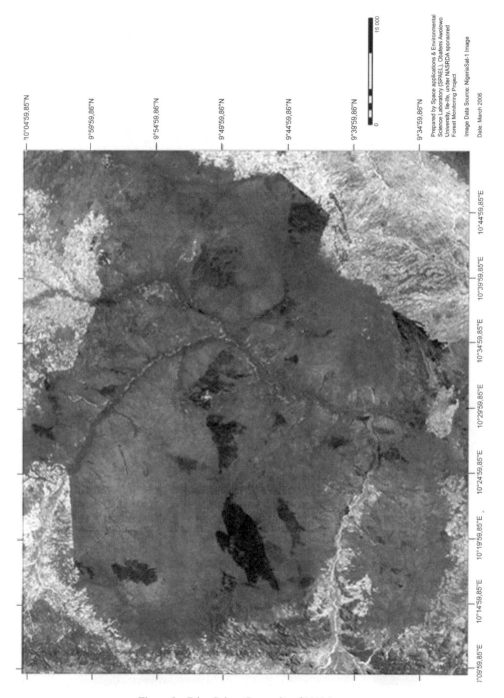

Prepared by Space applications & Environmental
Science Laboratory (SPAEL), Obafemi Awolowo
University, Ile-Ife, under NASRDA sponsored
Forest Monitoring Project

Image Data Source: NigeriaSat-1 Image

Date: March 2006

Figure 3a. False Colour Composite of 2003 Image.

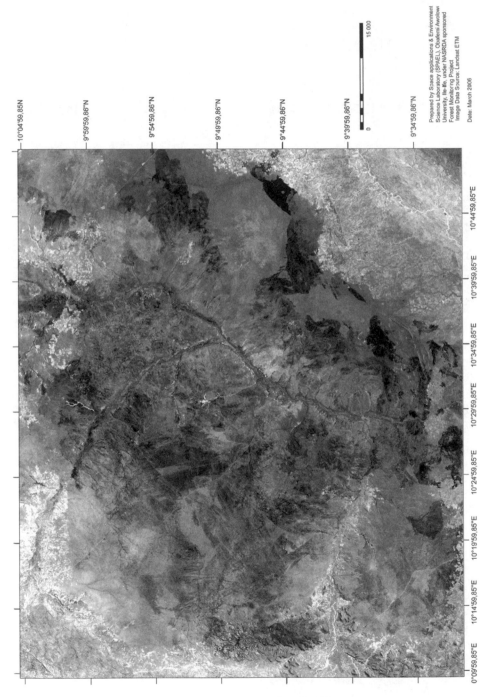

Figure 3b. False Colour Composite of 2001 Image.

Figure 4. Cover Types within and around Yankari National Park, Nigeria (2003).

Table 2. Classification accuracy of NigeriaSat-1 compared with other satellites (Salami and Balogun (2006).

Class Name	ASTER		NigeriaSat-1		LANDSAT ETM+		No. of points used for classification	No. of points used for accuracy assessment
	Producer Accuracy (%)	User	Producer Accuracy (%)	User	Producer Accuracy (%)	User		
Wetland								
Mangrove Swamp	82	81	80	81	75	76	27	45
Degraded Mangrove	86	75	83	71	67	100	21	42
Dryland								
High Forest	75	60	100	71	100	80	12	30
Light Forest	60	75	57	67	83	67	15	30
Other								
Mudflat/ Bare Surface	100	100	67	100	100	67	12	24
Water Body	50	66	50	100	60	100	09	24
Overall Accuracy	**75%**		**72%**		**78%**		**96**	**195**

assessment of these classes was then carried out. To do this, the x and y coordinates of the ground observation points recorded using the GPS, were dropped onto each image. Each point represents the centre coordinates of a plot of 35 × 35 m (i.e. about 0.1 ha) on the ground. Each plot was demarcated using narrow-cut lines. The selection of this plot size was governed by the spatial resolution of the images used. Theoretically, each plot consists of at least 1 pixel (30 × 30 m) of LANDSAT ETM+, 1 pixel (32 × 32 m) of NigeriaSat-1 and 4 pixels of ASTER (each pixel being 15 × 15 m). At least 6 observation points were selected from the surrounding 8 pixels of each field observation point for accuracy assessment. In all, a total of 195 points were used for the assessment, following the procedure outlined by Congalton and Green (1993). This yielded the producer and user accuracies for each cover type as well as overall accuracy for each image. The results are as shown by table 2. The results show that NigeriaSat-1 is of comparable accuracy levels for land use monitoring when compared with ASTER and LANDSAT. The overall accuracy obtained for ASTER is 75% and that of NigeriaSat-1 is 72% while that of LANDSAT ETM+ is 78%. This has obvious implications for forest monitoring and management in Nigeria, and possibly for entire Africa given the strategic position of Nigeria within the West Africa sub-region.

Table 3 shows the cost of NigeriaSat-1 images. A scene of LANDSAT is $400 compared to NigeriaSat-1 which is $175 for archived image and $205 for Real Time image. In terms of affordability, it is clear that it is cheaper for the poor countries of Africa. This implies that this satellite combines the advantages of cost lowering, improved temporal resolution and better accessibility with comparable accuracy level as LANDSAT. These potentials therefore need to be fully exploited for forest monitoring in Africa.

Table 3. Price List of NigeriaSat-1 Images (scene = 80 km × 80 km). Product description: LO = Raw image (individual band); LOR = radiometric corrected image; L1 = band registered image; L1R = radiometric corrected L1; LIG = eometric corrected L1R. High processing level: Stage I = precision geo-reference image; Stage II = ortho-rectified image.

Image Specific	Standard programming (one scene)	$	Priority programme (one scene)	$
ARCHIVE (Government Institutions)				
LO	N6.000	$41,96	N7.000	$48,95
LOR	N9.000	$62,94	N10.000	$69,93
L1	N12.000	$83,92	N13.000	$90,90
L1R	N15.000	$104,89	N16.000	$111,88
LIG	N18.000	$125,87	N19.000	$132,86
HPL Stage I	N25.000	$174,82	N29.000	$202,79
HPL Stage II	N30.000	$209,79	N34.000	$237,76
ARCHIVE (Commercial Institutions)				
LO	N12.000	$83,92	N14.000	$97,90
LOR	N18.000	$125,87	N20.000	$139,86
L1	N24.000	$167,83	N26.000	$181,82
L1R	N30.000	$209,79	N32.000	$223,78
LIG	N36.000	$251,75	N38.000	$265,73
HPL Stage I	N50.000	$349,65	N58.000	$405,60
HPL Stage II	N60.000	$419,58	N68.000	$475,52
REAL TIME (Government Institutions)				
LO	N8.000	$55,94	N10.000	$69,93
LOR	N12.000	$83,91	N13.000	$90,90
L1	N15.000	$104,89	N16.000	$111,88
L1R	N18.000	$125,87	N19.000	$132,86
LIG	N21.000	$146,85	N22.000	$153,84
HPL Stage I	N30.000	$209,79	N35.000	$244,75
HPL Stage II	N60.000	$419,58	N40.000	$279,72
REAL TIME (Commercial Institutions)				
LO	N16.000	$111,88	N20.000	$139,86
LOR	N24.000	$167,83	N26.000	$181,82
L1	N30.000	$209,79	N32.000	$223,78
L1R	N36.000	$251,75	N38.000	$265,73
LIG	N42.000	$293,70	N44.000	$307,69
HPL Stage I	N60.000	$419,58	N70.000	$489,51
HPL Stage II	N70.000	$489,51	N80.000	$559,44

13.4 CONCLUSIONS

Forest degradation has continued to attract the attention of researchers, probably because of its far reaching environmental implications, such as its potential influence on surface albedo, reduction of timber and fuel wood as well as declining genetic diversity (Salami, 1998). Paeth and Hense (2006) noted that reduction in freshwater supply over sub-Saharan

Africa is partly man-made in the context of massive deforestation currently going on while climatologists argue that the decrease in precipitation in the Sahel and Sudan savanna zones of Nigeria is linked with the disturbance of atmospheric circulation south of them, which is engendered largely by the declining vegetation cover (Fricke, 2004). There is therefore a need to have a reliable monitoring system which for obvious reasons must rely largely on remotely sensed data.

As at now, Africa depends on the rates of deforestation produced by international organizations such as FAO and World Bank, which are sometimes conflicting (depending on the source) and generally do not reflect the reality on ground in different parts of Africa. With the advent of the low cost satellites (like NigeriaSat-1) and their potentials for environmental management, African countries are now in a better position to monitor the trend of deforestation over the continent. This could be done through South-South collaboration involving the owner countries within the continent (i.e. Nigeria and Algeria) and other countries through a framework provided by New Partnership for Africa's Development (NEPAD). This is to ensure that Africa's environmental management initiatives reap the maximum benefits from the advent of low cost earth observation satellites.

REFERENCES

Akinbami, J-F.K, Salami, A.T. and Siyanbola, W.O. 2003, An Integrated strategy for sustainable forest-energy-environment interactions in Nigeria. *Journal of Environmental Management,* **69 (2)**, pp. 115–128.

Christenson, E., Jenson, J., Ramsey, E. and Mackey, H. Jr., 1988, *Aircraft MSS data registration and vegetation.*

Congalton, R. and Green, K., 1993, A practical look at the sources of confusing in error matrix generation. *Photogrametric Engineering and Remote Sensing,* **59**, pp. 641–644.

Jha, C.S. and Unni, N.V.M., 1994, Digital change detection of forest conversion of a dry tropical Indian forest region. *International Journal of Remote Sensing,* **15**, pp. 2543–2552.

FAO, 1990, Forest resources assessment, 1990, Tropical Countries, *FAO Forestry Paper 112,* Roma, Italy.

Fricke, W., 2004, Population shifts and migration in the Sahel and sub-Sudan zone of West Africa reviewed from the aspect of human carrying capacity. In *Biological Resources and Migration,* edited by D. Werner (Berlin, Heidelberg: Springer-Verlag), pp. 297–316.

Monat, D.A., Mahin, G.C. and Lancaster, J., 1993, Remote sensing techniques in the Analysis of change detection, *Geocarto Internatonal,* **2**, pp. 39–50.

Oyebo, M.A., 2006, History of Forest Management in Nigeria from 19th Century to Date. A Paper presented at the International Stakeholders' Workshop on *Geo-Information System-Based Forest Monitoring in Nigeria* (GEOFORMIN), Abuja, March 27–30, 2006.

Saadi, M. and Abolfazi, A., 2003, Analysis and estimation of deforestation using satellite imagery and GIS. Map India conference, *Forestry and Biodiversity,* GIS development. net

Salami, A.T., 1998, Vegetation Modification and Man-induced Environmental Change in Rural Southwestern Nigeria. *Agriculture, Ecosystem and Environment,* **70**, pp. 159–167.

Salami, A.T., 1999, Vegetation Dynamics on the Fringes of Lowland Tropical Rainforest of South-western Nigeria—An Assessment of Environmental Change with Air Photos and LANDSAT TM, *International Journal of Remote Sensing,* **20 (6)**, pp. 1169–1182.

Salami, A.T., 2000, Vegetation Mapping of a Part of Dry Tropical Rainforest of Southern Nigeria From LANDSAT TM. *International Archives of Photogrammetry and Remote Sensing,* **XXXIII**, (B7), pp. 1301–1308.

Salami, A.T., 2004, *Deforestation in Nigeria*, A Report Submitted to International START Secretariat, 2000 Florida , Avenue NW, Washington, DC 2009, U.S.A.

Salami, A.T., 2006, Monitoring Nigerian Forests with NigeriaSat-1 and other Satellites. In *Imperatives of Space Technology for Sustainable Forest Management in Nigeria*, edited by Salami, A.T., Space Applications and Environmental Science Laboratory, Obafemi Awolowo University, Ile-Ife, pp. 28–61.

Salami, A.T. and Balogun, E.E., 2005, *Comparative Assessment of NigeriaSat-1 and LANDSAT ETM+ for Deforestation Monitoring in the Niger Delta.* Proceedings of Africa GIS Conference, Pretoria, South Africa, pp. 1175–1180.

Salami, A.T. and Balogun, E.E., 2006, Utilization of NigeriaSat-1 and other satellites for forest and Biodiversity monitoring in Nigeria, A Monograph by National Space Research and Development Agency (NASRDA), Federal Ministry of Science and Technology, Abuja.

Paeth, H. and Hense, A., 2006, "On linear response of tropical African climate to SST changes Deduced from regional climate model simulations". *Theoretical and Applied Climatology*, **83**: 1–19.

World Resources Institute, 2003, *Forests, Grasslands, and Drylands—Nigeria*, EarthTrends.

CHAPTER 14

Landscape and vegetation patterns studied by remotely sensed data analysis in rain forest ecosystems near Ebolowa (Southern Cameroon)

Thorsten Herold

Institute of Physical Geography, Johann Wolfgang Goethe University, Frankfurt am Main, Germany

Jürgen Runge

Centre for Interdisciplinary Research on Africa (CIRA/ZIAF), Johann Wolfgang Goethe University, Frankfurt am Main, Germany

ABSTRACT: Central African landscapes have been subject to striking environmental changes induced either by climatic changes since the LGM or by the growing influence of humans since the Holocene. Within these tropical areas, today spatio-temporal evidence of forest reduction during arid and colder periods can sometimes still be recognized by enclosed, sometimes also edaphically determined savanna patches located inside the rain forest. The objective of this geographical study was the investigation of such 'suspicious' savanna islands and patches in the today's rain forest surrounding the town of Ebolowa in Southern Cameroon by applying different remote sensing approaches. It was focused on the topic whether some palaeoenvironmental traces can be evidenced by the recent forest savanna distribution. A LANDSAT ETM+ imagery recorded in March 2001 was interpreted by using the ENVI image processing software which rendered possible to give an overview of the different geographical conditions within the 42 km × 42 km large research area. However, the unsupervised and supervised classification of the LANDSAT data in connection with ground truthing led to the conclusion that the landscape and vegetation patterns found in the environs of Ebolowa are mainly induced by human action and that no evidence for edaphic nor palaeo-savanna patches ('savane incluse') can be shown.

14.1 INTRODUCTION

The humid tropics are dominated by a great number of different forest and savanna ecosystems which are characterized by a strong latitudinal modification if one moves north- or southwards from the equator. Typical transitional ecosystems from evergreen and semi-deciduous forests to semi-humid savannas are recognizable in many regions by complex forest-savannas mosaics (Favier *et al.*, 2004). Within evergreen and semi-deciduous forests often open patches appear, whose plant formations are typical for open savanna vegetation (Maley and Brenac, 1998).

Investigations on the spatio-temporal dynamics of these highly sensitive borders between forests and savannas in tropical Africa show that anthropogenic as well as natural factors have been influencing this zone (Favier *et al.*, 2004).

In this paper the spatial occurrence of enclosed savannas in actual rain forests by remote sensing techniques is examined, followed up by some ground truthing in the environs of Ebolowa in Southern Cameroon. It shall be clarified whether the shape of these physio-geographical units are mainly either of a natural or of an anthropogenic origin or of both.

14.2 STATE OF THE ART

Concepts relating to the sensitivity of landscapes emphasize that many ecosystems can experience drastic modifications in size and shape by environmental changes, even during relatively short periods, which led to visible changes of the appearance and the structure of the landscape (Thomas, 2004). Many research studying landscape dynamics in Central Africa document far-reaching changes in the prevailing vegetation, which were probably caused by abrupt climatic changes and/or human influence (Vincens *et al.*, 1999).

The current findings essentially rely on the following groups of indicators: Soil and sediment indicators, such as stable carbon isotopes ratios (C3/C4 plant composition) and palynological evidence, permit to reconstruct the palaeoenvironmental conditions in the investigated landscape area. Investigations of the current vegetation and its spreading structure in the area permit a characterization of the variety of forest and help to understand the proceedings which contributed to the spreading of forest. Hereby new methods are also applied, like investigation of seeds of heliophile Taxa, which help to reconstruct the regeneration processes in the forest. Investigations of charcoal in soils at archaeological sites permit to seize the influence of anthropogenic and natural fires on the ecological system (Servant, 2000).

In order to describe recent changes in the vegetation covering, methods of remote sensing are applied which are getting more and more effective and useful (Youta Happi *et al.*, 2000; Runge and Neumer, 2000).

There is already a lot of evidence that tropical rain forests have not been, as assumed for a long time, a stable ecological system for the last 10.000 years, but that it was subject to numerous deeply seizing changes (Servant, 2000). Palaeoecological studies carried out in the humid tropics from the 1970ies onwards, show that the tropical rain forest did not exhibit a continuous vegetation formation but was probably limited to shorter sections of the Holocene (Runge, 2001). Today it is assumed that the equatorial forests were subject, like many other regions of the world, to numerous floristic, structural and palaeogeographic changes in the past (Vincens *et al.*, 1999). Isotopic dating methods of morphological indicators show that the times of largest humidity correlates with interglacials (Runge, 2001), whereas the last Glacial Maximum (LGM) around 20–18 ka was strongly characterized by severe aridity (Thomas, 1994). Climatic changes can be detected within the range of the transition zone between forest and savanna, an ecologically unstable zone (Servant, 2000). Modifications in this transient area express themselves by the shifting of the forest/savanna border, by the occurrence or disappearing of large open surfaces within in the forests (Servant, 2000). So larger parts of today's wooded regions had lost their forest cover during the LGM and the Late Quaternary for several times (Maley and Brenac, 1998). There are still different opinions concerning the amount of reduction of tropical rain forest in South America in comparison to Africa due to unavailable data. Whereas the impact of LGM climate extremes was less severe in South America in comparison to Africa. During the LGM the area of humid forest in Africa was probably reduced by 84% in comparison to South America where the forest expansion was less reduced to 54% of their present-day extension (Anhuf *et al.*, 2006). Enclosed savannas could be interpreted as leftovers of drier periods, which prevailed ultimately in Africa between 4.000 and 1.200 B.P and in South America between 7.500 and 4.500 BP (Servant, 2000; Favier *et al.*, 2004). Schwartz (1992) proposed that the influence of humans played an important role for the formation of savanna patches. He suggested that certain enclosed savannas are related to the propagation of Bantu speaking groups to the south which took place around 3.000 BP.

Under today's climate conditions forests naturally can spread on savannas when the influence of humans is reduced or lacking as Favier *et al.* (2004) has shown.

14.3 METHODS

Before field work in Cameroon started, a satellite based overview of the study area was set up by mapping the physiogeographical ground patterns around Ebolowa. Older IGN topographic maps (1960–1970, scale 1:200.000), the Vegetation Map of Cameroon by Letouzey (1985) with a scale of 1:200.000 and a LANDSAT 7 ETM+ satellite imagery with a ground resolution of $28,5 \times 28,5$ meters per pixel from March 2001 were used as a basic geographical information. In addition SRTM elevation data (90×90 meter spatial resolution) from 2000 were downloaded from the Global Land Cover Facility of the University of Maryland (USA). The satellite images were then processed in the combination of bands 457 (RGB) using an ENVI 4.2 software package. The digital image processing mainly allowed a visual separation between different vegetation units and bare, unvegetated areas.

14.4 STUDY AREA

The study area covers the environs of approx. 15 km around the town of Ebolowa, which represents the administrative centre of the southern province of Cameroon. The town lies in a shallow geomorphic depression (580 m asl) surrounded by dome-shaped inselbergs.

The study area is located south of 3°N and shows a tropical climate that corresponds to the Equatorial Guinea type: e.g. with four seasons and an average temperature of 24 °C with a precipitation of 1.876 mm a year distributed over 164 days (Suchel, 1995) (Figure 1).

This zone is belonging to the geomorphological unit of the interior plateau, which forms the largest connected planation surface within Cameroon and spreads over the entire south and southeast of Cameroon (Segalen, 1969). Crystalline rocks (granites, gneiss and mica slates) of the southern plateau are of Precambrian age. During the Earth's history this area was subject to strong folding and subsequent continental denudation that formed a mainly planated landscape (Runge *et al.*, 2005). The long lasting denudational processes conferred the surface to a relatively uniform, ondulating Peneplain. Within this range, numerous groups of isolated inselbergs and mountain ridges occur, which can reach heights of over 700 m asl (Santoir, 1995).

The hot and humid equatorial climate caused the evolution of different types of strongly weathered ferralitic soils. Basement rocks were hydrolyzed and most of the bases were exhausted so that only residual elements such as kaolinit, quartz and ferrohydroxids remained. On the peneplain surfaces ferrialitic soils are strongly distinct whose horizons can exhibit more or less strong incrustations. The common lateritic crusts mainly developed within the deeper sections of the saprolite, however by strong denudation that caused a relief inversion, today the often can be observed either at mountain summits or at slope slips (Vallerie, 1995).

Rain fall occurs over the entire year with a maximum in September during the large rainy season and in March/April during the small rainy season. The 'driest' period of the year normally is from December to January during the large dry season and from July to

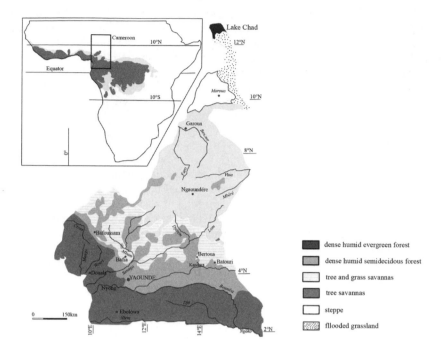

Figure 1. Location and vegetation map of the wider study area (Youta Happi *et al.*, 2000, adapted).

August during the short dry season (Etia, 1979). Under such climatic frame conditions, the dominating vegetation units in the area are represented by semi-deciduous and evergreen rain forests. The evergreen rain forest appears in the climatically most humid areas outside of occasionally flooded or continuously wetter locations (Knapp, 1973). The evergreen rain forest was and still is subject to strong human influences such as wood exploitation, cultivation of cocoa, oil palm, banana and coffee and clearings for settlement (Letouzey, 1979). The semi-deciduous forest extends from Yaoundé to the south. In the north it forms a distinct vegetation border to the savanna ecosystems (Villiers, 1995). Degraded forest formations resulted particularly from human influence within the range of the former cultivation of cocoa and in the environment of high population densities; such surfaces are mainly covered by savannas, fallows and secondary forests (Villiers, 1995).

Savannas of the 'Guinea Sudan Type' appear within semi-deciduous forests, either as large surrounding 'markers' close to the forest or as small enclosed islands inside. Trees and shrubs of savannas which there are in the boundary region of the forest are meagre, whereas the grasses form a continuous layer (Mayaux *et al.*, 1999). They are covered either with trees, shrubs or more or less dense herbs (Villiers, 1995). One reason for the occurrence of such savannas is due to edaphic factors: they can be caused by poor in nutrients sandy or incrusted lateritic soils. Another explanation can be the influence of recent and former settlements which mostly goes ahead with slash and burn agriculture (Mayaux *et al.*, 1999).

Further areas covered by savannas are swampy valleys, *Raphia* palms spotted inside savanna areas, grass-covered swamp areas in the forest, inundation 'meadows,' fraying formations at the banks of rivers and herb dominated vegetation cover on ferralitic crusts inside forested regions.

14.5 ANALYSIS OF REMOTELY SENSED DATA (LANDSAT ETM+)

Digital image processing of the ETM satellite image aimed to generate the mapping and classification of land cover classes using the ENVI 4.2 software. For this purpose a subset of the LANDSAT scene was generated for using the 42 km × 42 km investigation area as a smaller test site (Figure 2).

For the classification of the LANDSAT scene and for the other preparing steps, a methodology for mapping of forest areas was applied, proposed by Wulder *et al.* (2004). Before conducting the real classification of the land cover, it is necessary to carry out preprocessing steps, in order to achieve a good classification result and to eliminate disturbing effects.

14.5.1 Masking

Because of the location in the humid tropics, bright cloud patches and their shades on the ground may cause classification errors. Therefore it was important to mask out these ranges for the subsequent classification. Having identified all the clouds by using the tasselled cap algorithm, a mask was built which permits to fade out the clouds. The received cloud mask was afterwards moved around by a 3 pixels window to the west, in order to mask out the black areas of the cloud shade (Wulder *et al.*, 2004) (Figure 3).

14.5.2 Classification

In order to simplify the classification process, a vegetation index was computed, which permits to separate areas with high vegetation portion from unvegetated and bare areas (Wulder *et al.*, 2004).

The common NDVI vegetation index was generated, due to the relatively high vegetation degree of coverage within the investigation area. As NDVI is widely applied, it also allows an easy comparison of the accomplished results with other works carried

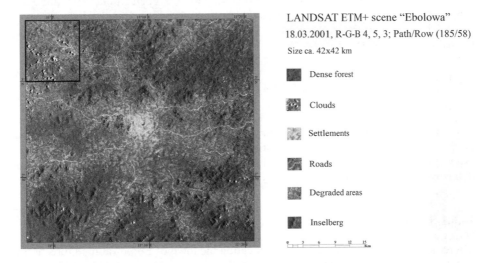

LANDSAT ETM+ scene "Ebolowa"

18.03.2001, R-G-B 4, 5, 3; Path/Row (185/58)

Size ca. 42x42 km

- Dense forest
- Clouds
- Settlements
- Roads
- Degraded areas
- Inselberg

Figure 2. The default subset of the LANDSAT-ETM satellite scenery (Path 185/Row 58) of the 42 km × 42 km large research area processed in RGB (457). The bright spot in the middle is the town of Ebolowa.

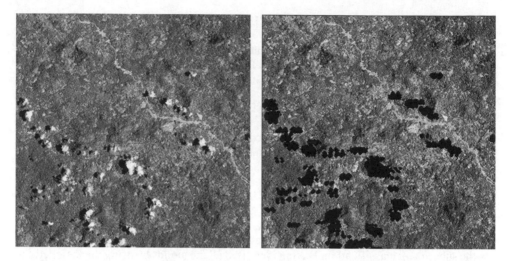

Figure 3. Comparison between the original imagery on the left and the scene with masked out cloud patches on the right, cut-out of 15 × 15 km.

out in the tropics (Jensen, 2000). The computation of the NDVI supplies dimensionless values between −1 and 1. With the help of a threshold value and a mask vegetation-free areas can then be separated from vegetation-covered ranges (Wulder *et al.*, 2004). A first overview of the vegetation patterns was generated by unsupervised classifications with the ISODATA algorithm, which were accomplished with different numbers of classes. It permitted to combine an optical interpretation and the comparison of spectral profiles of similar classes (Jensen, 1996). Thanks to collected information in the field about the vegetation formations and ground truthing it was possible to conduct later a supervised classification with the MAXIMUM LIKELIHOOD algorithm for 11 different classes. The last step was to evaluate the classification results with a CONFUSION MATRIX in order to meet a statement about the accuracy of the received classification (Jensen, 1996). With the help of the SRTM digital elevation model it was possible to rework areas where the classification algorithm failed, as for example in the shadows of inselbergs (Wulder *et al.*, 2004).

14.6 GEOGRAPHICAL PATTERNS

The ground truthing field campaign and the classification of the satellite data supplied a good overview of the individual physical geographical units within the study area. In addition, empirically collected information permits statements about land use and historical aspects of the former transformation of a site. 11 clearly distinctive coverage classes were determined by the remote sensing analysis (Table 1).

14.6.1 Settlements

There is a star-shaped road network connecting the town of Ebolowa in all directions. The setting up of this infrastructure network mainly goes back to colonial times when Ebolowa was an important trade base for cocoa, coffee and timber. The smaller villages are not equally distributed over the area, but they follow quite obvious the road network. However,

Table 1. Overview of the 11 geographical patterns deduced by classification and ground-truthing in the environs of Ebolowa, Southern Cameroon.

Type of geographical class/unit	Percent of the total area	Area in km^2
Settlements	0,5%	9,5
Roads and small villages	0,6%	10,6
Agricultural areas	5,8%	108,7
Burned surfaces	0,3%	6,3
Fallow	3,8%	71,5
Secondary forest	10,8%	202,2
Raphia palms	1,3%	24,5
Moist/swamp forest	31,1%	580,4
Evergreen and semi-deciduous forest	42,3%	789,4
Open lateritic areas (quarries)	1,9%	36,1
Water	0,1%	0,2
Masked out pixels	1,5%	27,3
Total	**100%**	**1.866,7**

some isolated, island-like concentrations of smaller settlements can be evidenced within the closed forest. Usually the villages are located in the proximity of the main traffic axes which also is a heritage of the colonial age (Santoir, 1995).

14.6.2 Agriculture

The effective areas for agriculture are relatively small plotted and have average sizes of 1–2 hectares (Loung, 1979). These surfaces are gained from the forest by slash and burn clearing methods. The small fields are then cultivated by a rotational system with fallow periods depending on soil fertility and groundwater availability. Shifting cultivation by using a hoe is the most frequently used method within the range of the forest. Inside the forests, land use is dominated by mixed cultures or small pure cultures, which positively affects the yield (Santoir, 1995).

The main crops cultivated within the study area are corn, manioc, yams and peanut. In the proximity to the settlements or in the lower vegetation layers of the forest also grow cocoa plants (Figure 4). The cultivation of cocoa plants was introduced to Cameroon during colonial times (Santoir, 1995).

Depending on the local conditions differences in cultivation practices appear. Within the research area there is a varying degree of vegetation coverage by primary- or secondary forest. This vegetation structure shows that not the complete cultivation area is cleared and primary forest remains inside the cultivation area so some fields are agriculturally used at only 50%. The distinctness of this transition zone to the primary forest in the field gives hints whether the area was formerly or recently cleared. Concerning land covers mapping of the satellite image all agricultural areas were not further specified as for crops since this was not feasible due to sensor resolution.

Figure 4. Cleared and cultivated small manioc field with about 50 cm high plants surrounded by primary rain forest located in the south western outskirts of Ebolowa (02.03.2006).

Figure 5. Ash covered surface inside the rain forest around Ebolowa: patterns of fresh slash and burned areas can be easily identified on satellite imageries (01.03.2006).

14.6.3 Burned surface pattern

Slash and burn agriculture destroys a lot of fresh biomass producing dark ash layers on the surface (Figure 5). These 1–2 hectares plots formerly covered by forest vegetation are highly significant on remotely sensed data. Due to their colour, their structure and spectral characteristics, fire surfaces can easily be separated from other physio-geographical units.

14.6.4 Fallow

When an agricultural field is no longer profitable, the surface becomes abandoned by the farmer and it is subsequently transformed into a fallow where a vegetation succession cycle takes place. Within the low land rain forest fallow areas or places, where the trees were felled, are quickly overgrown by thicket. On these areas early pioneer formations species develop, which are facultative bushy or lianic (Knapp, 1973) (Figure 6).

If the growing up of vegetation is more or less undisturbed, the shrubs are rapidly overgrown by colonizing pioneer trees. However, game and fire activity can prevent the ongoing succession, so that bush coverage can remain for a long time. Moreover these formations are very common in densely settled areas (Knapp, 1973).

In the study area such locations exhibited close homogeneous herb-like bushy vegetation, which reached heights between 1,5 to 2,5 m. Both on the agriculturally used areas and on fallow surfaces isolated rainforest trees, oilpalms or secondary forest can appear, which, however, have a portion of only 20% of the total vegetation cover.

14.6.5 Secondary forest

The longer a surface is not under cultivation the stronger the stock of this surface is covered by fast growing and light demanding pioneer formations. These plant species are sometimes similar to a preliminary forest and can very fast overgrow initial tree formations, if the vegetation development is an undisturbed one. The pioneer species can reach only quite a small age (Knapp, 1973). The shape of the upper layer of secondary forest is often relatively homogeneous and is dominated by certain species with large leaves such as *Musanga cecropioides* and *Fagara macrophylla* (Knapp, 1973; Lebrun and Gilbert, 1954; Trochain, 1980 in Mayaux *et al.*, 1999) (Figure 7).

Furthermore Euphorbiaceae trees are far spread within secondary forests (Knapp, 1973). Within the clearings these pioneer species, which are very light-demanding, show a fast growth in height of up to 15 m in three years and have a permanently high photosynthesis activity. On satellite images they can be identified by a signal of high reflection in the near infrared (Mayaux *et al.*, 1999).

Figure 6. An around two years old fallow area south of Ebolowa (28.02.2006).

Figure 7. Densely vegetated transition between an abandoned agricultural field and the fast growing, up to 5–25 m high secondary forest in the background (27.02.2006).

In comparison to mature forests much light can still reach the ground so that a relatively dense thicket and herb layer can develop. These forests can prevail on fallow areas, impact surfaces or along traffic routes during a period of usually 3 to 30 years after beginning of succession on fallow areas (Knapp, 1973; Mayaux *et al.*, 1999). The prevailing vegetation along roads consist of such a secondary regrowth, fallow areas, patterns of arable crops, back gardens and village plantings which form a very complex mosaic (Vandenput, 1981 in Mayaux *et al.*, 1999).

14.6.6 Moist forest/swamp forest

Geomorphological depressions and channel sections in the proximity of rivers and brooks show within the rain forest often moist loving vegetation These are caused by regular and frequent flooding and show clayed soils which are not sufficiently drained (Mayaux *et al.*, 1999). Different forests and also savanna types can develop according to the frequency of floodings and drainage conditions of the soil: bank forests, periodically swamped forests, constantly swamped forests (Evrard, 1968 in Mayaux *et al.*, 1999) (Figure 8).

In Cameroon these forests are mainly coverd by *Raphia* palm trees and bamboo (Letouzey, 1985). Swampy areas can be easily identified on the satellite image because of the striking physiognomy of the leaves.

14.6.7 *Raphia* palms

Along hydrological net within the range of depressions *Raphia* palms are spread in preferably pure formations. They prefer areas with flowing and not stagnating surface water which never dry up completely because of long continuous flooding. The prevailing soils exhibit high humus content and are nearly of a preat-like structure. The height of the palms remains usually low and exceeds only rarely 15 m (Knapp, 1973). Due to their leaf physiognomy they can be easily identified by satellite imagery.

Figure 8. Swamp forest with high ground water level and continuous flooding.
Covered by about 3 m high dense vegetation in background (27.02.2006).

14.6.8 Forest canopy with evergreen and semi-deciduous forest

Within the research area evergreen and semi-deciduous rain forest is dominating (Letouzey, 1985). The tropical evergreen forest which is characterized by Cesalpiniaceae spreads over the most humid areas and its upper stratum (35–45 m) is composed of a few shadow evergreen species as *Julbernadia secretii, Brachystegia laurentii or Anthonotha* (Letouzey, 1979; Mayaux *et al.*, 1999). The canopy density of this forest type prevents the development of shrub and herbaceous strata and favours epiphytes (Mayaux *et al.*, 1999).

The semi-deciduous forest usually prevails in less humid regions and characterized by up 70% decidous species mixed with evergreen species in the upper stream. This forest type can reach heights between 40–60 m and is characterized by Sterculiceae and Ulmaceae (Knapp, 1973; Letouzey, 1979). The lower canopy density in comparison to evergreen forest promotes the development of a continuous shrub stratum (Mayaux *et al.*, 1999).

A clear separation between a semi-deciduous forest of the Guinea/Sudan type and evergreen forest of the Biafra type is however difficult to make, because these two forest formations merge into one another. Furthermore the semi-deciduous forest sometimes replaces the evergreen forest (Letouzey, 1979). Due to physiognomic similarity and sensor resolution a distinction in satellite imagery between the two forest types was not possible

14.6.9 Lateritic soils and crusts

Several locations within the study area show hardened lateritic crusts were rain forest vegetation is often lacking due to soil properties. Vegetation establishment is mainly restricted to a sparse and open, sometimes savanna like vegetation (Kamagang Beyala *et al.*, 2000) (Figure 9).

Such pedologic conditions are of an interest for the palaeoenvironmental discussion. As far as they are not resulting from human influence, they can be interpreted in a climate controlled process of the past, meaning that these patterns are mainly caused by edaphic frame conditions. Edaphic determined lateritic occurrence may not have supported a forest

Figure 9. Former lateritic quarry used for road construction near the international road connecting Ebolowa with Gabon (27.02.2006).

development, as investigations on soil conditions in East Cameroon in the range of forest-savanna boundary suggest (compare Kamagang Beyala *et al.,* 2000).

These soil and crust properties may be relicts of former arid climatic phases which prevailed in Africa during the LGM (18.000 BP) and also between 4.000 and 1.200 BP and thus have prevented a settlement of the forest (Favier *et al.,* 2004). These open savanna-like structures within the forest are typically described as enclosed savannas or 'savanne incluse' (Mayaux *et al.,* 1999).

14.7 RESULTS AND DISCUSSION

Remote sensing techniques helped to give an overview of the different physiogeographical units and landscape patterns in a hardly accessible rain forest area around the town of Ebolowa, Southern Cameroon. Additional ground thruthing information collected during 7 days of field work rendered possible a labelling of land coverage classes and locally also of land use classes.

Concerning the 'savanna incluse' topic, the analysis of satellite images showed that no edaphic savannas can be identified as remnants of an assumed former 'catastrophic' rain forest destruction by a shift to climate much drier than today. Field observations, however, showed that other environmental factors might have caused some deforested and open savanna like features within the rain forest. Some locations with increased soil moisture and swampy conditions did not support a dense forest cover, therefore it was interrupted at such sites. These observations correspond to Letouzey (1979) who also describes open patches within the forest due to sandy soils at strongly hydromorphic influenced areas.

Suspending of forest vegetation is also typical around inselbergs where the rock crops out at the surface and the lack of fine material does not support the growth of bigger trees (compare Letouzey, 1979).

Recent researches in eastern Cameroon give evidence that there is also a correlationship between the occurrence of lateritic crusts and savanna patches. Within the forests these crusts are often covered by finer soil textures which allow the growth of larger trees (Youta

Happi, 2000). Therefore in the Ebolowa study special attention was drawn to areas which show a lateritic spectral signal on the LANDSAT image. As in most of these areas the influence of humans is obviously prevailing (villages, roads). Ground thruthing indeed showed that the lateritic spectral signals were dominating on sites strongly influenced by humans. Often in quarry sites within the rain forest, topsoil was spaciously removed to get lateritic clays for road construction. In conclusion it can be stated that the 'lateritic' spectral pixel signal in the satellite data that these open sites within the rain forest (savannas) are of an anthropogenic origin.

Former disturbances of the rain forest in the Ebolowa region are still not well documented due to the lack of proxy data and to their rough resolution in space (Höhn *et al.*, 2007, this volume). This leads to unequivocal research results: Maley (2004) postulates that Holocene forest refugia and forest savanna mosaics spread in the region between the Atlantic coast and Ebolowa/Ambam; however Höhn *et al.* (2007, this volume) could only get evidence for persisting forest vegetation. Although some archaeological and pollen data indicate a disturbance of primary forest in the described region leading to an expansion of secondary forest from 2.800 BP onwards, it is assumed that these disturbances took place only in a small scale with local openings in the rain forest (Höhn *et al.*, 2007 this volume).

Consequently a re-expansion of rain forest at the beginning of more humid climatic conditions in the Late Holocene was possible because of the small size of the savannas (Höhn *et al.*, 2007, this volume). If forest refugia within savanna environments had existed as assumed by Maley (2004), a re-colonisation of the forest would have been quite easy because the studied region is in proximity to a former LGM refugia region. Therefore it is very probable that the edaphic savanna regions, if they really had existed, have been closed by forest up to now, when referring to forest expansion rates on savannas areas observed during the twentieth century (Blanc-Pamard and Peltre, 1984; Maley, 1990, 1996; Fairhead and Leach, 1995, 1998; Servant, 1996 in Maley, 2002). Under nowadays humid climatic conditions savannas are not stable ecosystems, which can only persist, if they are kept open by human action (Schwartz *et al.*, 2000).

Investigations in the eastern part of Cameroon and in the Sanaga River region describe the existence of enclosed savannas in the proximity to the forest savanna contact zone. The further one moves to the south away from this transition zone, the smaller gets the probability of the occurrence of enclosed savannas (Schwartz *et al.*, 2000). A further factor in the Ebolowa research area is that wide ranges of the forests are influenced by settlements and agricultural use. This contributed to a considerable degradation of forest, particularly in the closer surrounding fields of the city (Santoir 1995).

Former IGN maps of the 1960ies which were worked out on the basis of aerial photographs shows a forest-savanna belt surrounding the city apart from the western range. During the field campaign in 2006 some of these areas were visited which showed that these areas were used for agricultural purposes. Due to the observed hoe culture, applied in this area individual trees remain within the agricultural fields. Therefore it is possible that these areas might have been mapped as wooded savannas. This assumption corresponds to Santoir (1995) and others showing that many savanna regions have been developed due to agriculture.

ACKNOWLEDGEMENTS

We would like to thank DFG (Deutsche Forschungsgemeinschaft) for funding and financing the project RU 555/14-2. Further thanks to members of RESAKO research group for fruitful discussions and supporting fieldwork.

REFERENCES

Anhuf, D., Ledru, M.P., Behling, H., Dacruz Jr., F.W., Cordeiro, R.C., Van der Hammen, T., Karmann, I., Marengo, J.A., De Oliveira, P.E., Pessenda, L., Siffedine, A., Albuquerque, A.L. and Da Silva Dias, P.L., 2006, Paleo-environmental change in Amazonian and African rainforest during the LGM. *Palaeogeography, Palaeoclimatology, Palaeoecology*, **239**, pp. 510–527

Etia, P.M., 1979, Climat. In *Atlas de la République Unie du Cameroun*, edited by Laclavère, G., (Paris: Groupe J.A.), pp. 16–19.

Favier, C., Chave, J., Fabing A., Schwartz D. and Dubois, M.A., 2004, Modelling forest-savanna mosaic dynamics in man-influenced environments: effects of fire, climate and soil heterogeneity. *Ecological Modelling*, **171**, pp. 85–102.

Höhn, A., Kahlheber, S., Neumann. K. and Schweitzer, A., 2007, Settling the rainforest- the environment of farming communities in southern Cameroon during the first millennium BC. *Palaeoecology of Africa*, **28**.

Jensen, J.R., 1996, *Introductory Digital Image Processing*, 2nd., edited, (New Jersey: Prentice-Hall Inc.).

Jensen, J.R., 2000, *Remote Sensing of the Environment, An Earth Resource Perspective* (New Jersey: Prentice-Hall Inc.).

Kamagang Beyala V., Ekodeck, G.E. and Achoundong, G., 2000, Essai d'interprétation de la dynamique de la mosaïque forestière dans la zone de contact forêt-savane du sud-est Cameron. In *Dynamique à long terme des écosystèmes forestiers intertropicaux (ECOFIT)*, Paris, edited by Servant, M. and Servant-Vildary, S. (Paris: UNESCO, IRD), pp. 175–182.

Knapp, R., 1973, *Die Vegetation von Afrika unter Berücksichtigung von Umwelt, Entwicklung, Wirtschaft, Agrar- und Forstgeographie*, (Stuttgart: Gustav Fischer).

Lebrun, J. and Gilbert, G., 1954, *Une classification écologique des forêts du Congo*, (Bruxelles: Institut National pour l'`Etude Agronomique du Congo Belge, série scientifique n°63).

Letouzey, R., 1979, Vegetation, In *Atlas de la République Unie du Cameroun*, London, edited by Laclavère, G., (Paris: Groupe J.A.), pp. 20–24.

Letouzey, R , 1985, *Carte phytogeographique du Cameroun, 1/50.0000, six feuilles 66 × 94 cm en couleurs, deux feuilles de légende et cinq fasiules*, (Yaoundé: Institut de la recherche agronomique, Toulouse: Herbier national, Institut de la carte internationale de la végétation).

Loung, J.F., 1979, Agriculture-Conditions générales des activités agricoles. In *Atlas de la République Unie du Cameroun*, London, edited by Laclavère, G., (Paris: Groupe J.A.), pp. 44.

Maley, J., 1990, Histoire récente de la forêt dense humide Africaine et dynamisme actuel de quelques formations forestières. In *Paysages Quaternaires de l'Afrique centrale Atlantique,* edited by Lanfranchi, R. and Schwartz, D. Editions de l'ORSTOM, Paris, pp. 367–382.

Maley, J., 2002, A catastrophic destruction of African forests about 2.500 years ago still exerts a major influence on present vegetation formations. In *Science and and policy process: perspectives from the forest*, edited by Leach, M., Fairhead, J. an Amanor, K. IDS bulletin, **33/1**, pp. 13–30.

Maley, J., 2004, Les variations de la végétation et des paléoenvironnements du domaine forestier africain au cours du Quaternaire récent. In Sémah, A.-M. and Renault-Miskovsky, J. (eds.), *L'Evolution de la végétation depuis deux millions d'années*. Editions Artcom/Errance, Paris, pp. 143–178.

Maley, J. and Brenac, P., 1998, Vegetation dynamics, palaeoenvironments and climatic changes in the forests of western Cameroon during the last 28.000 years BP. *Review of Palaeobotany and Palynology*, **99**, pp. 157–187.

Mayaux, P., Richards, T. and Janodet, E., 1999, A vegetation map of Central Africa derived from satellite imagery. *Journal of Biogeography*, **25**, pp. 353–366.

Runge, J. and Neumer, M, 2000, Dynamique du paysage entre 1955 et 1990 à la limite forêt-savane dans le nord du Zaïre, par l'étude de photographies aériennes et de données LANDSAT-TM. In *Dynamique à long terme des écosystèmes forestiers intertropicaux (ECOFIT)*, Paris, edited by Servant, M. and S. Servant-Vildary, (Paris: UNESCO, IRD), pp. 311–317.

Runge, J., 2001, Landschaftsgenese und Paläoklima in Zentralafrika. Physiogeographische Untersuchungen zur klimagesteuerten quartären Vegetations- und Geomorphodynamik in Kongo-Zaire (Kivu, Kasai, Oberkongo) und der Zentralafrikanischen Republik (Mbomou), *Relief, Boden, Paläoklima*, **17**, pp. 1–294.

Runge, J., Eisenberg J. and Sangen, M., 2005, Ökologischer Wandel und kulturelle Umbrüche in West- und Zentralafrika, Prospektionsreise nach Südwestkamerun vom 05.03.–03.04.2004 im Rahmen der DFG-Forschergruppe 510: Teilprojekt, Regenwald-Savannen-Kontakt (ReSaKo). *Geoöko* **26**, pp. 135–154.

Santoir, C., 1995, L'oro-hydrographie. In *Atlas régional du Sud- Cameroun*, Paris, edited by Santoir, C. and Bopda, A. (Paris: Orstom éditions et Minrest), pp. 4–5.

Schwartz, D., Mariotti, A., Trouve, C., Van den Borg, K. and Guillet, B., 1992, Etude des profiles isotopiques 13C et 14C d'un sol ferralitique sableux du littoral congolais. Implications sur la dynamique de la matière organique et l'histoire de la végétation. *Coptes Rendus de l'Académie des Sciences, Paris, série 2.*, **315**, pp. 1411–1417.

Schwartz, D., Elenga, H., Vincens, A., Bertaux J., Mariotti, A., Achoundong, G., Alexandre, A., Belingard, C., Girardin C., Guillet, B., Maley, J., De Namur, C., Reynaud-Farrera, I and Youta Happi, J., 2000, Origine et évolution des savanes des marges forestières en Afrique centrale Atlantique (Cameron, Gabon, Congo): approche aux échelles millénaires et séculaires. In *Dynamique à long terme des écosystèmes forestiers intertropicaux (ECOFIT)*, Paris, edited by Servant, M. and S. Servant-Vildary, (Paris: UNESCO, IRD), pp. 325–338.

Segalen, P., 1969, *Les sols et la géomorpholgie du Cameroun*. Cahiers ORSTOM, **7**, pp. 137–187.

Servant, M., 2000, Diversité actuelle de la forêt tropicale et changements passés du climat: le programme Écosystèmes et paléoécosystèmes des forêts intertropicales (ECOFIT). Bilan et perspectives. In *Dynamique à long terme des écosystèmes forestiers intertropicaux (ECOFIT)*, Paris, edited by Servant, M. and S. Servant-Vildary, (Paris: UNESCO, IRD), pp. 13–18.

Suchel, J.B., 1995, La climatologie. In *Atlas régional du Sud- Cameroun*, Paris, edited by Santoir, C. and Bopda, A. (Paris: Orstom éditions et Minrest), pp. 8–9.

Thomas, M.F., 1994, *Geomorphology in the Tropics. A study of weathering and denudation in low latitudes*. (New York, Brisbane, Toronto: Wiley & Sons).

Thomas, M.F., 2004, Landscape sensitivity to rapid environmental change, a Quaternary perspective with examples from tropical areas. *Catena*, **55**, pp. 107–124

Trochain J.L., 1980, *Ecologie végétale de la zone intertropicale non désertique*. (Toulouse: Université Paul Sabatier).

Vallerie, M., 1995, La pédologie. In *Atlas régional du Sud-Cameroun*, Paris, edited by Santoir, C. and Bopda, A. (Paris: Orstom éditions et Minrest), pp. 6–7.

Vandenput, R., 1981, Les principales cultures en Afrique Centrale. (Bruxelles: Administration Générale de la Coopération du Développement).

Vincens, A., Schwartz, D., Elenga, H., Reynaud-Farrera, I., Alexandre, A., Bertaux, J., Mariotti, A., Martin, L., Meunier, J.D., Nguetsop, F., Servant, M., Servant- Vildary, S.

and Wirrmann, J.D., 1999, Forest response to climate changes in Atlantic Equatorial Africa during the last 4.000 years BP and inheritance on the modern landscapes. *Journal of Biogeography*, **26**, pp. 879–885.

Wulder, M., Cranny, M., Dechka, J. and White, J., 2004, An Illustrated Methodology for Land Cover Mapping of Forests with Landsat-7 ETM+ Data: Methods in Support of EOSD Land Cover, Version 3, Natural Resources Canada, Canadian Forest Service, Pacific Forestry Centre, Victoria BC, Canada, pp. 1–35. http://www.pfc.forestry.ca/eosd/cover/methods_vers3_e.pdf

Youta Happi, J., Hotyat, M. and Bonvallot, J., 2000, La colonisation des savanes par la forêt à l'est du Cameroun. In *Dynamique à long terme des écosystèmes forestiers intertropicaux (ECOFIT)*, Paris, edited by Servant, M. and S. Servant-Vildary, (Paris: UNESCO, IRD), pp. 423–427.

CHAPTER 15

Remote sensing based forest assessment: recent dynamics (1973–2002) of forest-savanna boundaries at Ngotto Forest, Central African Republic (CAR)

Jürgen Runge

Centre for Interdisciplinary Research on Africa (CIRA/ZIAF), Johann Wolfgang Goethe University, Frankfurt am Main, Germany

ABSTRACT: Interpretation of remotely sensed data was carried out for an assessment of the Ngotto Forest using six LANDSAT MSS, TM and ETM sceneries dating back to February 1973, March 1979, December 1986, November 1990, January 1995, and April 2002. A multitemporal comparison between the 1995 and 2002 situation—while the ECOFAC/IFB management plan has already been set up—evidenced for the Ngotto region an increase in higher woody vegetation ('forest') from 66,2% to 71,8% while the area with savanna ecosystems shrinked slightly from 33,8% to 28,2%. Except of these positive effects of the ECOFAC conservation policies on the tropical forest, the partly economic collapse in the CAR during the last 10 years forced many people from the rural areas to leave their villages and move to bigger towns like the capital Bangui. As a consequence, regular bush fires for hunting and clearing are now less prominent; this socio-economic factor also contributes to the significant regeneration and extension of forests against savannas.

15.1 INTRODUCTION

Tropical forestry mainly in developing countries is nowadys an economic sector undergoing a dynamic evolution in a rapidly-changing world. The state of natural forests and the forestry sector in subsaharan Africa is shaped as much by external economic, political, demographic and social trends as they are by forces working within the sector itself. Since the United Nations Conference on Environment and Development (UNCED), held in Rio de Janeiro 1992, impetus and commitment was provided to international activity focused on the world forests (Harcharik, 1997). Today, fifteen years after Rio, there is widespread recognition and acceptance of the importance of ensuring that forests are sustainably managed. There is an ongoing strong need to reconcile the productive functions with the protective, environmental and social roles of tropical forests (FAO, 1997, 2001).

As an important tool, remotely sensed data from different earth observation satellites are available to monitor the spatial evolution of tropical forest ecosystems. Aside of the widespread LANDSAT imageries, there was a new international program proposed in 1996 to create the so-called Disaster Monitoring Constellation (DMC) led by Surrey Satellite Technology Ltd. (SSLT) in the UK. The objective was to provide a daily global imaging capability at medium resolution (30–40 m), in 3–4 spectral bands, for rapid-response disaster monitoring and mitigation (Sweeting and Chen, 1996). Aside of the United Kingdom, China, Turkey and Algeria, Nigeria launched its first remote sensing microsatellite

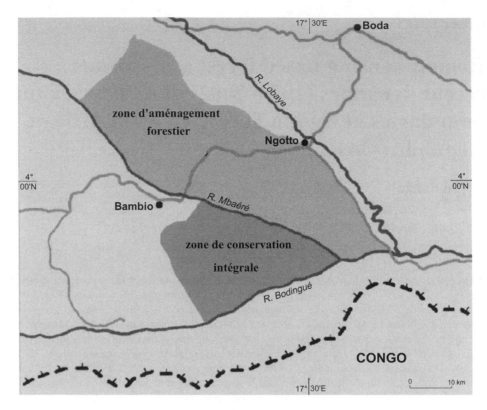

Figure 1. Location of the Ngotto Forest (zone d'aménagement forestier) in the south–west of the Central African Republic. The sustainable used forest zone between the Lobaye and the Mbaéré River also acts as a buffer zone for the Parc National Mbaére-Bodingué (zone de conservation intégrale) where no logging at all is allowed. (http://www.ecofac.org/Composantes/CentrafriqueNgotto.htm, modified, 30.04.2006).

NigeriaSat-1 succesfully on September 27, 2003 operated by the National Space Research and Development Agency (NASRDA, Abuja). Aside of disaster monitoring, this remotely sensed data can also be applied for the survey of forest resources in Africa and elsewhere in the world.

The example of sustainable forest use and monitoring of the spatial evolution in the Ngotto Forest in the Central African Republic was undertaken by using conventional LANDSAT (MSS, TM and ETM) data, however, it must be underlined that NigeriaSat-1 has almost the same capacity for this purpose, and could therefore be an interesting alternative to the LANDSAT sensor (compare chapter 13 in this volume).

15.2 STUDY AREA

The Ngotto Forest is located in the southwestern part of the Central African Republic (4°00'N, 17°30'E) covering an area of 195.000 ha in the Lobaye and Sangha–Mbaéré Prefectures (Figure 1).

The study region gets an average of 1.700–1.800 mm annual precipitation (distributed over 110–120 days) and is mainly covered by closed semi-evergreen to evergreen forest (60% of the total area), degraded forest (17%), temporarily flooded forest located along the watercourses crossing the area, and an extended swamp forest along the Mbaéré River.

Economically valuable red woods are Sapelli (*Entandrophragma cylindrium*) and Sipo (*Entandrophragma utile*); in the case of white wood Ayous (*Triplochiton scleroxilon*) is dominant. There are also more open savanna ecosystems bordering the northeastern forest fringe, with crop and fallow land, and a typical patchwork of savanna and gallery forest.

15.3 OUTLINE AND BACKGROUND OF THE ECOFAC PROJECT

Forest management by the European funded ECOFAC project (ECOsystèmes Forestiers d'Afrique Centrale) in the Ngotto area is the first pilot experiment in co-management with a logging company (IFB, Industrie Forestière de Batalimo) in CAR. This integrated management with timber production as its goal is part of a more comprehensive rural development program (Koko and Runge, 2004), that includes other components like conservation and rural development. Implementation of this project for a sustainable tropical forest harvesting (30 years cycles) in collaboration with a logging company has been carried out on the basis of a management plan and schedule of terms and conditions since 1992 (Figure 2).

Forestry harvesting in CAR was a small affair for a long time, the production of logs only reaching 100.000 m³ from 1960 onwards. Following that, it grew to reach a peak of 670.687 m³ in 1974. It was with the beginning of harvesting in the Sangha region (south–west) towards the end of the 1960s that statistics on log production began to be kept. In 1966 log production reached 128.000 m³ then 260.000 m³ in 1984 with a peak of 400.000 m³ in 1973. Nevertheless, this production dropped to its lowest level in 1991 to 114.000 m³. It has continued to grow over the past number of years, culminating with

Figure 2. Ecological damage of tropical forests and landscape degradation or sustainable forest management with regularly, every 30 years felling cycles according to the ECOFAC project concept as an alternative? The image shows an almost totally cleared forest within a log of the Industrie Forestièrs de Batalimo (IFB) company near the village of Ngotto in March 2005 (Photo: M. Neumer).

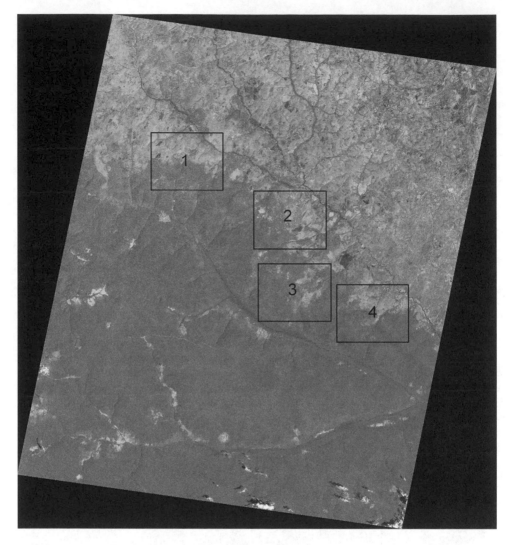

Figure 3. Location of the multitemporal subsets 1–4 within the LANDSAT scene (182–57, RGB 457, approx. 190 km × 190 km wide) from January 16th, 1995 (see also table 1, figure 4) located on the transition between rain forest (dark grey patterns in the lower part) and savanna environments (light gray mosaics in the upper part of image).

553.000 m³ in 1999. Log production in the Central African Republic is subject to many ups and downs of all kinds. It is conditioned not only by the international market situation, but even the distance of this market limits the number of species for export, thus obliging companies just to concern themselves with species of great value in economic terms, such as species from the Meliaceae. The percentage of log exported, around 25–30% of the volume fell, is on the decrease since 1980, the foreign market essentially being Europe with almost 65% in volume (Yalibanda, 2004).

Until very recently, the forest sector in CAR had received no attention from authorities or donors, and only in the last ten years special interest has been directed to it. Forest regulations, previously based on the 1962 postcolonial Forest Code, were modified by a new law promulgated on 9. June 1990 with an application decree on 9. February 1991. The aim of the new code was to harmonize the demands of making the forest heritage profitable with those of conservation, using forest management to do so. It defines different types of forest: State forests (complete nature reserves, national parks, wildlife reserves, recreational forests, protected areas, reforestation areas and production forests) community and private forests.

15.4 METHODS

Mostly cloud cover free, good quality LANDSAT MSS, TM and ETM sceneries were downloaded for the Ngotto Forest from the internet at the University Maryland NASA Science Server (http://svs.gsfc.nasa.gov). Image processing in the infrared mode (RGB 457) and pixel classification followed up with an ERDAS Imagine 8.5 software package. Unsupervised ISODATA classification with only five classes allowed a good distinction of open savanna against closed forest and higher woody biomass ecosystems. Subsequently, only two qualitative pixel classes ("forest" and "savanna") were integrated and processed for the complete LANDSAT imageries for the years 1973, 1979, 1986, 1995 and 2002 to illustrate the natural and human induced dynamics of the vegetation cover. In addition the effectiveness of the taken measures of the ECOFAC project against uncontrolled deforestation towards a sustainable forest management should be monitored as well. As it turned out that it was difficult to show clear trends for the complete LANDSAT sceneries, it was decided to create four smaller subsets of the complete imageries that gave a better representation of the different environmental context in the surroundings of Ngotto Forest (Figure 3).

15.5 RESULTS: SPATIAL DYNAMICS OF THE FOREST–SAVANNA BOUNDARY IN THE NGOTTO FOREST 1973–2002

Interpretation of remotely sensed data was carried out for an assessment of the Ngotto Forest using five LANDSAT MSS, TM and ETM sceneries dating back to February 1973, March 1979, December 1986, January 1995 and April 2002. A multitemporal comparison between the 1995 and 2002 situation—while the new ECOFAC/IFB management plan has already been set up—evidenced for the total Ngotto region an increase in higher woody vegetation ('forest') from 66,2% to 71,8% while the area with savanna ecosystems shrinked slightly from 33,8% to 28,2%. Except of these positive effects of the ECOFAC conservation policies on the tropical forest, the partly economic collapse in the CAR during the last 10 years forced many people from the rural areas to leave their villages and to move to bigger towns like the capital Bangui.

As a consequence, regular bush fires for hunting and clearing are now less prominent; this socio-economic factor also contributes to the regeneration and extension of forests against savannas in more recent years. Larger scaled data was obtained by four subsets taken from the LANDSAT imagery.

This spatial development of forest against savanna areas for subsets 1–4 (see figure 3) is shown in table 1 and figure 4.

15.6 CONCLUSIONS

Forestry management of the Ngotto Forest in Central African Republic undertaken by the European funded ECOFAC project was the first pilot experiment in co-management with the logging company Industrie Forestière de Batalimo (IFB), and therefore reflected the strong need for effective concepts of sustainable use of tropical wood. It is of importance to mention that this project became a success because other local stakeholders were included into the program by general conservation efforts and integrated rural development measures.

According to Cirad Forêt (1996) and Cossocim (1996) the following factors helped IFB, ECOFAC and the supervising ministry to prepare the management and working plans together:

— IFB's commitment to participating regularly in the various discussion meetings that led to a document with the endorsement of all the parties involved;
— IFB's undertaking to implement this plan for the sustainable management of timber resources and the sustainable supply of its Ngotto industrial plant;
— the support of the government department in charge of forest management;
— the plan does not so far stipulate any silvicultural measures other than harvesting and pre- and post-harvest monitoring of the dynamics of the stand on monitoring plots.

In addition to vegetation type classification of the harvesting area, other elements were agreed on together with IFB: the definition of harvest potential; an establishment of the division into felling plots, and the setting up of harvesting rules (e.g. 30 year felling cycles).

Table 1. Vegetation and landscape changes within selected Ngotto subsets 1–4 (see figure 3) in hectar (ha) and percent (%). It is obvious that beside of the socio-economic and political aspects (see above), there is no significant trend of a severe shrinking nor a strong expansion of the forest (statistics by E. Becker, 2006).

Date and Subset	14.02.1973		18.03.1979		09.12.1986		16.01.1995		01.04.2002	
Subset 1	Area (ha)	**Area (%)**	Area (ha)	**Area (%)**	Area (ha)	**Area (%)**	Area (ha)	**Area (%)**	Area (ha)	**Area (%)**
'Forest'	21.229	**59,1**	24.417	**67,8**	22.608	**62,2**	19.661	**54,7**	22.625	**63,0**
'Savanna'	14.667	**40,9**	11.575	**32,2**	13.713	**37,8**	16.307	**45,3**	13.303	**37**
Total	35.896	**100**	35.992	**100**	36.321	**100**	35.968	**100**	35.932	**100**
Subset 2										
'Forest'	17.951	**50,0**	22.497	**62,7**	20.130	**56,3**	20.791	**57,7**	23.824	**66,3**
'Savanna'	17.945	**50,0**	13.398	**37,3**	15,632	**43,7**	15.231	**42,3**	12.107	**33,7**
Total	35.896	**100**	35.896	**100**	35.763	**100**	36.022	**100**	35.932	**100**
Subset 3										
'Forest'	25.374	**70,5**	23.959	**66,7**	25.930	**71,5**	26.028	**73,4**	28.209	**78,5**
'Savanna'	10.618	**29,5**	11.937	**33,3**	10.329	**28,5**	9.926	**27,6**	7.722	**21,5**
Total	35.992	**100**	35.896	**100**	36.259	**100**	35.954	**100**	35.932	**100**
Subset 4										
'Forest'	26.156	**72,9**	27.291	**76,0**	26.742	**74,8**	26.962	**75,0**	25.928	**72,1**
'Savanna'	9.740	**27,1**	8.604	**24,0**	9.018	**25,2**	8.922	**25,0**	10.064	**27,9**
Total	35.896	**100**	35.896	**100**	35.760	**100**	35.954	**100**	35.992	**100**

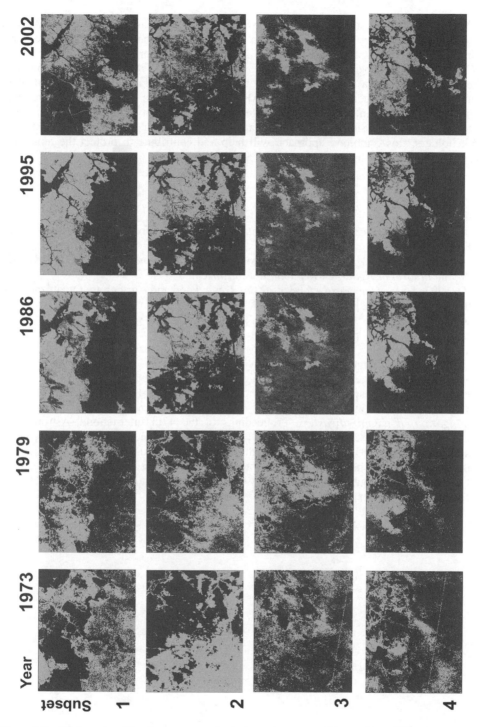

Figure 4. Spatial dynamics of forest-savanna mosaics in the environments of Ngotto (for position of subsets see figure 3; forest areas are in dark grey, savanna and land used areas in light grey).

The Ngotto Forest experiment clearly provides a 'showcase for the Central African Republic's forest policy' (Georges N'Gasse, ECOFAC). It has been promoted by the forest administration and attracted the attention of donors. The experiment owes its success to ECOFAC's methodological and financial contribution and IFB's undertaking to implement the plan and ensure a sustainable supply for its industrial plant. It is hoped that other commercial logging companies in the CAR will adopt to this sustainable approach that point the way ahead in tropical forestry. Remotely sensed data can contribute to monitor the changes of ecosystems and produce an independent data base for use by different stakeholders; this combined approach will help and contribute to protect the remaining forests resources in subsaharan Africa.

REFERENCES

Cirad Forêt, 1996, *Plan d'aménagement du PEA 169 de la forêt de Ngotto*. Montpellier, France.

Cossocim, B., 1996, *Mission d'appui au suivi du plan d'aménagement forestier du PEA 169 de la forêt de Ngotto en République centrafricaine.*

FAO, 1997, *State of the world's forests 2001*. (Rome: Food and Agriculture Organization of the United Nations), pp. 1–200.

FAO, 2001, *State of the world's forests 2001*. (Rome: Food and Agriculture Organization of the United Nations), pp. 1–181.

Harcharik, D.A., 1997, *Foreword in State of the world's forests 1997*, (Rome: Food and Agriculture Organization of the United Nations).

Koko, M. and Runge, J., 2004, *La dégradation du milieu naturel en République Centrafricaine*. Zeitschrift für Geomorphologie, N.F., Suppl.-Bd. **133**, pp. 9–47.

Sweeting, M. and Chen, F., 1996, *Network of low cost small satellites for monitoring and mitigation of natural disasters*. Proceedings of the 47th International Astronautical Congress, Beijing, China, October 7–11, 1996, IAF-96-C.1.09 (http://www.ee.surrey.ac.uk/SSC/CSER/UOSAT/papers/iaf96/disnet/disnet.html, 30.04.2006).

Yalibanda, Y., 2001, The forest revenue system and government expenditure on forestry in Central African Republic. Forestry Policy and Planning Division, Rome, Regional Office for Africa, *Accra*, pp. 1–54.

CHAPTER 16

Late Neoproterozoic Palaeogeography of Central Africa: relations with Holocene geological and geomorphological setting

Boniface Kankeu

Institut de Recherches Géologiques et Minières (IRGM), Yaoundé, Cameroon

Jürgen Runge

Centre for Interdisciplinary Research on Africa (CIRA/ZIAF), Johann Wolfgang Goethe University, Frankfurt am Main, Germany

ABSTRACT: The long and complex crustal evolution of Central Africa during the Late Neoproterozoic is presented, in order to advance the knowledge on former environmental changes taken place in and around the rain forest zone. A close relationship between the Precambrian heritage and the Early Phanerozoic intra-plate tectonics and sedimentation is sketched out. We suggested that this may also be the case before, around and after the Holocene.

16.1 INTRODUCTION

The majority of the rain forest–savanna mosaics in Central Africa are situated on planation surfaces formed over Gondwana basement rocks and covered by extended lateritic crusts (Runge, 2002). Therefore, amalgamation of Gondwana and also a number of other, possibly related, phenomena such as the final break up of Rodinia, the Pan-African orogeny and the last known period of equatorial glaciation which occurred approximately 600 Ma during the Late Neoproterozoic, one of the most enigmatic time periods in geological history, can improved the knowledge on former environmental changes taken place in and around the rain forest zone in Central Africa, by giving suitable regional to local information for the interpretation of climate, vegetation and morphodynamic processes before, around and after the Holocene. In this study, we present the long and complex crustal evolution of Central Africa during the Late Neoproterozoic with special paid to relationship between the Precambrian heritage and the Early Phanerozoic intra-plate tectonics.

16.2 PRECAMBRIAN EVOLUTION OF WEST AND CENTRAL AFRICAN BASEMENT

In all continents, Proterozoic orogenic belts occur as sinuous zones forming complex nets around older Cratons. Proterozoic activity occurred through continental accretion, island-arc and continental-margin magmatism, and possibly accretion of displaced terranes (Black *et al.*, 1979; Condie, 1982; Stoesser and Camp, 1985). In Africa, well-defined linear belts are the result of Pan-African tectonic activity.

16.2.1 The Palaeoproterozoic

The Precambrian basement of West and Central Africa has a long and complex crustal evolution, beginning with the Archaen (2.900–3.000 Ma), followed by two successive orogenies: (1) the Eburnean (Africa) –Transamazonian (Brazil) orogeny, at 2.100 Ma which joined the Congo Craton (CC) of Africa and the Sao Francisco Craton (SFC) of Brazil about 2,05 ± Ga, possibly during formation of a Palaeoproterozoic supercontinent. Evidence for this fusion event is found both on the east coast of Brazil (Atlantic granulite belt of Bathia and Sergipe) and on the west coast of Africa (Nyong series in southern Cameroon) and (2) the Pan-African at 600 Ma, which have extensively altered the Early Precambrian basement in Central Africa.

16.2.2 The Neoproterozoic and the Pan-African

The Neoproterozoic was a period of major plate re-organizations including the breakup of a Mesoproterozoic supercontinent and the subsequent dispersal and later reassembly of its constituent fragments into the Late Neoproterozoic/Early Palaeozoic Gondwana supercontinent during the Pan-African orogeny.

The Pan-African event characterizes the structural differentiation of Africa into Cratons and mobile belts during the Late Precambrian/Early Paleozoic (Figure1).

In terms of plate tectonics this differentiation was accomplished during a multi-episode process (by ensialic processes, Wilson-cycle-type evolution or thermal reactivation of old terranes) that started some 950–900 Ma ago with incipient or complete continental breakup (i.e. a partial or total continental fragmentation) in continental Africa. Basin closure and orogeny in the Pan-African mobile belts took place during discrete tectono-metamorphic episodes that ended with uplift and cooling of the crystalline basement of the African continent as part of a Late Neoproterozoic supercontinent Gondwana, some 450 Ma ago. The Pan-African belt of Central Africa which is the main topic of our study is best interpreted in the framework of the large scale/regional Pan-African/Brasiliano chain (Figure 2).

The general organization of this major structure is the result of the confrontation (convergence and collision) between three major continental domains. These were the West African and Sao Francisco/Congo Cratons and a Late Proterozoic mobile domain composed in its inner part, of the polycyclic basement provinces (Central-East Hoggar, Nigeria, Borborema, Central Africa provinces) and bounded, opposite to the Cratons, by a fringe of metasedimentary thrust belts (Pharusian, Atacora, Sergipano, Central Africann fold belts) (Castaing *et al.*, 1994; Trompette, 1994). Two segments can be recognized. These are the N–S trending Trans-Saharan belt against the West African Craton and the SW–NE trending Central African belt against the Congo Craton.

The Pan-African Trans-Saharan belt of West Africa extends from the Hoggar Mountains in the north to the Gulf of Benin in the south, and further south into northeastern Brazil, which was linked with Africa during Proterozoic. It has been established as a typical example of a collisional orogen. The Pan-African evolution is interpreted in terms of transpressive tectonics and terrane accretion (Black *et al.*, 1994; Ferré *et al.*, 1996). The Pan-African orogenic cycle began ca. 780 Ma ago with a partial or total continental fragmentation involving, to the west, a passive margin bording the West African Craton and, to the east, an active margin facing the Touareg shield and the Nigerian province

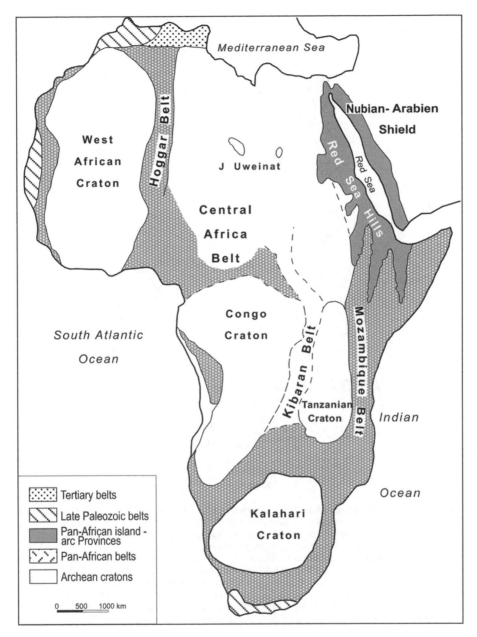

Figure 1. The main structural elements of the Africa basement, with Pan-African belts forming complex nets around older/Archean Cratons (modified after Clifford, 1970; Kröner, 1979; Schandelmeir and Harms, 1987; Vail, 1990; Abu Fatima, 1992; Babiker and Gudmundsson, 2004).

(Caby *et al.*, 1981). The genesis of the belt can be divided into three main periods: (1) a first period characterized by oblique collision inducing oblique SW thrusting, (2) a second period with anatectic doming; (3) a third period of wrench faulting with associated N–S post-foliation folding under late E–W shortening.

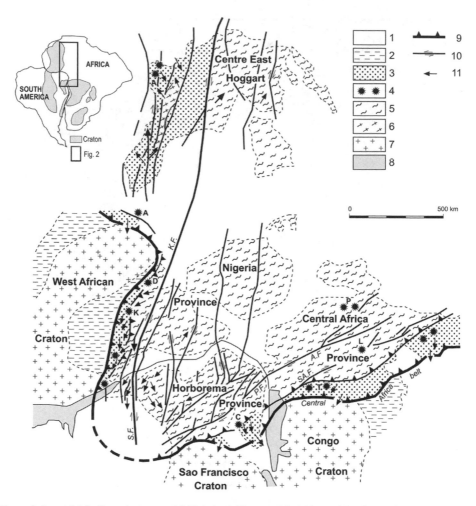

Figure 2. Insert: Main Cratonic areas and folds belts (white space) in Africa and South America and localization of Figure 2. General organization of the Late Proterozoic Pan-African/Brasiliano belt as the result of the confrontation between three major continental domains during formation of Gondwana (modified from Castaing *et al.,* 1993). Make a note of the Late Proterozoic mobile domain composed of a central part, bordered, opposite to the Cratons, by a fringe of metasedimentary thrust belts. 1 = Phanerozoic cover, 2 = Proterozoic cover, 3 = Pan-African/Brasiliano metasedimentary fold belts including basement granulite (6) in the western Hoggar, 4 = main basic and ultrabasic massifs underlining the suture zone, 5 = undifferentiated mono-and polycyclic gneiss and metasediment, migmatites, granite and charnockite, 6 = Ouzzal and Iforas Eburnian, 7 = 2 Ga Cratons, 8 = sea, 9 = major shear zone, 10 = shear zones, 11 = tectonic nappe-transport direction. A.F. = Anaga-Adamaoua fault, P.F = Pernambuco fault, S.F = Sobral fault, SA.F = Sanaga fault, K.F. = Kandi fault.

The Pan-African belt of Central Africa is a more or less E–W trending belt. The location of Cameroon in the Pan-African network is of prime importance as it covers the opposite margin of the Pan-African belt against the Congo Craton. Southern Cameroon represented the orogenic margin of the belt where marginal thrusts transported Pan-African orogenic material southwards onto the Congo Craton orogenic foreland (e.g. Castaing *et al.*, 1994) whereas, Central and northern Cameroon, represent the more internal domains of the orogen, dominated by strike-slip tectonics.

16.2.3 Nature and location of terrain boundaries

The nature and the location of terrain boundaries relative to the various Pan-African domains in Central Africa are still in debate. On the basis of field, petrographic, structural, and isotopic studies (Toteu *et al.*, 2001; Penaye, 1988; Soba, 1987; Ngako *et al.*, 2003) several domains can be recognized in Cameroon (Figure 3).

① The northern edge of the Congo Craton corresponds to the Ntem complex which outcrops in Cameroon, Gabon and Congo. The Ntem complex is an Archean core that formed about 2,9 Ga as juvenile crust (Toteu *et al.*, 1994). It is dominated by massive and banded plutonic rocks of charnockitic and/or non charnockitic composition. They display large xenolith of supracrustal rocks interpreted as remnants of greenstone belts.

② The Palaeoproterozoic polycyclic basement (ca. 2.100 Ma) in the west corresponds to the Nyong and Lokundje series. Rock units attributed to the Nyong and Lokundje series are metasediments and metavolcanosediments which include metaquartzites, metaarkoses, Banded Iron Formations of Mamelles type associated with migmatitic grey gneisses and syn- to late-tectonic granitoid rocks (charnockites, granites, granodiorites and syenites). It has first been regarded as a reactivated part of the Ntem complex (Bessoles and Trompette, 1980), and interpreted as an Early Proterozoic nappe, thrust to the south onto the Congo Craton (Feybesse *et al.*, 1986) or as a basement nappe of Pan-African age (Toteu *et al.*, 1991). The structures associated with the Nyong series, is now attributed to intense Palaeoproterozoic orogenic activity ca. 2.100 Ma (Eburnian orogen) (Toteu *et al.*, 2001) with recycling of Archean crust and newly accreted material, deformation, metamorphism, crustal melting and emplacement of plutonic rocks. The structures associated with this Eburnian orogeny are preserved only in the Nyong series, where Pan-African activity was less important. Unfortunately, it is not easy in most cases to discriminate between Neoproterozoic (Pan-African) and Palaeoproterozoic (Eburnean) high-grade rocks in the field.

③ The Mesoproterozoic to Neoproterozoic monocyclic units (1.000–700 Ma) forming volcano-sedimentary basins, are now deformed and metamorphed into schist and gneisses of the Lom and Yaoundé Mbalmayo series. The latter is an association of metasediments and metaplutonics that underwent regional metamorphism culmining in granulitic facies. It is interpreted (Nedelec *et al.*, 1986; Nzenti *et al.*, 1996) as shallow-water sedimentary sequences composed mainly of shallow and volcanogenic greywackes with dolomitic

Figure 3. (A) The Pangaea at upper Palaeozoic/Permian (after Aubouin J. *et al.*, 1979). (B). Fragmentation/ break-up of Gondwana; after Waterlot (1985).

marls, dolomites, evaporitic sediments, quartzites, and iron-rich beds. The question of the origin of the series (autochthonous or allochthonous on the Craton) and the significance of granulitic rocks (roof of collision zone or intracontinental?) is still on debate. (4) The Pan-African granitoids emplacement ages range from the early stage of the deformation to the late uplift stages of the belt.

16.2.4 Tectonic evolution

The tectonic evolution of the Pan-African belt in Central Africa is not yet well documented. The chronology of deformation is similar in the entire fold belt north of the Congo Craton (Nzenti *et al.*, 1988, Ngako *et al.*, 1989), with D1 (older, compressive deformation) during the peak of Pan-African collision, associated with an early medium- to high-pressure regional metamorphism followed by intense migmatization and or without emplacement of plutonic bodies during the deformation. It is followed by D2 deformation, dominated by N–S to NE–SW trending, steeply dipping foliations (in the more internal domain), or E–W trending flat-laying foliations (in the orogenic margin/frontal belts). The age of this D2 event is estimated at 565 Ma (Lasserre and Soba, 1979). The subsequent episode resulted in the development of a complex network of continental-scale transcurrent shear zones. As a consequence, the older structural elements are cut by late, regional-scale strike-slip faults (D3).

16.3 LATE NEOPROTEROZOIC PALAEOGEOGRAPHY OF CENTRAL AFRICA

The Late Neoproterozoic Palaeogeography of Central Africa that emerges is characterized by the presence of a sinuous and narrow Pan-African mobile belt (of Central Africa) forming complex nets around the Archean Congo Craton. The external domain of the belt (or frontal belt) thrust over the Congo Craton, derived from sedimentary deposits and displaying low grade schists to high-grade metamorphic rocks. More internal domains composed of orthogneiss, monocyclic metasedimentary rocks, reworked Eburnean rocks/ gneiss and orthogneiss and intrusive granite and charnockite are dominated by NE trending large-scale shear zone. These are the Adamaoua line and the so called "Sanaga wrench fault". The later is now interpreted as a large-scale transpressional structure (Kankeu and Greiling, 2006). Transpression may be the mechanism which thickens the crust and structurally inverts sedimentary basins, and is responsible for heating, metamorphism, and generation of granitoid magmas. The shear zone network in southern Cameroon consists of major NE trending movement zones and numerous, dominantly N–S trending, "subsidiary" shear zones.

16.4 EARLY PHANEROZOIC EVOLUTION OF CENTRAL AFRICA

The Phanerozoic evolution of Africa reflects both the assembly of Pangea, and the polyphased breack-up of the Gondwana supercontinent (Geraud *et al.*, 2006). With (almost) no subduction zones in Central Africa during the Phanerozoic, global tectonic processes are confined to combinations of rifting and transcurrent motions (transpression and transtension). The Congo Craton has remained relatively rigid since Pan-African times, surrounded by narrow Pan-African mobile zones that accommodated these dynamic processes. Only gentle deformation affected Central Africa during the Palaeozoic. By upper

Palaeozoic/Permian, the PANGAEA (Figure 4.A) was a simple supercontinent surrounded by the universal ocean Panthalassa. Tethys was a minor arm of this ocean. In the Early Mesozoic, initial rifting of Pangaea along the line of the North Atlantic Ocean and Tethys produced two large continental masses: northern LAURASIA and southern GONDWANA. Cretaceous times correspond to a crucial period for Central Africa. Active rifting episodes over large domains led to the break-up of Western Gondwana and the opening of the South and Equatorial Atlantic oceans. Magmatism often accompanied rifting. From Aptian times and peaking in the Late Cretaceous, strong warming in global climate resulted in wide transgression over large continental subsiding basins, followed by the inversion of many ~E–W trending basins in Late Cretaceous.

16.4.1 The fragmentation (or break up) of Gondwana

The Gondwanian assemblage was reworked under the effect of a generalized distension which led to the opening of the equatorial domain of the South Atlantic. The fragmentation of Gondwana (Figure 3.B and 4.A) started with ① the separation of West Gondwana (South America-Africa) from the East (Antartica, India, Australia) during the Lower Mesozoic/ Trias or Jurassic, followed by ② the separation of Africa-South America during upper Mesozoic (end of Jurassic and Lower Cretaceous). The opening of the Gulf of Guinea ③ during the lower Cretaceous represented the final Gondwanian link between the Central and South Atlantic (Figure 4.B). The present features of African and South American continents appear as from the uppermost Albian to Lower Cenomanian, times of final crustal disruption followed by a progressive separation of the land masses along the strike-slip equatorial margin (Popoff, 1988).

16.4.2 Rework of NE–SW ductile Gondwanian shear zone (Intra-plate tectonics)

At the same time as the opening of the Gulf of Guinea (Figure 4.A), a transcurrent shear occurred along the transform faults of the Gulf of Guinea (Figure 4.B), and was transmitted to their continental extensions. The dominant NE–SW ductile Gondwanian shear zone of Central Africa was reworked leading to the installation of the Central African lineament also called Central African rift system which corresponds to major mylonitic accidents dissecting granitised Precambrian basement of Central Africa and extending without discontinuity over more than 2.500 km from Western Cameroon to Central Sudan (Figure 5). Brittle tectonics giving rise to fractures, faults or joints were most prominent along these inherited Gondwanian mega-discontinuities with mainly vertical movements (Ngangom, 1983). We shall summarize here after the tectonic evolution of Central Africa during Late Mesozoic and Early Cenozoic.

A rapid change in the intra-plate stress-field occurred in the Early Cretaceous/ Early Aptian. The extension direction, formerly 160° NNW–SSE to N–S, moved to NE–SW (Guiraud and Maurin, 1992). Dextral transtension initiated along the Central African fault Zone (Browne and Fairhead, 1983) activating minor NW–SE trending rifts or pull-apart basins.

A tectonic event in response to ~160° NNW–SSE oriented shortening occurred in the Upper Cretaceous/Late-Santonian. This event corresponds to the first general compressional episode registered by the African-Arabian plates. 70° to E–W trending segments of the Central African Rift System experienced dextral transpressional deformation resulting in large folds and positive flower structures, and the inversion of most of the E–W to ENE–WSW trending cretaceous troughs (Giraud et al., 2005).

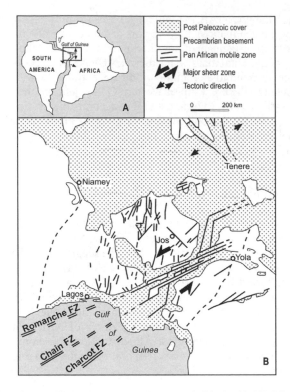

Figure 4. Opening of the Gulf of Guinea (Lower Cretaceous/Albian) with (A) Schematic separation of Africa from South America and (B) Sinistral transcurrent movement along the Benue Trough in the prolongation of the oceanic transform faults (tf) and related structures (modified from Grant, 1971; Benkhelil and Robineau, 1983).

By the Early Tertiary/Late Eocene (~37 Ma) a brief, major compressional event occurred (Giraud *et al.*, 2006). This event was another major stage in collision of the Africa-Arabian and Eurasian plates. On a plate scale, stress field analysis shows a general NNW–SSE (~160°) to N–S shortening (Guiraud and Bellion, 1995). E–W trending dextral Central African fault zones were rejuvenated as strike-slip faults with associated drag folds. Like the Late Santonian event, it resulted from changes in the rates and directions of opening of the Central, South, and North Atlantic oceans. This compressional phase was rapidly followed by the development of the NNW–SSE to NNW–SSE trending Oligocene rifts which registered active more or less synchronously magmatic provinces amongst which the Cameroon Line, the Tibesti and Hoggar. Present tectonic readjustments in Central Africa can be inferred from the present seismicity of certain regions such as the 2005 Monatélé earthquake north of Yaoundé region.

16.4.3 Sedimentation

Although originated during the Pan-African, the more or less E–W oriented zone of latent weakness have been reworked several times since the Pan-African and dissected the Precambrian basement into a number of grabens and half-grabens where sediments

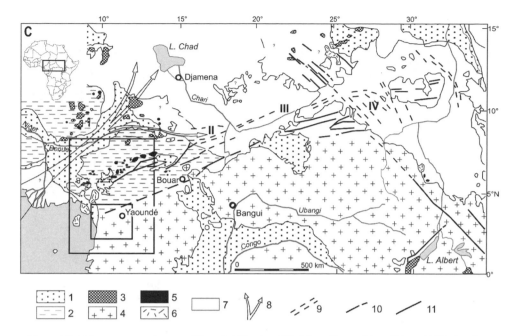

Figure 5A. Schematic tectonic map of Central Africa (modified from Cornacchia and Dars, 1983) showing major faults/mylonitic accidents extending without discontinuity from Western Cameroon to Central Sudan and locally hidden by superficial deposits belonging to the Tertiary and the Quaternary. 1 = Tertiary and Quaternary. 2 = undifferentiated Mesozoic, Cretaceous 3 − Cretaceous, 4 − Precambrian basements (undifferentiated), 5 = Tertiary/Eocene granites of Cameroon, 6 = Volcanic rocks, 7 = Quaternary, 8 = Cretaceous transgressions, 9 = sedimentary basins of M'Bere (I), Dobo (II), Birao (III) and Abu Gabra (IV), 10 = Fault, 11 = Shear zones.

accumulated up to more recent geological times (Neogene ?). It has influenced the transgression which has progressed from the Atlantic Ocean into the heart of the continent during the Cretaceous (Figure 5A).

In fact, Middle to Late Cretaceous (Latest Albian to Middle Santonian) is characterized by decrease in the tectonic activity and strong warming in the global climate resulting in a wide marine transgression over continental subsiding basins. For Central Africa the subsidence is considered to be due to NE–SW directed extension (Guiraud, 1993; Janssen, 1996), sometimes associated with thermal relaxation. With regard to the Palaeogeography, a major change occurred as the sea invaded the intra-continental basins from the Atlantic via the Benue Trough. Consequently Early Phanerozoic basins became oriented according to the inherited Gondwanian mega-discontinuities. The Late Eocene compressional event generated large uplifts in the Central Africa intraplate domain. As a result, the size of the remaining intraplate sedimentary basins was reduced and this tendency intensified later in response to younger compressional events and development of large uplifted magmatic provinces. Seas no longer invaded the intra-continental basins of Central Africa that underwent fluviatile-lacustrine sedimentation.

16.4.4 Magmatism

The Cameroon Volcanic Line (Figure 5C) is an alignment of Tertiary to recent alkaline oceanic and continental volcanoes and anorogenic pluton trending averagely 30°N and stretching for over 1.500 km. Volcanism along the CVL dates from 65 Ma to present day.

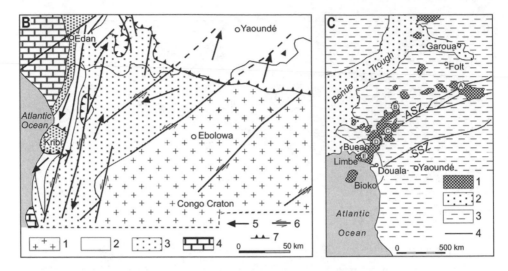

Figure 5B. Sketch map of southern Cameroon (data from Nzenti *et al.* (1988) and Feybesse *et al.* (1986)), showing major structural units (after amalgamation of the Gondwana supercontinent) and a complex network of N-S to NE trending transcurrent shear zone. (1) Ntem complex (Archean Congo Craton), (2) Nyong series (Palaeoproterozoic polycyclic basement, ca.2.100 Ma), (3) Yaoundé and Mbalmayo series (Mesoproterozoic to Neoproterozoic monocyclic units, between 1.000 Ma and 700 Ma), (4) Cretaceous to recent formations, (5) Plunge direction of stretching lineation, (6) shear zones, (7) major shear zone.

Figure 5C. 1 = Main volcanic centres of the Cameroon Volcanic Line (A = Ngaoundéré plateau, B = Biu plateau, C = Oku, D = Bambouto, E = Manengouba, F = Cameroon.) and relation with Precambrian structures (SSZ = Sanaga shear zone, ASZ = Adamawa shear zone); 2 = Creataceous (undifferentiated), 3 = Precambrian basements (undifferentiated), 4 = shear zones.

The most widely accepted structural explanation for the origin of Cameroon Volcanic Line, is that it is the product of reactivation of Pan-African strike-slip faults trending 70° E–W. This large-magmatic province resulted in uplifts that influenced the Palaeogeography. Suh *et al.* (2001) shown that the 1999 eruption of Cameroon Mountain was controlled by NE–SW trending fissures and that brittle failure and vertical/horizontal dilatational ground movements were the main styles of deformation accompanying the eruption.

16.5 DISCUSSION AND CONCLUSIONS

One axiom of the plate tectonics paradigm is that, plates are rigid. However, in the Himalaya orogen, major deformation is documented thousands of kilometers far from the suture zone, in the core of the Asia plate, after closure of all oceans (Liegeois, 2004). This underscores the importance of post-collisional events inside the plates themselves.

In Central Africa, dominant NE–SW ductile shear zones acted as limits of mobile crustal blocks on both sides of the Gulf of Guinea during the Late Precambrian collision stages in between the West African and Sao Francisco/Congo Cratons. Although originated during the Pan-African, these major mylonitic accidents have been reworked during the generalized distension which led to the opening (Lower Cretaceous) of the equatorial domain of the South Atlantic with the development of the Central African Rift System, the plate scale Late Santonian transpressional event and the early Late Eocene major compressional event. This frequent rejuvenation of the Pan-African fault net also influenced the development of magmatic provinces and possibly the present tectonic readjustment.

These phenomena which resulted in frequent changes in palaeoenvironments, reveals the importance of intra-plate Phanerozoic tectonics in Central Africa and furthermore a close relationship between Early Phanerozoic intra-plate tectonics and Gondwana megadiscontinuities in Central Africa.

As sedimentary systems over geological times ensure the erosion of continental relief, the transportation of the resulting sediments and their preservation within sedimentary basins, sediment transfer which controls the topographic evolution may depend of several parameters such as the deformation history of the region and the climatic situation of the drainage system. This may also be the case in Central Africa, where the zone of latent weakness dissected the Precambrian basement into a number of grabens and half-grabens where sediments accumulated up to more recent geological times, influenced the transgression during the Cretaceous and the genesis of the Tertiary plutonic ring complexes and volcanoes satured along the Cameroon Line. We believe that (neotectonic?) deformation along Precambrian structures modified rivers and the drainage net and controlled the geomorphic evolution of the Holocene basin in Central Africa.

REFERENCES

Abu Fatima, M., 1992, Magmatic and tectonic evolution of the granite-greenstone sequences of the Sinkat area, Red Sea Province, NE Sudan: *unpublished M. Phil.,* department of Geology, University of Portsmouth, UK, p. 276.

Aubouin, J., Brousse, R. and Lehman, J.P., 1979, *Precis de géologie.3. Tectonique, tectonophysique, morphologie,* **4**, (Paris: Dunod Université).

Babiker, M. and Gudmundsson, A., 2004, Geometry, structure and emplacement of mafic dykes in the Red Sea Hills, Sudan. *Journal of African Earth Sciences,* **38**, pp. 279–292.

Benkhelil, J., and Robinau, B., 1983, Le fosse de la Bénoué est – il un rift ?. *Bulletin Centre Recherches et Exploration-Production Elf Aquitaine,* **7**, 1, pp. 315–321.

Bessoles, B. and Trompette, R., 1980, Géologie de l' Afrique. La chaine panafricaine «zone mobile d' Afrique Centrale (partie sud) et zone mobile soudanaise». *Memoire du Bureau de Recherches Géologiques et Minières,* Orleans , **92**, p. 394.

Black, R., Caby, R., Moussine-Pouchkine, A., Bayer, R., Bertrand, J.M., Boulier, A.M. and Fabre, J., 1979, Evidence for late Precambrian plate tectonics in West Africa. *Nature,* **278**, pp. 223–227.

Black, R., Latouche, L., Liegeois, J.P., Caby, R. and Bertrand, J.M., 1994, Pan-African displaced terranes in the Tuareg shield (Central Sahara). *Geology ,* **22**, pp. 641–644.

Browne, S.E. and Fairhead, J.D., 1983, Gravity study of the Central African Rift System: a model of continental disruption. 1. The Ngaoundere and Abu Gabra Rifts. *Tectonophysics,* **94**, pp. 187–203.

Caby, R., Bertrand, J.M.L. and Black, R., 1981, Pan-African ocean closure and continental collision in the Hoggar-Iforas segment, central Sahara. In *Precambrian Plate Tectonics,* edited by Kroner, A., (Amsterdam: Elsevier), pp. 407–434.

Castaing, C., Triboulet, C., Feybesse, J.L. and Chevremont, P., 1993, Tectonometamorphic evolution of Ghana, Togo and Benin in the light of Pan-African/Brasiliano orogeny. *Tectonophysics,* **218**, pp. 323–342.

Castaing, C., Feybesse, J.L., Thieblemont, D., Triboulet, C. and Chevremont, P., 1994, Paleogeographical reconstructions of the Pan-African/Brasiliano orogen: closure of an oceanic domain or intracontinental convergence between major blocks. *Precambrian Research,* **69**, pp. 327–344.

Clifford, T.N., 1970, The structural framework of Africa. In *African Magmatism and tectonics,* edited by Clifford, T.N. and Gass, I.G. (Edinburgh: Olivier and Boyd), pp.1–26.

Condie, K.C., 1982, Plate tectonic models for Proterozoic continental accretion in the SW United States. *Geology,* **10**, pp. 37–42.

Cornacchia, M. and Dars, R., 1983, Un trait structural majeur du continent africain. Les lineaments centrafricains du Cameroun au golphe d'Aden. *Bulletin de la Societé Géologique Francaise* , (7), **XXV 1**, pp.101–109.

Ferré, E., Déléris, J., Bouchez, J.L., Lar, A.U. and Peucat, J.J., 1996, The Pan-African reactivation of contrasted Eburnean and Archaen provinces in Nigeria. *Tectonics,* **145**, pp. 1205–1219.

Feybesse, J.L., Johan, V., Maurizot, P. and Abessolo, A., 1986, Mise en évidence d'une nappe synmétamorphe d'âge Eburnéen dans la partie NW du Craton zairois (SW Cameroon). In *Les formations birrimiennes en Afrique de l'Ouest,* (Paris: Centre International de formation et d' Etudes Géologiques, publication occasionnelle), **10**, pp. 105–111.

Guiraud, M., 1993, Late Jurassic rifting–Early Cretaceous rifting and Late Cretaceous transpressional inversion in the upper Benue Basin (NE-Nigeria). *Bulletin Centre Recherches et Exploration-Production Elf Aquitaine,* **17**, pp, 371–383.

Guiraud, R. and Bellion, Y., 1995, Late Carboniferous to Recent geodynamic evolution of the West Gondwanian Cratonic Tethyan margins. In *The Ocean Basins and Margins 8, the Tethys Ocean,* edited by Nairn, A., Ricou, L.E., Vrielynck, B. and Dercourt, J. (New York, Plenum Press), pp. 101–124.

Guiraud, R., Binks, R.M., Fairhead, J.D. and Wilson, M., 1992, Chronology and geodynamic setting of Cretaceous—Cenozoic rifting in West and Central Africa. *Tectonophysics,* **213**, pp. 227–234.

Guiraud, R., Bosworth, W., Thierry, J. and Delplaque, A., 2006, Phanerozoic geological evolution of Northern and Central Africa: an overview. *Journal of African Earth Sciences,* **43**, pp. 83–143.

Janssen, M.E., 1996, *Intraplate deformation in Africa as a consequence of plate boundary changes. Inferences from subsidence analysis and tectonic modelling of the Early and Middle Cretaceous period.* Ph.D. Thesis, Free University Amsterdam, Holland, p. 161.

Kankeu, B. and Greiling, R.O., 2006, Magnetic fabrics (AMS) and Transpression in the Neoproterozoic basement of eastern Cameroon (Garga-Sarali area). *Neues Jahrbuch für Geologie und Paläontologie,* **239**, pp. 263–287.

Kröner, A., 1979, Pan-African plate tectonics and its repercussion on the crust of Northeast Africa. *Geologische Rundschau,* **68**, pp. 565–583.

Lassere, M. and Soba, D., 1979, Migmatisation d'âge panafricain au sein des formations camerounaises appartenant à la zone mobile d'Afrique centrale. *Comptes Rendus Societé Géologique Francaise,* **2**, pp.64–68.

Liégeois, J.P., 2004, Cratons, meta Cratons and juvenile terranes in the Pan-African Tuareg shield (Central Sahara) and generation of granitoids batholiths. In *Abstracts of 20th colloqium on African Geology,* Orleans, 2–7 June (Orleans: BRGM).

Nédélec, A., Macaudiere, J., Nzenti, J.P. and Barbey, P., 1986, Evolution structurale et métamorphique des schistes de Mbalmayo (Cameroun). Implications pour la structure de la zone mobile Pan-Africaine d'Afrique Centrale au contact du Craton du Congo. *Comptes Rendus Académie Sciences,* Paris, **303**, pp. 75–80.

Nédélec, A., Minyem, D. and Barbey, P., 1993, High P-high T anatexis of Archean tonalite grey gneisses: the Eseka migmatites, Cameroon. *Precambrian Research,* **62**, pp. 191–205.

Ngangom, E., 1983, Etude du Fosse Crétacé de la Mbéré et du Djerem, Sud–Adamaoua, Cameroun. *Bulletin Centre Recherches et Exploration-Production Elf Aquitaine,*7,1, pp. 339–347.

Ngako, V., Jegouzo, P. and Djallo, S., 1989, Déformation et métamorphisme dans la chaîne Pan-Africaine de Poli (Nord-Cameroun): implications géodynamiques et paléogégraphiques. *Journal of African earth Sciences,* **9**, pp. 541–55.

Ngako, V., Affaton, P. Nnangé. J.M. and Njanko, T., 2003, Pan-African tectonic in Central and Southern Cameroon: transpression and transtension during sinistral shear movements. *Journal of African Earth Sciences*, **36**, pp. 207–214.

Nzenti, J.P., Barbey, P., Macaudière, J. and Soba, D., 1988, Origin and evolution of the late Precambrian high-grade Yaoundé gneisses (Cameroon). *Precambrian Research,* **38**, pp. 91–109.

Nzenti, J.P. and Tchoua, F., 1996, Les roches à scapolites de la chaine Pan-Africaine d'âge proterozoique en marge du Craton du Congo. *Comptes Rendus Académie Sciences,* Paris, **323**, pp. 289–295.

Popoff, M., 1988, Du Gondwana à l'Atlantique Sud: les connexions du fossé de la Bénoué avec les bassins du Nord-Est brésiliens jusqu'à l'ouverture du golfe de Guinée au Crétacée inférieur. *Journal of African Earth Sciences*, 7, 2, pp. 409–431.

Popoff, M., Benkhelil, J., Simon, J. and Motte, J.J.,1983, Approche geodynamique du fosse de la Bénoué (NE Nigéria) à partir des donnees de terrain et télédétection. *Bulletin Centre de Recherches Exploration-Production Elf Aquitaine*, 7, pp. 323–337.

Runge, J., 2002, Holocene landscape history and palaeohydrology evidenced by stable carbon isotope (δ^{13}C) analysis of alluvial sediments in the Mbari valley (5°N/23°E), Central African Republic. *Catena*, **48**, pp. 67–87.

Schandelmeier, H. and Harms, U., 1987, The Northern Zalingei fold zones an intercontinental mobile belt of mid-Proterozoic age in Northeast Africa. In *Current Research in African Earth Sciences*, edited by Matheis, G. and Schandelmeir, H. (Rotterdam: Balkema), pp. 45–48.

Stoeser, D.B. and Camp, V.E., 1985, Pan-African microplate accretion of the Arabian shield. *Geological Society of America Bulletin*, **97**, pp. 817–826.

Suh, C., E., Ayonghe, S.N. and Njumbe, E.S, 2001, Neotectonic earth movement related to the 1999 eruption of Cameroon mountain, west Africa. *Episodes*, **24**, pp. 9–12.

Vail, J.R., 1990, Geochronology of the Sudan. In *Overseas geological Mineral Resources*, **66**.

Toteu, S.F., Bertrand, J.M., Penaye, J., Macaudiere, J., Angoua, S. and Barbey, P., 1991, Cameroon. A tectonic keystone in the Pan-african network. In *The Early Proterozoic Trans-Hudson Orogen of North America*, edited by Lewrey, J.F. and Stauffer, M.R., (London: Geological Association of Canada, Special Paper, **37**), pp. 483–496.

Toteu, S.F., Van Schmus, W.R., Penaye, J. and Nyobe, J.B., 1994, U-Pb and Sm-Nd evidence for Eburnean and Pan-African high-grade metamorphism in Cratonic rocks of Southern Cameroon. *Precambrian Research*, **67**, pp. 321–347.

Toteu, S.F., Van Schmus, W.R., Penaye, J. and Michard, A., 2001, New U-Pb and Sm-Nd data from north-central Cameroon and its bearing on the pre-Pan-African history of central Africa. *Precambrian Research,* **108**, pp. 45–73.

Trompette, R., 1994, Geology of western Gondwana (2000-500Ma). *Pan-African-Brasiliano aggregation of South America and Africa*, (Rotterdam: A.A. Balkema Press).

Waterlot, M., 1985, La fragmentation du Gondwana. In *Evolution geologique de l'Afrique: Seminaire de formtion*. CIFEG. Publication occasionnelle 1985/4, pp. 249–263.

CHAPTER 17

A palaeoecological approach to neotectonics: the geomorphic evolution of the Ntem River in and below its interior delta, SW Cameroon

Joachim Eisenberg

Institute of Physical Geography, Johann Wolfgang Goethe University, Frankfurt am Main, Germany

ABSTRACT: The Late Tertiary to Quaternary evolution of the Ntem interior delta in SW Cameroon shall be modelled. A step fault was formed along neotectonically remobilized Precambrian structures. Uncalibrated ^{14}C-datations in this 'sediment trap' show Pleistocene to Holocene ages. Both within and below the interior delta pebbles and clasts which are cemented in an iron and manganese matrix were found. These 'fanglomerates' are used to discuss different processes of the younger evolution also concerning climatic fluctuations in the study area.

17.1 INTRODUCTION

In the framework of the subproject 'ReSaKo' (Rainforest-Savanna-Contact) of the DFG-Research Unit 510 'Environmental and cultural change in West and Central Africa' the palaeoenvironmental significance of alluvia of the Ntem interior delta in SW Cameroon will be investigated (see Sangen, this publication). To understand the sedimentation processes in the study area, it is important to comprehend the evolution of the interior delta in the context of neotectonics during the past 5–6 Ma.

17.1.1 State of the art

After the progressive opening of the southern Atlantic at about 95 Ma (Reyment and Tait, 1972; Robert, 1987) the drainage network was rejuvenated along the newly exposed continental margin (Summerfield, 1996; Moore and Larkin, 2001). Initially the drainage system was oriented towards the old cratons away from the centres of uplift, which triggered the opening. Along small passages (Congo River) or aulacogenes (Benue) the streams made their way into the proto-Atlantic Ocean. Consequent drainage systems flowed directly into the southern Atlantic. As a result, Summerfield (1991) distinguishes between basinward- and oceanward-oriented rivers. The majority of river catchment area on the African continent has a basinward orientation.

The collision of the African plate with Eurasia 30 Ma ago (Burke and Whiteman, 1973; Burke, 1996) initiated uplift and formed the typical African basin and swell structure, as well as increasing volcanic activity along the East African rift after a long period of inactivity (Burke, 1996). In Cameroon the uplift induced the formation of the 'surface côtière', the coastal surface which is recently separated from the 'surface intérieur' in the hinterland by a peneplain step (Segalen, 1967).

On the African continent, neotectonic impulses are generally associated with the young East African rift structures (Summerfield, 1991; McCarthy *et al.*, 1993) although a minority are associated with the passive Central African continental margin. Suh *et al.* (2001) investigated the neotectonic forms on the slopes of Mount Cameroon triggered by the most recent eruption of this active volcano in 1999. They interpret the Cameroon Volcanic Line (CVL) as an eastward offset of the panafricanic West African rift system which has been reactivated since the Tertiary by hot plumes originating in the asthenosphere (cf. Fairhead, 1985; Déruelle *et al.*, 1987). For southern Cameroon a seismologic map by Krenkel (1921, cited after Fairhead, 1985) states an 'area within which earthquakes commonly occur' (1–5 per year). Fairhead (1985) confirms the accuracy of this map by his own actual seismologic measurements.

Lucazeau *et al.* (2003) used thick sediment layers accumulated inside the Congo drainage system (cf. Karner *et al.*, 1997) to calculate the isostatic uplift of the Congo craton's swells by up to 500 m since the Miocene. Summerfield (1996) refers to the recent uplift in sub-Saharan Africa manifested by massive denudation and accumulation processes at the continental margin, which originate from the remobilization of Precambrian structures since the Phanerozoic (Summerfield, 1985; cf. Daly *et al.*, 1989).

Concerning African interior deltas McCarthy *et al.* (1993; cf. McCarthy, 1993) point out to neotectonic influences on the water dispersal of the Okavango delta in Botswana which is related to the southwestern prolongation of the East African Rift. A comparison with the Ntem interior delta is difficult due to different recent climatic and vegetation habitats. However, the Ntem study area is situated close to the Cameroonian northeast trending Pan-African belt which is described as 'a line of weakness' (McCarthy *et al.*, 1993).

17.1.2 Methods

Analysis and interpretation of topographic and geologic maps as well as Landsat 7 ETM+ scenes (21.02.2001; EarthSat), SRTM and JERS-1 radar data revealed several sedimentation areas in different catchments. These data layers were combined in a Geographic Information System (GIS) with basic information about infrastructure and physiogeographical characteristics (DCW—Digital Chart of the World). Outline maps were created for the field trips.

On-site, waypoints and exploration routes were edited with Magellan ProMark 2-GPS and OziExplorer software. Geologic lineaments were documented in the field. Bedding dips and inclination were measured using a Breithaupt clinometer, and altitudes were recorded with the Thommen Classic altimeter.

Bedrock samples and iron oxide-cemented river pebbles were collected along the Ntem River for petrographical and geochemical analyses. The mineralogical laboratory of *Institut des Recherches Géologiques et Minières (IRGM)* in Yaoundé prepared rock samples for thin section manufacture. The Institute of Geosciences in Frankfurt analysed four samples of ferruginous rock patina from different locations in the interior delta using X-ray diffractometry (RFA). In addition, the Department of Mineralogy produced several thin sections of a gravel fanglomerate for geochemical analysis (polarizing microscopy, electronic microprobe: element distribution map).

17.2 THE NTEM INTERIOR DELTA

The Ntem River catchment covers an area of about 31.000 km² (Olivry, 1986). The river has its source in Gabon at an altitude of about 700 m a.s.l. and flows into the southern Atlantic

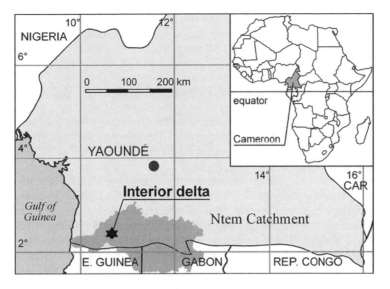

Figure 1. Location of the study area in SW Cameroon.

close to Campo/Cameroon; hence, according to Summerfield's (1991) definition, it is an oceanward draining river. At this location, the Ntem River represents the border between Equatorial Guinea and Cameroon (Figure 1).

In the sub-prefecture of Ma'an (prefecture Ambam/Vallée du Ntem, 2°22'N, 10°37'E), the Ntem River forms an interior delta, covering a surface of 210 km².

The river fans out into a multi-branched system over a length of ~25 km from SE to NW and a maximum width of up to 10 km. Near the settlement Akom (2°26'N, 10°29'E) the multiple branches rejoin again to a single channel due to the influence of a SSW–NNE striking inselberg ridge at this point. The river crosses this section in a sharp bow and subsequently flows in SW direction for a distance of ~10 km. Just downstream of the

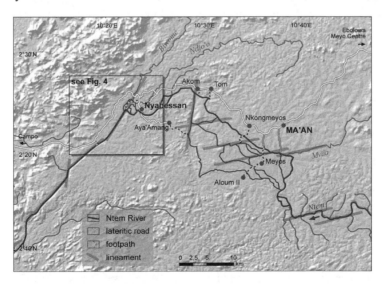

Figure 2. The Ntem interior delta with generalized lineaments interpreted as influencing the river course.

village of Nyabessan (2°24'N, 10°24'E) the river fans out once again. Here the Ntem flows into a SW orientated rectilinear V-shaped gorge, cascading down the ~10 m waterfalls of 'Chutes de Menvé'élé' (Figure 2 and 4).

17.2.1 Physiogeography: Climate, geology and geomorphology

The climate within the Ntem catchment is tropical to semi-humid with a short (April–June), and a longer (September–December) rainy season. The mean annual rainfall amounts 1.675 mm at Nyabessan and 1.695 mm in the entire Ntem catchment (Olivry, 1986). Mean annual temperature is around 25°C.

The interior delta is situated on the northwestern swell of the Congo Craton just above the peneplain step slanting down towards the Atlantic Ocean. The Ntem crosses the transition from the higher and lower peneplain levels over several cataracts. The river drops 200 m in altitude over a distance of 40 km. Below the peneplain step the Ntem crosses the coastal surface ('surface côtière') which is about 150–200 km wide (Segalen, 1967).

The peneplain step is characterized by an escarpment ('l'escarpement du Ntem', Kuete, 1990) at an altitude of about 1.000 m a.s.l., formed by charnockites of the Precambrian basement. The escarpment subdivides the migmatitic gneisses of the 'surface intérieur' from those of the coastal plain (Kuete, 1990; Maurizot, 2000).

Three main strike directions of geomorphologic linear structures were identified using radar data: NNE–SSW, NE–SW and ENE–WSW. These show parallels to the southern Atlantic opening direction, the CVL, and the West African rift system, which is represented by the Sanaga Fault north of the study area (cf. Ngako *et al.*, 2003).

17.2.2 Fluvial lineaments

The river network is influenced, and locally oriented parallel to structures in the bedrock.

Within the interior delta the main NE flow direction of the Ntem River deviates at regular distances to a WSW flow direction. Extrapolation from the linear river reaches highlights lineaments with a length of up to several kilometres. The regular spacing suggests the existence of a step fault system which interrupts and deviates river flow, resulting in a change in river gradient and the formation of an interior delta defined by channel splitting (see figure 2 and 3). The occurrence of thick gravel layers in and below the river channel within this 'interior delta' suggests the influence of neotectonic controls on the present river morphology. The orientation of assumed lineaments could be the result of remobilization of Precambrian structures (Eisenberg, 2007; Runge *et al.,* 2006).

The rose diagram of the river channel orientations (see figure 3) highlights the subordinate influence of the structural lineaments discussed in 17.2.1 and the Ntem's wide arc of channel orientations around the main north-westerly flow direction. The V-shaped valley, within which the Ntem River is confined below the interior delta where it crosses the peneplain step, is linear over a distance of about 30 km, with only two deviations. Neotectonic structural control is assumed to have influenced the river along this reach.

17.3 RESULTS

Downstream of the interior delta, three sites with gravel layers were identified. The matrix and clasts of these gravels are cemented by iron and manganese hydroxides. The expression *gravel fanglomerate* will be used in the description because the matrix contains both fluvial

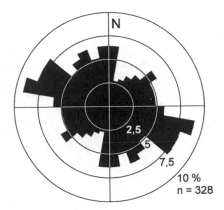

Figure 3. Rose diagram showing channel orientations within the Ntem River network in the study area.

rounded clasts and angular locally-derived detrital material. Three transverse valley profiles are presented to define the different characters of the gravel occurrences (Figure 4).

They reflect the situation directly below the waterfalls of Menvé'élé (A–B), at the beginning of the V-shaped valley after all branches flow into it (C–D), and about 7 km in a southwest-ward direction (E–F). The sketch is based on SRTM data in combination with GPS tracks and detailed field observations (see figures 4 and 10).

17.3.1 Transect A–B

The 'Chutes de Menvé'élé' location can be reached via a pathway from Nyabessan along a deep gorge-like valley (increasing up to 20 m in depth). Four waterfalls lie on the opposite

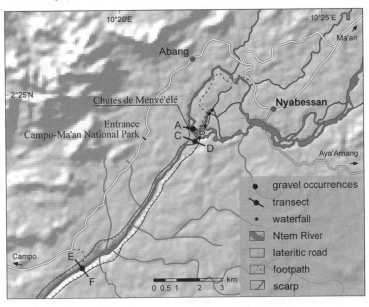

Figure 4. Detailed map of the V-shaped valley situated below the waterfalls of (Chutes de) Menvé'élé showing the location of the three valley profile transects, also shown in Figure 10.

Figure 5. Situation below the 'Chutes de Menvé'élé' waterfalls showing the gravel bed (thickness indicated), recent gravel cover, and the location of the fanglomerate which have been sampled. The Ntem River flow direction is outlined by the curved white arrow.

slope, flowing into this valley (see figure 4). The pathway branches off towards the WSW and leads to a large gravel plain below the Menvé'élé waterfalls. Directly below the falls, the Ntem has incised into the gravels and formed an inner and outer-bank. The gravels comprise rounded to subangular clasts, mixed with angular rock fragments, coated by a dark brown to black patina. Examination of the patina composition using X-ray diffractometry (RFA) identified the coating as amorphous, presumably organogene material. The occurrence of goethite and manganese hydroxide was verified as the likely source of the black surface colouring.

At two sites gravel samples were taken: (1) at the inner bank, influenced by the ever drenching humidity of the waterfall, and (2) below a small tributary fan reaching into the Ntem's valley (see figure 5 for exact locations).

(1) A small hollow space below the sample permitted the removal of the fanglomerate from the gravel bed. The sectioned sample shows rounded clasts which are overlain by poorly sorted subangular to angular quartz grains. Underneath the clasts a ~2 cm reddish to dark brown, iron and manganese crust has formed. The coarse grains of the upper layer do not occur in the crust.

(2) The second sample contains subangular to rounded pebble clasts although the primary components of the fanglomerate are angular to subangular fragments of up to 10 mm in diameter. Two areas of angular quartz particles cemented by a bright brown matrix occur within the hardened gravel sample (see figure 6).

17.3.2 Transect C–D

At the northeastern border of the V-shaped valley a basin was formed where all branches of the Ntem channel amalgamate. The basin is surrounded by linear stretching steep valley sides of the outcropping bedrock with NNE and ENE striking, along which one branch of the Ntem flows into the basin.

At the western flank, the tributary, which is fed by the waterfalls of Menvé'élé, flows into the basin across several bedrock steps, each a maximum of 2 m in height. At the

Figure 6. Cut sections through samples of the gravel fanglomerates taken below the waterfalls of Menvé'élé.

junction of the NNE and ENE aligned outcrops, the southernmost Ntem channel branch crosses a 5 m step to reach the basin (see figure 4).

About 150 cm above the dry season water level, a fringe of cemented gravels was observed at the ENE flank. At a similar elevation, gravels crop out on a small island within the basin. At two sites samples of the fanglomerate were taken: (1) from a small 50 cm wide cavity in the outcropping basement where the gravel was protected from erosional processes, and (2) directly from the gravel fringe.

(1) The sample has a breccia-like character (Figure 7 (1)). Most components are subangular, and are cemented by a fine-grained, brown to dark brown material. The parent rock is unitary weathered and it seems to originate from the same region.

(2) The fanglomerate comprises well rounded pebble clasts and angular rock fragments. The spaces are characteristically unsorted and coarse grained (Figure 7 (2)).

Figure 7. Cut sections through the gravel fanglomerates which were sampled from the basin at the northeastern end of the V-shaped valley (transect C–D).

17.3.3 Transect E–F

The Ntem river bed within the steep sided valley is marked by irregular beds of cemented gravel matrix and basement outcrops. In a few sections the gravel matrix is superficially eroded. Moreover, in several erosional hollows loose gravels have accumulated. On those gravels remnants of a dark patina give a hint to their origin from the hardened gravel matrix. Apart from those loose gravels, red to brown coloured sand banks exist behind large blocks of outcropping basement.

The fanglomerate sample which was collected at this site is characterized by two different areas (Figure 8: divided by the white line). The right side comprises strongly corroded, subangular to well rounded clasts which are cemented by a silty matrix. On the left side of the sample, the clasts are primary pebble clast fragments, cemented in a hardened matrix with coarse grained particles of quarzitic rock and some ferruginous pisoliths up to three millimetres in size. In this area there is also a clast fragment which is partially coated by a ferruginous coating or patina (see figure 8 inset). The photomicrograph from a thin section of this patina shows an alternating stratification of bright and dark brown coloured layers.

17.4 DISCUSSION

A fanglomerate is a cemented accumulation of unsorted material which contains rounded as well as clastic pebbles which formed during an arid climate. During excessive precipitation periods, the material originating from slopes will be loosed and accumulated at the valley bottom as a mud deposit (Murawski, 1992). This definition has to be adjusted to the study area. A tectonically triggered process has to be considered. The term 'alluvial fan' seems to be also suitable for the unsorted and cemented deposits. Blair and McPherson (1994) refer to all global climatic regimes in which the alluvial fans occur. However, some morphologic features which are related to this term beneath the texture of a fan deposit, particularly the V-shaped valley downstream the interior delta can not be applied to the study area

Figure 8. Cut section through the gravel fanglomerate from the Ntem River bed within the steep-sided valley (transect E–F).

This valley is defined by a distinct linear structure, offset by a fault which intersects the valley at a 40° angle, some 20 km SW of the waterfalls (Figure 2). The structure of this deep valley presumes a former fault or graben which was probably neotectonically remobilized below the Ntem escarpment in the undulating peneplain. The gravel layers, blocks of unweathered basement rocks and residuals of a ferricretic crust of up to 2 m in diameter below several waterfalls flowing into the northern end of the gorge (transect A–B, figure 9(1)), on the margin of the basin (transect C–D), and also downstream within the Ntem River gorge (Figure 9(2)) are indicators of a sedimentary regime that differs from the present one. This is probably the result of the neotectonic movement on fault planes.

Nearly all fanglomerates are composed of well rounded clasts, subangular rock fragments, and angular quartzitic particles. The rounded gravel indicates long-distance fluvial transport whereas the rock fragments and quartzitic clasts probably represent local debris flow deposits. The gravel composition suggests that an initial alluvial gravel layer accumulated along the NE–SW oriented, structurally controlled river gorge. Landslides triggered by the tectonic activity related to transported angular rock fragments into the valley. These deposits were admixed with alluvial gravels prior to cementation by the ferruginous matrix. These processes are documented by the different clast shape and textural composition exhibited in the cut rock section (transect E–F, figure 8).

The former extent of the gravel layer in the basin (transect C–D) is marked by small gravel remnants cemented to the outcropping rock. The gravel is also preserved in cavities and on an island within the basin. Directly below the waterfalls of Menvé'élé (Figure 5) and at the valley site the distribution of the incised gravel remnants suggests a wider distribution within the valley. The tectonic forces responsible for the formation of the linear gorge and the presumed former distribution of the gravel level are sketched in figure 10. The figure shows the position of the present Ntem river bed, incised into the cemented gravels (transect E–F).

The interpretation of the phases of gravel deposition and cementation can be interpreted in the context of climatic changes; the cementation of the fanglomerates occurred during

Figure 9. Boulders (marked) below several waterfalls across the pathway leading to the waterfalls of Menvé'élé (1) and close to some cataracts at the southernmost branch of the Ntem River (2).

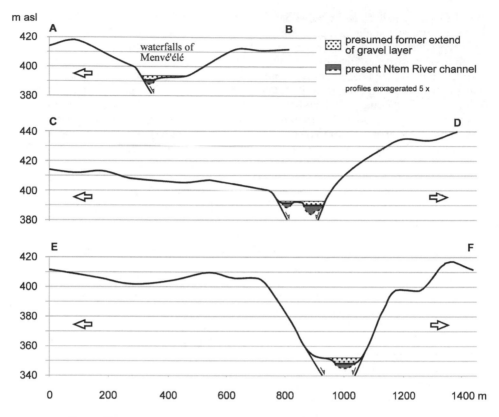

Figure 10. Valley profiles at transects A–B, C–D, and E–F based on SRTM DEM data (obtained from the GLCF server of the University of Maryland) and detailed field observations. The positions of the transects are shown in figure 4.

recurrent moistening and desiccation of the gravel layer (personal communication with Prof. G. Brey, Department of Mineralogy, Frankfurt; cf. Alexandre and Lequarré, 1975). This process can be triggered by a different seasonality with long lasting dry seasons and only short but intensive rainy seasons.

The patina which partially coats a gravel fragment (see figure 8) also suggests cementation under climatic conditions that differed from that experience by the area today. The stratified ferruginous coating could point to longer dry and shorter rainy seasons and to distinct changes between arid and humid climatic cycles. The bright layers indicate a phreatic groundwater regime, whereas the darker layers are presumed to have formed under vadose groundwater conditions (personal communication with Dr. G. Ries, Department of Mineralogy, Hamburg). It is speculated that this stratification could have formed in response to the short-term alternation between arid and humid phases which were postulated by Taylor *et al.* (1993) for the Late Pleistocene interglacials.

In this context the quartz particles found below the waterfall also have to be discussed. They have been fixed in a bright brown matrix and hardened within it (Figure 6: second sample). The preservation of ferruginous coatings on some clasts within the cemented gravel points to polyphase cementation and reworking of gravel beds in the river valley.

The first accumulation of pebbles along the structurally controlled course of the Ntem River occurred >50 kyrs BP—due to the sediment ages in the interior delta—and was associated with the formation of the step fault which lowered the river gradient and formed the interior delta (see section 17.2.2). There are no absolute age determinations on the gravels or cement although Colin *et al.* (2005) have shown that such materials can be dated using the ^{40}Ar/^{39}Ar technique. The age of the hardening of the deposits is assumed to be Late Pleistocene.

17.5 CONCLUSIONS

In the rain forest of SW Cameroon the Ntem River has formed an interior delta above a peneplain step slanting down towards the South Atlantic. It is assumed that the delta was formed by neotectonic activity due to the accumulation of pebbles along the structurally controlled Ntem River course. Organic horizons found in the sediments of the interior delta were dated to (uncalibrated ^{14}C-) ages of up to nearly 50 kyrs BP (see Sangen, this publication).

The clasts provided by the evolution of the study area were accumulated and cemented both within and below the interior delta along the V-shaped valley. The cut sections of the samples show indications of landslide processes which can be related to tectonic activity. Besides, the deposits are marked by fluvial deposition, and also cementation processes of a climate different from todays. By finding datable material in the fanglomerates, different cementation phases could be identified and correlated to the palaeoenvironmental work which was done on the lacustrine sediments and deep sea fan deposits of Central Africa (Giresse *et al.*, 1994; Marret *et al.*, 1998; Nguetsop *et al.*, 2004; Zabel *et al.*, 2001; Zogning *et al.*, 1994).

ACKNOWLEDGEMENTS

This research was supported by grants of the German Research Foundation (Deutsche Forschungsgemeinschaft, DFG), Bonn, within the DFG-Research Unit 510: 'Ecological and cultural change in West and Central Africa' (sub-project RU 555/14-3; Prof. J. Runge). I am grateful to the Cameroonian colleagues from the 'Département de Géographie' at the 'Université de Yaoundé I' (Prof. Dr. M. Tsalefac, Dr. M. Tchindjang) for assisting and supporting fieldworks in 2004, 2005, and 2006, and to Dr. B. Kankeu from the MINRESI/ IRGM (Ministère de la Recherche Scientifique et l'Innovation/Institut de Recherches Géologiques et Minières). I thank Prof. G. Brey and Dr. H. Höfer from the Department of Mineralogy at the University of Frankfurt and Dr. G. Ries from the Department of Mineralogy at the University of Hamburg for the inspiring discussion, and finally Dr. G.A. Botha for several critical and constructive remarks on this text.

REFERENCES

Alexandre, J. and Lequarré, A., 1975, Essai de datation des formes d'érosion dans les chutes et les rapides du Shaba. In *Geomorphologie dynamique dans les regions intertropicales. Colloque de Géomorphologie de l'Environnement dans les régions intertropicales, Lumbumbashi*, edited by Alexandre, J. (Presses universitaires du Zaire), pp. 279–286.

Bessoles, B. and Trompette, R., 1980, *Géologie de l'Afrique. La chaine panafricaine «Zone mobile d'Afrique Centrale (partie sud) et zone mobile Soudanaise».* (Orléans: Mémoires du Bureau de recherches géologiques et minières, Èditions B.R.G.M.).

Blair, T.C. and McPherson, J.G., 1994, Alluvial Fan Processes and Forms. In *Geomorphology of Desert Environments*, edited by Abrahams, A.D. and Parsons, A.J. (London: Chapman & Hall), pp. 354–402.

Burke, K., 1996, The African Plate. *South African Journal of Geology*, **99 (4),** pp. 341–409.

Burke, K. and Whiteman, A.J., 1973, Uplift, Rifting and the Break-up of Africa. In *Implications of continental rift to the Earth Sciences*, edited by Tarling, D.H. and Runcorn, S.K. (London and New York: Academic Press), pp. 735–755.

Colin, F., Beauvais, A., Ruffet, G. and Hénoque, O., 2005, First ^{40}Ar/^{39}Ar geochronology of lateritic manganiferous pisolites: Implications for the Palaeogene history of a West African landscape. *Earth and Planetary Science Letters,* **238**, pp. 172–188.

Daly, M.C., Chorowicz, J. and Fairhead, J.D., 1989, Rift basin evolution in Africa: the influence of reactivated steep basement shear zones. In *Inversion tectonics*, edited by Cooper, M.A. and Williams, G.D. (London: Special Publication of the Geological Society of London), **44**, pp. 309–334.

Déruelle, B., N'ni, J. and Kambou, R., 1987, Mount Cameroun: an active volcano of the Cameroon line. *Journal of. African Earth Sciences,* Vol. **6**, No. **2**, pp. 197–214.

Eisenberg, J., 2007, Neotektonische Prozesse und geomorphologische Entwicklung des Ntem-Binnendeltas, SW-Kamerun. *Zentralblatt für Geologie und Paläontologie, Teil 1.*

Fairhead, J.D., 1985, Preliminary study of the seismicity associated with the Cameroon Volcanic Province during the volcanic eruption of Mt. Cameroon in 1982. *Journal of African Earth Sciences*, Vol. **3**, No. **3**, pp. 297–301.

Giresse, P., Maley, J. and Brenac, P., 1994, Late Quaternary palaeoenvironments in the Lake Barombi Mbo (West Cameroon) deduced from pollen and carbon isotopes of organic matter. *Palaeogeography, Palaeoclimatology, Palaeoecology*, **107**, pp. 65–78.

Goudie, A.S., 2005, The drainage of Africa since the Cretaceous. *Geomorphology,* **67**, pp. 437–456.

Hori, N., 1977, Landforms and superficial deposits in the coastal region of Cameroon. In *Geomorphological studies in the forest and savanna areas of Cameroon. An interim report of the Tropical African Geomorphology Research Project 1975/76,* edited by Kadamoura, H. (Hokkaido), pp. 37–52.

Kadomura, H., 1977, Some aspects of Geomorphology in the Forest and Savanna Areas of Cameroon, with Special Reference to South–North Variation. In *Geomorphological studies in the forest and savanna areas of Cameroon. An interim report of the Tropical African Geomorphology Research Project 1975/76,* edited by Kadamoura, H. (Hokkaido), pp. 7–36.

Karner, G.D., Driscoll, N.W., McGinnis, J.P., Brumbaugh, W.D. and Cameron, N.R., 1997, Tectonic significance of syn-rift sediment packages across the Gabon-Cabinda continental margin. *Marine and Petroleum Geology*, Vol. **14**, No. **7/8**, pp. 973–1000.

Kuete, M., 1990, *Géomorphologie du plateau sud-Camerounais à l'ouest du 13°E.* Thèse de Doctorat d'Etat, Université de Yaoundé 1.

Latrubesse, E.M., Stevaux, J.C. and Sinha, R., 2005, Tropical rivers. *Geomorphology,* **70**, pp. 187–206.

Leturmy, P., Lucazeau, F. and Brigaud, F., 2003, Dynamic interactions between the Gulf of Guinea passive margin and the Congo River drainage basin: 1. Morphology and mass balance. *Journal of Geophysical Research*, Vol. **108 (B8)**, **2383**, pp. 8/1–13.

Lucazeau, F., Brigaud, F. and Leturmy, P., 2003, Dynamic interactions between the Gulf of Guinea passive margin and the Congo River drainage basin: 2. Isostasy and uplift. *Journal of Geophysical Research*, Vol. **108 (B8)**, **2384**, pp. 9/1–19.

Marret, F., Scourse, J.D., Versteegh, G., Jansen, F.J.H. and Schneider, R., 1998, Integrated marine and terrestrial evidence for abrupt Congo River palaeodischarge fluctuations during the last deglaciation. *Quaternary Research*, **50**, pp. 34–45.

McCarthy, T.S., 1993, The great inland deltas of Africa. *Journal of African Earth Sciences*, Vol. **17**, No. **3**, pp. 275–291.

McCarthy, T.S., Green, R.W. and Franey, N.J., 1993, The influence of neo-tectonics on water dispersal in the northeastern regions of the Okavango swamps, Botswana. *Journal of African Earth Sciences*, Vol. **17**, No. **1**, pp. 23–32.

Moore, A.E. and Larkin, P.A., 2001, Drainage evolution in south–central Africa since the breakup of Gondwana. *South African Journal of Geology*, Vol. **104**, pp. 47–68.

Murawski, H., 1992, *Geologisches Wörterbuch*, (München: Enke).

Ngako, V., Affaton, P., Nnange, J.M. and Njanko, Th., 2003, Pan-African tectonic evolution in central and southern Cameroon: transpression and transtension during sinistral shear movements. *Journal of African Earth Sciences*, **36**, pp. 207–214.

Nguetsop, V.F., Servant-Vildary, S. and Servant, M., 2004, Late Holocene climatic changes in West Africa, a high resolution diatom record from equatorial Cameroon. *Quaternary Science Reviews*, **23**, pp. 591–609.

Olivry, J.-C., 1986, *Fleuves et rivières de Cameroun*, (Paris: Éditions de l'ORSTOM).

Runge, J., Eisenberg, J. and Sangen, M., 2006, Geomorphic evolution of the Ntem alluvial basin and physiogeographic evidence for Holocene environmental changes in the rain forest of SW Cameroon (Central Africa)—preliminary results. *Zeitschrift für Geomorphologie N.F.*, Suppl.-Vol. **145**, pp. 63–79.

Segalen, P., 1967, Les sols et la géomorphologie du Cameroun. *Cahier ORSTOM, Série Pédologie*, Vol. **V**, Nr. **2**, pp. 137–188.

Suh, C.E., Ayonghe, S.N. and Njumbe, E.S, 2001, Neotectonic earth movementd related to the 1999 eruption of Cameroon mountain, west Africa. *Episodes*, **24**, pp. 9–12.

Summerfield, M.A., 1985, Tectonic background to long-term landform development in tropical Africa. In *Environmental Change and Tropical Geomorphology*, edited by Douglas, I. and Spencer, T., pp. 281–294.

Summerfield, M.A., 1991, *Global geomorphology: an introduction to the study of landforms,* (Essex: Longman).

Summerfield, M.A., 1996, Tectonics, Geology, and Long-Term Landscape Development. In *The Physical Geography of Africa*, edited by Adams, W.W., Goudie, A.S. and Orme, A.R., (Oxford: University Press), pp. 1–17.

Taylor, K.C., Lamorey, G.W., Doyle, G.A., Alley, G.A., Grootes, P.M., Mayewski, P.A., White, J.W.C. and Barlow, L.K, 1993, The 'flickering switch' of late Pleistocene climate change. *Nature*, **361**, pp. 432–435.

Thomas, M.F., 1998, Lanscape sensitivity in the humid tropics—a geomorphologic appraisal. In *Human Activities and the tropical rainforest. Past, Present and Possible Future*, edited by Maloney, B.K. (Dordrecht: Kluwer), pp. 17–47.

Tooth, S. and McCarthy, T.S., 2004, Anabranching in mixed bedrock-alluvial rivers: the example of the Orange River above Augrabies Falls, Northern Cape Province, South Africa. *Geomorphology*, **57**, pp. 235–262.

Zabel, M., Schneider, R., Wagner, T., Adegbie, A. T., De Vries, U. and Kolonic, S., 2001, Late Quaternary Climate Changes in Central Africa as inferred from Terrigenous Input in the Niger Fan. *Quaternary Research*, **56**, pp. 207–217.

Zogning, A., Giresse, P., Maley, J. and Gadel, F., 1997, The Late Holocene palaeoenvironment in the Lake Njupi area, west Cameroon: implications regarding the history of Lake Nyos. *Journal of African Earth Sciences*, **24/3**, pp. 285–300.

CHAPTER 18

Effects of forest clearings around Bangui: urban floods in densely populated districts of the Central African capital

Cyriaque-Rufin Nguimalet

Laboratoire de Climatologie, de Cartographie et d'Etudes Géographiques (LACCEG), Département de Géographie, Faculté des Lettres et Sciences Humaines, Université de Bangui, Bangui, Central African Republic CNRS-Laboratoire de Géographie Physique « P. Birot », Meudon, France

ABSTRACT: Located on the right bank of the Oubangui River, the site of Bangui belongs to a typical transitional forest-savanna mosaic in Central Africa. The setting up of a French military station in 1889, that later became the capital of the Central African Republic, marked the beginning of severe ecosystem changes causing the degradation of 'urban' soils. The consequences of such changes, reinforced by the spatio-demographic growth and the process of an 'anarchistic urbanization', contributed to a modification of the hydrodynamic functioning of soils as well as to changed morphodynamic processes of the urban rivers. The study examines the spatio-temporal dynamics of landscapes in formerly forested, now urban areas, and highlights the effects of vegetation clearings and surface denudation which have contributed to urban floodings and inundations. Hydrological behaviour of urban rivers and channel morphology of streams was strongly modified within the urban areas making these drainage systems rather artificial. They collect rain fall water from roads and roofs and when certain threshold values of swelling are reached, it contributes to flooding disasters in Bangui. The uncontrolled setting up of houses in squatter-like districts with corrugated sheet roofs contributes to this process as a planned discharge system is lacking. The issue of the effect of rain risk in these districts was studied with regard to what had happened in 1973 and also most recently between 1998 and 2005. It shows that the extension of the urban areas with poor town planning and the lacking of regular cleaning of the main sewers, often blocked by the household's refuse, mineral remains and grass growing in the channels, increased significantly the risk of flooding at the town of Bangui.

18.1 INTRODUCTION

The impact of forest vegetation as a protection against the streaming of floods are beyond doubt although the results are sometimes mitigated. In fact, a forest can significantly reduce the effects of floods, but its effects may be unless in cases of exceptional downpours (Guilcher, 1979). Similarly, a controversial effect as far as streaming is concerned as streaming according to soil quality was studied by Bravard and Petit (1997). Quoting Cosandey (1995), these authors explained that a forest basin with a high evapotranspiration capacity would let less flows into the discharge system. Reversely, a significant deforestation would contribute to modifications of the rate flow in terms of not only the basic rate of flow but also of flow in times of flood. These various hydrological manifestations of the basins tend to bring light to the effects of forest clearings.

Clearings constitute a complex chain of processes induced by humans like agriculture, logging, earthworks and others which gradually strip the surface. The site of Bangui (Figure 1) belongs to the Congo–Guinean sector of humid dense forest. It was particularly covered by semi-evergreen forests surrounded by more open savannas (Boulvert, 1986). The establishment of the French colonial military post on the right bank of the Oubangui River in 1889 initially started the great clearings which led to the denudation of the urban ground. These were achieved because of the combination of two factors: one was related to the urbanization and the colonization of neighbouring areas of the former town centre, the second factor has to do with farming, wood cutting by sculptors and by charcoal burners as well as for the production of firewood.

This study analyses the effects of clearings around Bangui in terms of risks relating to urban floods. Because clearings change surface run-off conditions and induce the regularity of the rise of fast run-off, indeed, urban floods result of channels or other linear drainage flow to evacuate rain fall water. They display a lack of prevention of streaming floods. In developed countries these urban floods result from the extension of chain networks. It is a fact that in order to control rain water, one must develop the chain system with regard to the degree of exposure to these risks (Doulens, 1992). This problem basically affects

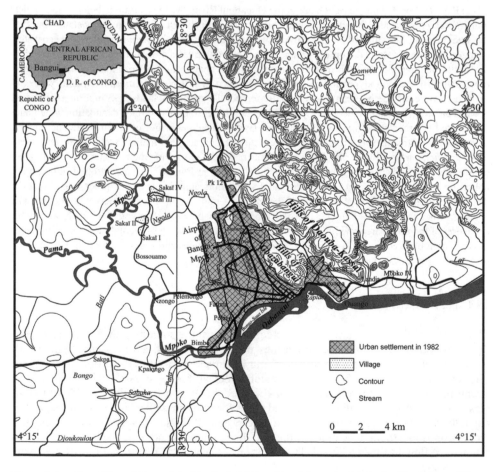

Figure 1. Location of Bangui, the capital of the Central African Republic, on the right bank of the Oubangui River.

the so called 'popular districts' or slums established in the old drained, shallow swamps which are often subjected to the concentration and the invasion of water. Is it possible to get more evidence in which way the loss of forest cover has modified the hydrology of the urban rivers? Can one explain how local hydropluviometric constraints affect socio-urban processes? Can it be admitted whether the problems of run-off in Bangui, that is, the rain risk from the clearings of the urban area or from an inefficient management of the territory by public authorities (see Nguimalet, 2004).

18.2 METHODS AND DATA

The former extension of the forest and the stream network at Bangui and its surroundings was reconstructed by interviews, oral reports and by the interpretation of aerial photographs, landscape photographs and other written reports and scientific references. A diachronic study based on aerial photo-interpretation and the use of the present day data on the development of the urban landscapes, in terms of space and demographic growth, approached many processes of land cover dynamics especially the clearings of the soil. Thus, the modification of landscapes was concerned to the way how humans shaped the Bangui area. Mainly, it was focused on the migration and installation of people into the urban basin-slopes at Nguitto, Ngoubagara and Ngongonon. In connection with urban streams Ngoubagara is the only one to be used for measurement for the rate of flow carried out by Kokamy–Yambéré (1994), but these measurements were not continuous due to the unpredictability of precipitation. This fact complicated the comparison modes of the early clearings and those of today. The data of flow were not continuous either and limited themselves only to the rainy seasons in the case of Ngoubagara. However, the daily rain fall records (amount and duration) were not to characterize the rain risk at Bangui, in relation to the process of 'anarchic planning', at the origin of populated districts. The data did not raise, however, of the pluviograms and showed the total time of a rain fall event. This made it difficult to correlate amount, duration and frequency of the rain fall, and the consistency of the floods in the vulnerable zones of the city (Table 1). Thus, the urban floods which result from this are systematically analysed on historical data (1973) and recent precipitation records (1998–2005). Field work together with such data, even the information was coming from the assessment of damage documented in official reports, of various known disasters led to the mapping of urban areas subjected to the frequence occurrence of these phenomena, giving evidence of the consequences of vegetation clearings of the urban surface related to the presence of people at Bangui since 117 years.

18.3 FOREST VEGETATION AND OUBANGUI RIVER
DYNAMICS BEFORE 1889

Dybowski (1891, in Boulvert 1989) noticed that the site of Bangui corresponds to the 'northern edge of the equatorial forest', (5°00' 45"N, 19°15' 45"E) located close to a big 'loop' of the Oubangui River. The site belongs to the Congo–Guinean sector with humid dense rain forest, which is mainly composed by semi-evergreen forest species like *Triplochiton scleroxylon*, *Terminalia superba* and *Celtis* spp. and inserted savanna patches mainly characterized by *Hymenocardia acida*, *Annona senegalensis* and *Bridelia ndellensis* (Boulvert, 1986). This forest belongs to the 'Lower Lobaye District' (Figure 2) which experienced recent clearings in the urban areas. There is evidence for this modification in vegetation cover at the Gbazabangui hills close to Bangui's downtown area: here one still can discover remnants of natural forest in the west SW and NW of the city beyond

Table 1. Rain fall characteristics causing severe floods at Bangui.

Date	Amount of rain fall (mm)	Duration (hours and minutes)	Intensity (mm/h)	Outline assessment	Affected sectors	Observations
1983	?	3. 00	?	100 destroyed houses	Districts Gbaya-Dombia, Lower Kotto, Ngouciment I and II	Overflow of river beds of Ngoubagara and Ngongonon
28.05.1998	37,2	2. 46	13,45			
12.08.1998	2,5	?	?	Stranding of the	?	
25.08.1998	2,2 0	? —	? —	piscicultural basins, floods		
18.08.2004	43,5	8. 16	5,26	12.096	Flood of the	
23.08.2004	13,5	5. 05	2,66	affected	bridge of the 8th	Flood
14.09.2004	28	2. 12	12,73	persons and	arrondissement.	
21.10.2004	70,7	4. 24	16,07	5 cases of		
01.11.2004	62,8	8. 02	7,81	death	'The most violent event'	
27.06.2005	90	5. 45	15,65			
06.08.2005	64,2	?	?	14.517		
07.08.2005	4,2	10. 22	0,41	victims and	Pétévo and	general
08.08.2005	0,5	?	?	2.607	Gbangouma	flooding of
11.08.2005	20,6	2. 42	7,63	collapsed	and others	Bangui
12.08.2005	9,3	2. 30	3,72	houses,		
13.08.2005	30,9	3. 30	8,83	1.082 totally destroyed and 1.525 damaged		

the Mpoko River and its tributaries (Pama River). However, often it is already degraded by slash and burn agriculture and logging of trees. Other relicts are still recognizable in the state-owned forest at Yangana located on the northern edge of the city. Riparian forests, developed under the influence of long lasting inundations with *Uapaca heudelotii* and *Cathormion altissium* can be observed west of the city centre in the surroundings of the tributaries Mpoko and Pama. Wooded river marshes with *Raphia* palms are typical for these swampy zones into which the city has moved. Gras dominated savanna patches caused by regular and long lasting flooding composed of *Paspalum commersonii*, *Jardinea congolensis* and *Vetiveria nigritana* are locally observed. Finally, on rocky outcrops and bare ground often *Euphorbia darbandensis*, *Sanseveria trifasciata* and *Aloe schweinfurthii* are located in the north eastern and mainly hilly areas of Bangui.

This spatial variety of flora already shows the exuberance of the former vegetation before the foundation of Bangui. The indigenous population settled in isolated villages

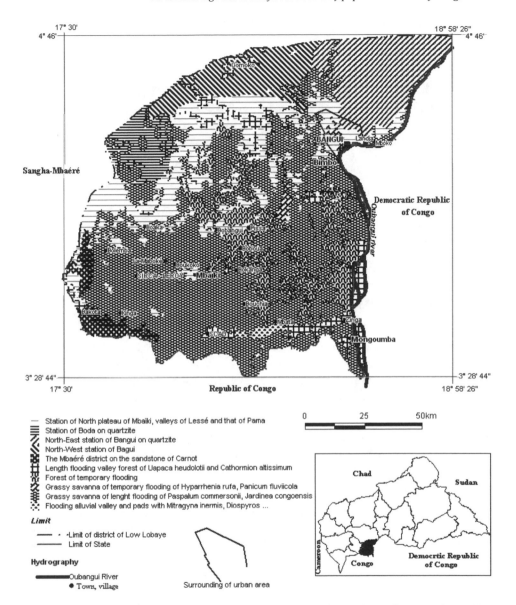

Figure 2. The forest district of the Low-Lobaye including the urban area of Bangui (according to Boulvert, 1986).

on the banks of the Oubangui and along the Mpoko and Pama River and they regularly 'visited' these forests. Even if this pre-colonial presence of humans at the site led to a local deforestation, it can't be compared to the transformations that followed up and especially to the level that it reached today in affecting fluvial dynamics. Thus, in the landscape context of before 1889, the streams of the site had been more or less in a 'symbiosis' with their environment (Figure 3), because the density of plant cover did not cause a deterioration of the general flow conditions. Owing to the fact that some streams are already naturally marshy on some of their sections of the river bed and having their springs uphills, the rising of floods should be synchronous, slow and progressive in both cases,

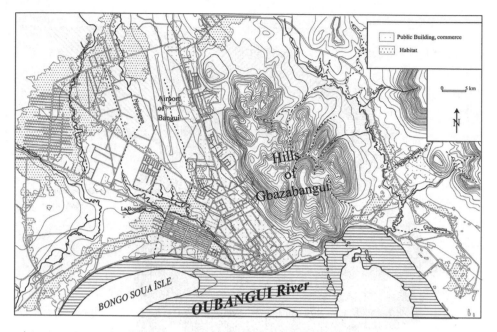

Figure 3. The initial drainage area network in the surroundings of Bangui (Nguimalet 2004, modified).

i.e. an intercommunication was set at the marshy river zones already, even if the presence and the direct influence of humans on the ecosystems was generally weak.

18.4 DEGRADATION OF RIVER FLOW CONDITIONS AND LANDSCAPE DYNAMICS AT BANGUI IN THE 20TH CENTURY

Bangui and its surroundings as it can be observed today is the result of recent clearings (Figure 2). According to oral reports (Nguimalet, 1999), in the 1940ies and 1950ies the Ngoubagara valley for example was covered by herbaceous vegetation (reed) from up to downstream. There were no trees at all. Only on the current entry of Dédengué V district (Fouh), there was a small forested 'island' where later, after deforestation a laterite quarry was set up. The Kalakpata marsh, located behind the former UCATEX factory, was similarly surrounded by a forest until 1982 which was evidenced by one remaining Kapok tree. Deforestation in and at the surroundings of Bangui was accelerated by a strong increase of charcoal production. In the area behind the police station of the 8th 'arrondissement' towards the recent district of the flooding disaster, a former savanna woodland coverage was obviously destroyed by wood cutting and for the making of charcoal. However, the sector where the international airport of Bangui–Mpoko was built in 1966–1968, had since long already been covered by an edaphic grass savanna with *Borassus* palms due to the unfertile ferrallitic soils with frequent outcrops of lateritic crusts.

One realizes that the clearings and the earthworks occurred at Bangui since then are mainly controlled by several factors: urbanization and domination of surrounding spaces, slash and burn agriculture and felling of trees that apply various activities as sculpturing, charcoal and firewood production. For setting up public buildings and roads the territory was largely cleared and formerly vegetated ground became bare surface subsequently.

The construction of the first airstrip of Bangui between 1920 and 1925, today located within the city on the Avenue of the Martyrs, between Koudoukou Avenue and the University of Bangui, caused a reduction in dense forest that is estimated in total up to 600 ha (Boulvert, 1989).

Parallel to this, the supplying of the city of Bangui with all kinds of staple caused a change of its hinterland, evidenced by the establishment of farms and plantations, timber logging and stone extraction on Gbazabangui Hills in the urban contact zone. This anarchic conquest, concerning the craggies of western sides of the Gbazabangui Hills, makes currently vulnerable the marshy sector in the city plain, with the strong streamings and the erosion whose matters concentrate there. The processes on these slopes engaged by the crushing of the quartzited pinnacles, then their fall, destroying the undergrowth and creating the corridors of evacuation for rain water (Nguimalet, 2004). This clearly underlines the deterioration of the natural flow conditions (rising) to which the urban rivers are 'subjugated'. In the local urban environment, indeed, the influence of humans is the most important factor of stream exacerbation and erosion (Nguimalet, 2004). Vegetation cover rather plays a stabilizing role for the environment, but its degradation amplifies the streaming and erosion in highly sensitive sites like that of Bangui. In the sequence of changing landscapes from forest to clearings and finally to bare ground, other factors are contributing to the current vulnerability of the city due to the precipitation risk are observable; these are land pressure on marshy grounds, easily flooded areas, and the run-off by sheet aluminium roofs acting like 'contributive surfaces' to the hydrological system.

In December 2003 Bangui's population was counted by a census up to 622.771 inhabitants with an average population density of 200 people per km² and probably even more in the slum-like 'popular districts'. Fast urban population growth creates a need for building plots, obliges the city-dwellers of all social groups to fold back itself on the insanitary grounds. On marshy, episodically flooded river beds or open 'free' spaces of river channels (Figure 4). This attitude highlights the non rational occupation of the land. Houses with corrugated sheet roofs act as impermeable surfaces, similar as the hydromorphic grounds, which amplify the concentration of precipitation in the old marshy dip, river branches and the collecting streams. These stream organizations have moreover a capacity of the channels reduced by the fact of the coming out of herbs and the rejection of the household's refuse. The combination of these facts enlarges the constraints of drainage when the density of drainage is low as in Bangui, where the surface waters compose of the temporary ponds as well in the inhabited courses in the not built spaces. These side-effects on the fluvial processes thus legitimate the regularity of the flood phenomena.

18.5 EFFECTS OF CHANGED LANDSCAPES ON THE HYDROLOGY OF URBAN RIVERS

Because of the absence of longer run-off records for the capital site, two case-studies (Tixier, 1953; Kokamy–Yambéré, 1994), however, allowed to determine the changes occurred in the streams as a direct response to the modification of flow at the Bangui site. The Ngola stream basin located in the periurban area (Figure 1), is partially covered with savanna woodland and shows a relatively accentuated relief, while the Ngoubagara stream drains a basin slope section, mainly bare of vegetation due to urban influence.

The tentative calibration of the gauging-station at the Ngola River (basin area of approx. 27 km²) is given in ten gaugings for flows from 0 to 6 m³ corresponding to heights of water of 0,50 m to 1,65 m (Tixier, 1953). The strongest increase was observed by Kokamy–Yambéré (1994) for 'weak' heights of the rains are approximately 15 m³ for 3 m height of water. But these gaugings were difficult because of the speed at the same

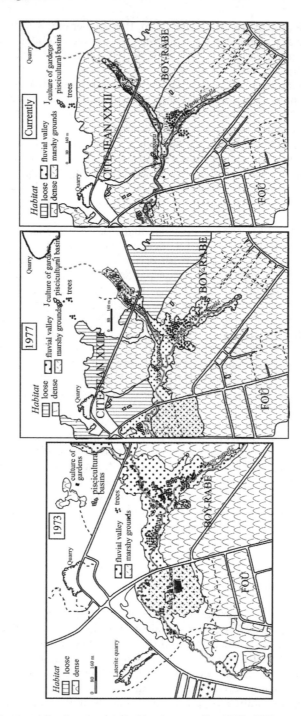

Figure 4. Multitemporal flooding events in the Ngoubagara valley (1973, 1977, recent) with traces of destruction (according to Nguimalet, 2004; modified).

of rising and the fall related to a stressing of the gradient of the stream in the area of the hills. It is in this suddenness rising-fall that the Ngola is connected to the Ngoubagara. In the other case when it occurred on August 22. 1967, a downpour of 171,5 mm measured at ORSTOM Bangui station (4°26' N, 18°32'43" E) an extraordinary rising of the Ngola stream followed, with an overflow of the bridge at km 12 (Figure 1). The downpour had a strong heterogeneity in space (Callède, 1969, 1970; Callède and Arquisou, 1972), unfortunately there are no data on the intensity of this downpour to get a better understanding of the event. One can suppose that precipitation of a secular frequency (Gumbel distribution) to a multi secular frequency (Gauss distribution) worked out by Nguimalet (2004), would generate more significant damage on an elementary catchment, stripped like that of the Ngoubagara stream. These also makes it possible to predict the behaviour of drainage basins with respect to exceptional rainy episodes covered by natural vegetation, in particular those of the site of Bangui during the early days of the city some hundred years ago. Moreover, if one considers the height of the bridge consecutively invaded by the Ngola River to the downpour of August 22, 1967, one would assess the flow of rising at 30, even 45 m³, that is to say 2 to 3 times more than that of the risings determined by Tixier (1953).

With regard to the Ngoubagara stream, the measurements carried out by Kokamy–Yambéré (1994) aimed to evaluate the effect of the degradation on slopes to fluvial dynamics in an urban environment. As the Ngongonon River is equally draining the center of Bangui (Figure 3), it is a genuine catchment of pluvial water.

However, precipitation measured during this period showed only small amounts in the mid 1990ies. Measurements proved downstream slightly higher run-off figures (station of SICA III: 11,7 km², with 11 m³/s and 40 cm water height) than upstream, respectively at the station of the Hôpital de l'Amitié (3,9 km²) and at the station at Miskine Market (6,8 km²). Thus for 40 cm on the scale, the flows do not reach, even 2 m³/s this appears normal, considering the principle of run-off increase in the flow to the downstream. The analysis of the work of Kokamy–Yambéré by Nguimalet (2004) noticed a very spontaneous response of the Ngoubagara basin to rainy episodes, posing a real hydrological problem as the river changed in time and space during the installation of the city. According to some hydrogrammes of discharge rising (Figures 5a, b), the hydrological mode of Ngoubagara would be currently closer to that of the torrents in mountain areas or that of wadis in semi-arid zones, although this river flows in a humid tropical environment.

Certainly, large floods are observed when precipitation is strong, however, even in the rainy season the river bed could be dry if there is no rain fall. This hydrological process results in the importance of the short time run-off, which is one of the main causes of severe urban floods posing cleansing issue.

18.6 URBAN FLOODING, RAIN RISK AND URBANIZATION

The recurrent phenomena of floods, classified as pluvial risk, occurs under local geomorphological conditions (marshy grounds, rise in the water level) triggered by the different types of ground occupation by the local population. The hydraulic 'vagaries' of floods on Bangui's surface is indeed mainly dependent on the pluviometric characteristics (e.g. height, duration and frequency of precipitation), but also on the hydromorphic grounds which developed in relation to the existence of a tectonic ditch directed to the north–south of which the depth with more than 200 m is not reached in the sector of the Bangui–Mpoko international airport. These variables associated with the shapes of ground occupation, in particular the installation of houses in the precarious zones as stream marshes, underline the vulnerability of urban slum districts. Two case studies of the Bangui flood disasters

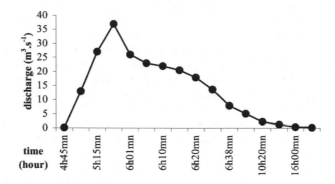

Figure 5a. Hydrograph of flood in the station of Miskine (14.10.94).

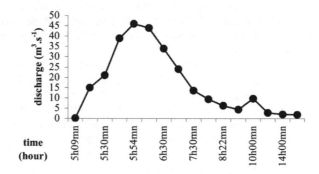

Figure 5b. Hydrograph of flood at the station of SICA III (14.10.94).

happened in 1973 and 2005 are used to illustrate the spatio-temporal effects of the inundations in these districts.

18.6.1 Flooding disasters cases studies: September 1973 and August 2005

The context in which the selected disasters occurred do not resemble, although they both occurred in the rainy season. In 1973 Bangui's urban population was of 277.815 inhabitants (Cursus, 1975), which is 2,5 times less than in 2005.

The first event took place in 1973 as the consequence of 64,4 mm of rain fall (probable frequency once a year) registered from September 12–13 on urban ground that already was saturated by earlier fallen precipitation in July, August and September (Figure 6).

Many districts were quickly flooded in relation to the medium basins and downstream of Ngoubagara and Ngongonon (Figure 7). The identification of the hazard victims in the different districts was difficult due to their fuzzy limits and their anachronic membership. By these floods 14.499 persons were affected corresponding 5,22% of the urban population at that time, including 2.905 homeless. 3.754 cabins were flooded, 733 of them destroyed. The main part of the results obtained underline a close correlation between flooded surfaces and the unplanned settling in marshy zones.

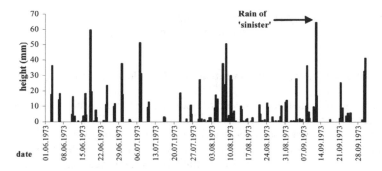

Figure 6. Daily rain fall records at Bangui between June and September 1973.

In 2005 several rainy days from August 6–12 with respective heights of precipitation reaching 64,2 mm, 4,2 mm, 0,5 mm, 1,8 mm, 20,6 mm, 9,3 mm and 30,9 mm (Table 1), generated severe floods in the sectors of Pétévo to the SW and Gbangouma in the east of the city (Figure 7). The flood of August 2005 was given publicity in the media and affected 14.517 city-dwellers and 2.607 houses, from which 1.082 houses were totally destroyed and 1.525 houses damaged according to the local Red-Cross report.

In comparing the two flood events, the last incident appears, however, minor. This could be due to the high level of the concentration of habitation and persons in these

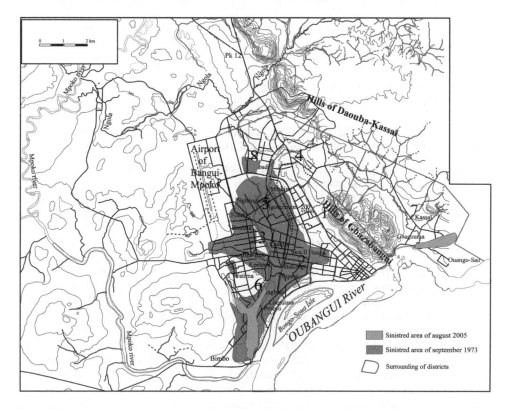

Figure 7. Spatial extent of the urban floods of September 1973 and August 2005 (according to Nguimalet, 2004, modified).

vulnerable sectors (marshy shoals and flooding zones) and because of the land pressure. According to oral sources the number of the victims was over-estimated.

18.6.2 Relationship between rain events and urban floods at Bangui

The establishing of clear trends and links between rain fall events, their duration, daily amounts and the process of a subsequent flooding event did not appear to be simple (Table 1). Run-off related floods seem to be frequent in Bangui, however, as there is no reliable data logging system available, it is still difficult to establish the link between various processes (run-off and rain fall) on the basis of empirical data. The data that show the balance of urban floods are scattered and difficult to obtain (Table 1), also as it lacks of a good collaboration with the appropriate services.

This does not forecast simplicity in the interpretation of the factors of causality of the flooding phenomena. Large difficulties in doing so is due to a strong spatial variability of rain fall in the urban areas of Bangui which has as a reference only one (!) pluviometric station. Thus, the highest variability of precipitation will play a significant role in the hydrological behaviour of streams, notably on a local scale (Cosandey and Robinson, 2000). The latter authors point out that the density of pluviometric stations generally will not contribute to a better understanding of the reaction of surfaces affected by storms or by heavy rain fall. This would show that excessive events such as those in 1998 or 2005 (Table 1) could be harmful. However, data on the total duration of a rain fall event which one used do not contribute in correlating height-duration-run-off events indeed, whereas the cases of overflows of the anthropic hydraulic gaps and the 'natural' channels are most notable in Bangui.

18.7 SIGNIFICANCE OF REOCCURRENCE OF URBAN FLOODS TO THE SLUM DISTRICTS OF BANGUI

The repetition of urban floods in the slum districts of Bangui are due to a combination of three types of physio-geographical, socio-economical and social vulnerability:

(1) The physio-geographically caused vulnerability is initially based on the geomorphological contrast between hill and plain (Figure 8). It forms a steep upstream zone consisting of two successive fault scarps, transferring discharge to an extended downstream plain where thick sediment layers had been accumulated over a Precambrian half-graben structure, where the city of Bangui is located. This geologically controlled graben plain has a slow drainage and is therefore mainly marshy with hydromorphic soils. Other contributive surfaces maintaining a seasonal increase of the ground water table, which are originated by the rising of rainy flows. Precipitation (e.g. amount and duration of rain fall) determines the regularity of the floods because Bangui is located in the fringe zone of the humid tropical climate with approximately 1.500 mm annual rain fall distributed over 120 days.

(2) Socio-economic vulnerability can be summarized with the insufficiencies concerning rational use of the occupied ground. The exposed housing areas and the low socio-economic level together with the fast increase of population density even triggers the degradational risk (Nguimalet and Boulvert, 2006). Land pressure exerted by the city-dwellers on these grounds acquired with water is a function of the level of their incomes. That explains the quality of the dwellings which is

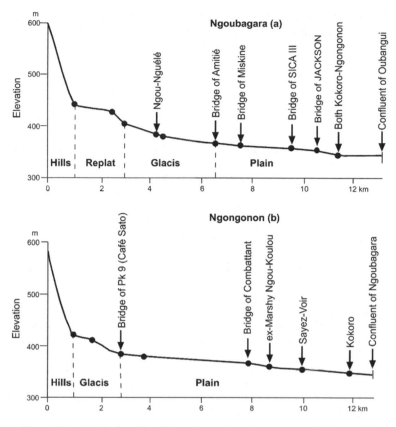

Figure 8. Longitudinal profiles of the urban streams Ngougagara and Ngonngonon (Nguimalet, 2004, modified).

generally built in adobes and exposed to the collapses because of a strong moisture of the grounds and water soaking in these districts with spontaneous population development. Whereas for these grounds, one would need more means to build a comfortable dwelling. Indeed, the limited financial resources do not allow many city-dwellers to offer healthy grounds, currently quite expensive in urban zones. Also in spite of the 'natural' weakness of the drainage, efforts were not made to clean the marshy zones strongly occupied by the housing and to equip them with the adequate works of rain fall drainage, nor of the structures of collection of the household refuse. These reasons attesting for the insufficiency of the adjustments and the non urbanization of the totality of Bangui city, constitute the real factors of vulnerabilities of the people living in these districts without infrastructure and of spontaneous extension. The major stakes in the populated districts (dwellings, drownings or losses in human lives, water pollution of the traditional wells, diseases of hydrous origin) are indeed the result of non urbanization of these areas subject to repetitive invasions of surface waters.

(3) The social vulnerability of the city-dwellers is in fact their aptitude to minimize the effects of an episode of flood, i.e. time, but of course the materials average and others, which they put to find a new balance if one regards the production of flood as a disturbing phenomenon of the urban system. This capacity of response

is at the time individual and collective owing to the fact that the consequences of floods will be felt individually and through what all the urban community records. Thus, the analysis of the insufficiency of installations and the non urbanization of the districts subjected to the passages of water of floods reveals the incapacity of urban populations as well as their basic structures to face a natural disaster.

18.8 CONCLUSIONS

Forest vegetation plays an undeniable role in protecting the ground against run-off. Its cut, even its grubbing on that or space scale of a basin-slope, shows a disturbance of the natural system or a change of the conditions of surface flow (denudation), of which the effects result in an increase in frequency of streaming directly interconnected with the stripping of formerly vegetated ground. The site of the city knew same processes in the sense that urban space became a suffering 'impluvium' of adequate discharge system from where regularity of the urban floods, in particular in the not urbanized districts. Thus, the specificity of the tackled questions is noticed in three fundamental points. The first refers to the precariousness of the site characterized by a geomorphological hill-plain contrast and the concentration and quasi-permanent stay of water in the densely populated areas of the city. The second point considers the grubbing of the urban surface according to the space growth and the activities of humans which again strip and waterproof the ground. Insufficient drainage networks in these 'sealed' areas and the dwellings built in the transitional zones to the marshy and often flooded zones do not support at all the circulation of water. The regularity of the urban floods is also the consequence of an ineffective water management by the authorities. This combination of facts and processes determines the risk of urban floods at Bangui whose consequences are noted in the slum districts established on the insanitary marshy grounds. Urban floods are common in Bangui and hold more with the absence of rationalization of the occupation of the land with the deficiency of the drainage systems and with the obstruction of linear flow by a rejection not controlled of the household's refuses by the population that with the morphodynamic energy of the rain fall generates them. The insufficiencies in terms of improvement and urban management justifies the dynamism of this hydrological risk in the districts of the city installed in the plain area where rain water concentrates, coming from the zones of hills, piedmont, glacis or slopes.

This risk should be less if the indicators of dense forest, covering the hills with the stepped slopes around Bangui, are not attacked by the forms of urban conquest and periurban development, which induced the degradation of the surface subjected to strong transfers of material (run-off water and suspensions) into the plain where the city is located and where the flood disasters are frequent.

ACKNOWLEDGEMENTS

The author wishes to express his appreciation for the work done in particular by Gabriel Danzi, Dean of the Faculty of Letters and Humanities for his rereading of the initial text in french, Hermann Jean-Michel Djimapo, for the english version, and Bertin Ouakanga who carried out some of the figures from the numerical data base on phytogeography.

REFERENCES

Boulvert, Y., 1986, Carte phytogéographique de la République centrafricaine à 1: 1000.000ᵉ. *ORSTOM éd., Coll. Notice Explicative*, **104**, (Paris: ORSTOM), p. 131.

Boulvert, Y., 1989, *Bangui 1889–1989. Points de vue et témoignages*. SEPIA, Saint-Maur (France), p. 311.

Bravard, J.-P. and Petit, F., 1997, *Les cours d'eau. Dynamique du système fluvial*, (Paris: Armand Colin), p. 222.

Bureau central du recensement (BCR), 2005, La République centrafricaine en chiffres. In *Résultats du Recensement Général de la Population et de l'Habitation (RGPH) de décembre 2003*, (Bangui: République centrafricaine, Ministère de l'Economie, du Plan et de la Coopération Internationale, Direction Générale de la Statistique, des Etudes Economiques et Sociales), p. 23.

Callède J., 1969, Premiers résultats des mesures effectuées à la station bioclimatologique de Bangui. Période 1964–1968. *Centre ORSTOM de Bangui* (RCA), p. 39.

Callède J., 1970, Note sur une averse d'intensité exceptionnelle observée à la station bioclimatologique de l'Orstom à Bangui. *Centre ORSTOM de Bangui (RCA)*, (in edit), p. 6.

Callède, J. and Arquisou, G., 1972, Données climatologiques recueillies à la station bioclimatologique de Bangui pendant la période 1963–1971. *Cahiers ORSTOM, Série Hydrologie*, IX, **4**, p. 26.

Cosandey, C., 1995, La forêt réduit-elle l'écoulement annuel? *Annales de Géographie*, 581–582, pp. 7–25.

Cosandey, C. and Robinson, M., 2000, Hydrologie continentale, (Paris: Armand Colin), p. 360.

Dourlens C., 1992, Le ruissellement pluvial et la ville. In *La ville au risque de l'eau. La sécurité dans les secteurs de distribution de l'eau et de l'assainissement pluvial*, edited by Dourlens, C. and Vidal-Naquet, P. A., (Paris: Editions l'Harmattan, Collection Logiques Sociales dirigée par D. Desjeux), p. 128.

Guilcher, A., 1979, *Précis d'hydrologie marine et continentale*, (Paris: Masson), p. 344.

Kokamy-Yambéré, S., 1994, Erosion et dégradation des collines de Bangui: impacts sur le milieu urbain. Bilan des trois années d'étude (1991, 1992, 1993). *Centre ORSTOM, Laboratoire de Géologie et d'Hydrologie*, Bangui (RCA), p. 76.

Nguimalet, C.-R., 2004, *Le cycle et la gestion de l'eau à Bangui (République centrafricaine). Approche hydrogéomorphologique du site d'une Capitale africaine*, Thèse doctorat, Géographie, Aménagement et Urbanisme, Université Lumière Lyon 2 (France), Lyon, p. 447.

Nguimalet, C.-R., 2005, Systèmes fluviaux et risques hydrologiques dans les quartiers (urbains) de Bangui. *Poster, Promotion de la culture scientifique et technique, «Caravane de la science» en Centrafrique*, 15 juin-27 août 2005, Bangui.

Nguimalet, C.-R. and Boulvert, Y., à paraître, Les crues historiques de l'Oubangui et leurs implications dans la gestion de l'eau du site de Bangui, capitale de la République centrafricaine. *Colloque international «Interactions Nature-Sociét: analyse et modèles», La Baule, France, 3–6 mai 2006.*

Tixier, J., 1953, Etude des crues sur un petit bassin de la région de Bangui (Oubangui-Chari). *Annuaires Hydrologiques de la FOM*, pp. 29–47.

CHAPTER 19

Non Woody Forest Products (NWFPs) and food safety: sustainable management in the Lobaye region (Central African Republic)

Cyriaque-Rufin Nguimalet, Marcel Koko and Félix Ngana

Laboratoire de Climatologie, de Cartographie et d'Etudes Géographiques (LACCEG), Département de Géographie, Faculté des Lettres et Sciences Humaines, Université de Bangui, Bangui, Central African Republic

Arsène Igor Kondayen

Institut Supérieur de Développement Rural (ISDR), Université de Bangui, Bangui, Central African Republic

ABSTRACT: This study highlights the importance of rain forest products aside of commercial timber exploitation summarized under the term 'Non Woody Forestry Products (NWFPs)' and their contribution for the local population's food safety. Processes of deforestation and the permanent use of these products by a rapidly growing number of people in and around the forests endanger the extinction risk of several non woody forest resources in the future. Stakeholders involved into these processes are local gatherers, hunters and fishermen (Pygmies), sometimes also autochthonous people migrating from outside into the forest, and small traders who buy, export and sell the non woody forest products on local and supraregional markets. Constraints relating to their exploitation are considered as well as the perspective of domestication of NWFPs. Thus, field surveys were conducted among the villagers of an area to inventory them and specify their interest and the role and function, NWFPs play for social balance within the Ngotto, Kenga and Kela Forests in the SW of the Central African Republic (Lobaye). Inventories of NWFPs allowed to classify two groups: (1) products of vegetable origin, and (2) products of animal origin. Their role as an important complement to food supply and an important source of income for the rain forest population was evidenced.

19.1 INTRODUCTION

The Lobaye River is a right bank tributary of the Oubangui River that runs from the northwest to the southwest. It is named after a densely forested area in the southwest of the Central African Republic. The Lobaye region belongs to the Mbaéré district located mainly on the Carnot sandstone and to the Lower Lobaye that belongs to the Congo–Guinean vegetation sector characterized by humid and dense rain forest vegetation, often more particularly composed by semi-evergreen forest (Boulvert, 1986). This forest vegetation is of an interest as far as the 'Non Woody Forest Products' (NWFPs) are concerned. The setting apart of the local population and recent processes of deforestation, interconnected with land degradation and the gradual disappearance of tropical biotopes, makes it necessary to study NWFPs in more detail. Several papers already lined out the great importance of

NWFPs and there is a broad focus on the topic by different scientific studies as shown by Falconer (1990), Ogden (1990), Wickens (1991) and FAO (1992).

According to FAO (1992), Non Woody Forest Products are all biological products and materials aside of timber which can be taken from the natural forest ecosystems. For example the collection of certain plants, insects, worms and hunting of small animals which can be directly used by the local population as food or to be economically marketed, or in addition, which also may have a social, cultural or religious background for humans. NWFPs can be plants for use either by humans or for animal consumption, or combustibles for the production of beverages, remedies, fibres, biochemical products and others. As Ogden (1990) and Falconer (1990) point out that the nutritional value of food produced by the forest is high, because it contains proteins, biogenic salts and vitamins. Their dietetic value was also observed. Thus, the purpose of this work was to determine the contribution of NWFPs in terms of food safety, and to identify the constraints relating to their durable exploitation in perspective to perpetuate the activity of gathering and collection by domestication.

Access to food coming from the rain forest becomes an important topic for the sustainable management of the forest resources by various stakeholders (gatherers, hunters, fishermen, and intermediates). The dynamic of deforestation concerned with the socio-demographic and economic pressure, will probably cause a gradual disappearance of NWFPs which earlier provided a supplementary amount of food and also generated considerable income for the local populations. It has to be examined how to conserve and to sustain these natural resources against the threats of deforestation.

19.2 METHODS

First some documentary sources were consulted to get quantified data on vegetation cover and species distribution in the study area. The phytogeographical map of Boulvert (1986) covers the district of the Lobaye and two other districts with rain forest which will be the showcase for this study: the Mbaéré district located on the Carnot sandstone and the Lower Lobaye district. Two types of surveys were jointly carried out among the villagers in particular those of Mongo, Mbata, Zomia, Bakota, Ngola, Ngotto, Bossabo, Kenga and Ibengué villages. The first type of survey consisted in an inventory of the non woody forest products, which allowed to establish a typology and reference collection, according to the relative range in local consumption compared to the agricultural production. This showed that NWFPs are not exclusively consumed locally, but that they are also marketed, e.g. bringing an income to the population. It was also tried to emphasize the importance of these food products for the 'social balance' in all points of view. Importance was also attached to the periodicity of collection or harvesting, of the frequency and the volume of the setting apart and re-entering in the local consumption as well as in marketing. A dividing space of the NWFPs, moreover, was accomplished by taking into account the local and respective ecological conditions of the different types of products.

A second examination focused on the prices and the economic values of the NWFPs in periods of abundance or in periods with shortages of certain products. This method resulted in estimating the profit margins in both cases which the actors of this commercial chain of the products could gain, according to their degree of importance concerning consumption. Data processing with EXCEL software allowed the extrapolation of the results, going into the direction of the clarification of the stakeholders' demands for food safety and the valorisation of NWFPs by a rational management of the forest resources that hopefully support the development of the local populations of Ngbaka, Issongo, Boffi, Gbaya, Ali, Monzombo and Aka Pygmies.

19.3 REGIONAL INVENTORY OF NWFPs

Even if large parts of the southern Central African Republic are covered by rain forest, forestry is not so well developed. Until recently the official administration regarded NWFPs to be of no importance, even if these resources had been subjected to an irrational exploitation. However, in less populated areas like in the Lobaye (12,8 inhabitants/km^2 in 2003), the gathering (picking/collecting), hunting and fishing still plays a great role in the subsistance economy of many locals. This should cause a collective conscious of the problem with the aim to manage properly these resources.

The inventoried species in the forests of Ngotto and Kéla are locally classified according to criterias of category such like usage, exploitation and availability. Beyond considering the role of certain NWFPs in consumption such as the liana *Gnetum* sp. and several caterpillars, one presents them according to their geo-ecological zones in the forest districts of Mbaéré and the Lower Lobaye. It could be shown that the trees with caterpillars are more frequent in the Lower Lobaye, whereas *Gnetum* sp. can be equally found and collected in both of the two forest districts in question, and also in the rest of the country when similar ecological conditions are prevailing.

19.3.1 Products of vegetable origin

The list of edible products from the forest is long (Table 1). Besides numerous fruits of trees and lianas there are many other exploitable small trees and certain herbaceous plants with a great variety of spontaneous tubers, edible leaves, and even certain lianas which, if once cut into sections of one meter length, let run out a sap with a pleasant flavour the peasants arc using for refreshing themselves. Some mushrooms and oleaginous nuts of *Elaeis guineensis* and others could be also added to table 1.

Apart from the forest block in the southern part of the country, there are the two edible liana species *Gnetum africanum* and *Gnetum bucholzianum*. The genus of *Gnetum* consists of about 30 species—the majority of them are tropical lianas, including the two ones in Central Africa (Bois, 1967). *Gnetum africanum* which is locally called 'kökö' has larger leaves than those of *Gnetum bucholzianum*, or 'kali' in Issongo dialect spoken at Mbaïki. *Gnetum africanum* occurs abundantly in dense and dry forest as well as in secondary forests and in gallery forests, whereas *Gnetum bucholzianum* grows preferably on the edges of the gallery forest and close to fallow areas. Both liana species are locally consumed as a vegetable food; *Gnetum bucholzianum* tastes a little bitter and the dark green leaves of *Gnetum africanum* are known for their great nutritional value.

There are also many wild tubers (yam) in the forest, however only *Dioscorea praehensilis*, called 'Esuma' by the Pygmies, is regularly used as a starch food supply. It preferably grows in sandy soils and furnish under dense forest and gives tubers in about 3 to 4 meters depth (Bahuchet, 1978). The oil palm (*Elaeis guineensis*), or 'Kaya' in Issongo dialect, occupies vast surfaces within the Lobaye. Their natural distribution and dissemination is probably caused by birds or cynocephali and bats. One also finds quite regularly many planted *Elaeis* species in the outskirts of the villages.

19.3.2 Products of animal origin

Snails, caterpillars, termites, honey and game constitute the main NWFPs of this category. Of course, the caterpillars are more valued of which these is a half-dozen of the two types in the area of Lobaye (Table 2).

Table 1. Principal fruit trees and wild fruits of the forest in the Lobaye (Ngana, 2004).

Family	Scientific name	Local name	Characteristics	Period of maturation
Annonaceae	*Anonidium manni*	Mobaï	Sweetened fruit, large globulous bay with maturity	June–September
Anacardiaceae	*Antrocaryon klaineanum*	Magnégné	Sweetened fruit	June–September
Burseraceae	*Canarium schweinfurthii*	Fatou	Fruit with the drawn taste on avocado	May–June
Burseraceae	*Dacryodes edulis*	Safou	Fruit with the drawn taste on avocado	May–August
Herbaceae	*Aframomum*	Modoko	Fruit with the acidulous taste	January–March
Apocynaceae	*Landilphia owariensis*	Gift (Ndri) Malo	Fruit composes globulous, sweetened taste	June–September
Moraceae	*Myrianthus arboreus*	Modiki	Fruit composes globulous with sweetened taste with maturity	June–September
Sapotaceae	*Aningeria robusta*	Mboulou (Gbidiguilis)	Fruit composes globulous, red with maturity and sweetened	July–August
Sapotaceae	*Aningeria altissima*	Monoungo-unoungou	—	—
Palmaceae	*Borassus aethiopum*	Kpô	Fibrous and sweetened fruit	January–March
Irvingiaceae	*Irvingia excelsa*	Payo	Almond	June–September
Irvingiaceae	*Irvingia gabonensis*	Mossombo	Almond	June–September
Annonaceae	*Monodosa myristica*	Mokpakpi	Edible fruit	April–July
Moraceae	*Treculia africana*	Pushed	Edible seeds	June–September

By the method of the specific abundance (sign, indication) or 'IPA' (Bastin, 1996), it was evidenced that many mammals and other species of tropical fauna from small antelopes to pachyderms and primates are present in the forests of Ngotto and Kéla (Table 3).

Table 2. Edible types of caterpillars in Lobaye (Ngana, 2004).

Scientific name	Local name	Wood specific	Characteristics
Imbrasia oyomins	Mboyo	*Entandraphragma cylindricum* (Sapelli)	Small black prickle thorns on the body, a little large is cut, green color. Collected between July and August
Imbrasia willobscura	Guéguéret	*Triplochiton scleroxlon* (Ayous)	Small pivots yellow cuts average, red–yellow color. Collected between July and August
Psendathea discription	Kangha	*Entandraphagma candollei* (Kossopo)	Big size, red color. Collected between July and August
Anaph sp.	Ndossi	*Triplochiton scleroxylon* (Ayous)	Small size, yellow color, small yellow hairs. Collected to leave September
Imbrasia epimethea	Soungan	*Ricinodendron hendeloti* (Essessang)	Cut average, black color, without pivots. Collected between July and August
Imbrasia tluncata	Bangha	*Petersianthus macrocorpus* (Essia)	Large, without prickles, color yellow–black cuts. Collected from August

Table 3. Some animal species inventoried in the Ngotto and Kéla Forest.

French/English name	Scientific name	Local name	Order
Gorilla	*Gorilla gorilla*	Sore	Primate
Chimpanzé	*Pantroplodyte*	Soumbou	Primate
(Dwarf) elephant	*Loxodonta africana*	Nzokou	Mammal
(Dwarf) buffalo	*Syncerus caffernaus*	Mboko	Mammal
Blue Céphalophe	*Philantomba caerula*	Dengbè	Mammal
Porc-épic	*Atherurus africanus*	Ngomba, Nguenzé	Rodent
Potamogale	*Atamogale velex*	Papassi	Rodent
Viper	*Bitis gaborica*	Pélé	Reptile
Goat-sucker	*Macrodypteryx*	Yeko	Bird

19.4 PERIODICITY OF COLLECTION

The Lobaye is rich in NWFPs which are collected by the local population all around the year without any concern of renewing them. Therefore, it is of importance to study the periodicity of collection of these resources. A 'picking' calendar of NWFPs in Kenga village was established for the years from 2001 to 2003 in order to show the sharing out of work of collection over these periods according to seasons (Figure 1). The production of edible plants and products (mushrooms, tubers and fruits), insects (termites, caterpillars, bees/honey) is related to several climatic factors of which rain fall, drought and photoperiodicity will have various effects on the inventoried species (Tables 1 and 2).

Obviously there does not exist a 'real' biological driven cycle (Figure 1) concerning the 'harvesting' of natural resources interconnected with the ongoing influence on the ecosystem. However, there are some rules as for example the picking up of termites is done the night after the first rain storms which normally mark the onset of the rainy season. The take-off of swarms is quite spontaneous, but it can be stimulated by a lamp-torch placed in a hole dug around the termite mound. As the gathering of the caterpillars is concerned it is carried out from mid-July to mid-September. During this period the peasants often leave their villages for staying in the forest.

This calendar also underlines (Figure 1) that there is practically no month without rain, even if there is an alternation of rainy and dry seasons over the year. Annual average precipitation comes up to 1.600 mm. The rainiest months are from July to October, whereas the shorter dry season is observed from November to February. The photoperiodicity is marked by a relative vegetative rhythm of the NWFPs, which is however not clearly seasonal. The changes of vegetation do not coincide with the changes of seasons when

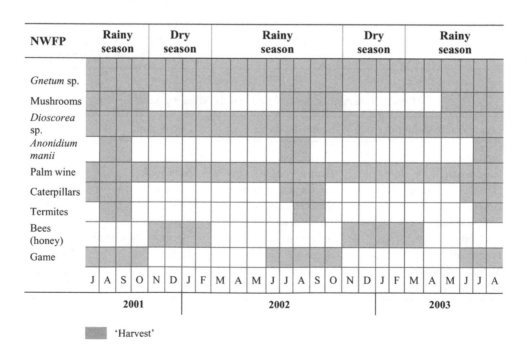

Figure 1. Calendar of gathering non woody forest products in the village of Kenga (2001–2003).

talking on the species flowering according to these changes. Because it is necessary that such observations cover a longer period of time, it is indeed difficult to catalogue species of which the rhythm of flowering, foliation, defoliation and repending of fruits corresponds to one or another season. In addition, one retains that the easily locatable and frequent species as *Gnetum africanum* and *Gentum bucholzianum* are those which are mainly collected by the villagers. *Gnetum africanum* is an evergreen plant with continuous foliation, therefore it is collected all the year round. *Gnetum africanum* and *Gentum bucholzianum* also develop tubers similar to those of yams some twenty centimetres below the surface. These tubers can survive over many years even if the predominant vegetation and the lianas already had been superficially cleared (Tisserand, 1961). According to the last author, the peasants of Kenga consume these tubers like spontaneous yams during times of food shortage which underlines the general importance of *Gentum* for food safety.

The wild yam is a climbing plant in forest and savanna environments with two main species: (1) the soft, directly edible, after short cooking, one, and (2) the bitter one, that has to be soaked longer in water and cooked because of its toxicity. As for the wild fruits (Table 1), *Irvingia gabonensis* and *Afrostyrax lepidophyllus* are among the most known. The main wild fruits are picked in the Lobaye area in the rainy or the dry season after the passage of fires. Because of deforestation some of the small trees and the crawling plants bearing these edible fruits, are in the process of disappearing around the towns of Mbaïki, Boda and Mongoumba. The trees with caterpillars (Table 2) are also affected. Thus the caterpillars are increasingly far away from the village ground. The pickers are occupied during all the period of the caterpillars corresponding to the rainy season (Figure 1). However, it is at this period that also corn is sown, groundnut and manioc. One thus understands why one doubts the agricultural abilities of the forest people.

The palm tree gives place to two kinds of oil (palm and cabbage tree), whose extraction can absorb half of the working day of the women. However, the wine drawn from the palm tree, locally called 'Kangoya', is consumed in the Lobaye and elsewhere, and it brings back money to its producers when it is sold. Its collection is done either by the uprooting of the tree, or by the incision of the base in the buds on the sharp tree.

The fauna of the region plays also a great role for food supply (bush meat). Leclerc (1999) pointed out that the forests and savannas covering this ground are obviously not 'only a vegetable space'. Apart from the totems that one avoids, the richness in fauna involves an organization of hunting for the food needed (Table 3). Wild fauna and game is regularly hunted by the population with the aim of nourishing themselves, but also to sell the meat to urban markets.

19.5 EXPLOITATION AND VALORIZATION OF NWFPs

As shown above, there are numerous types of NWFPs (Tables 1, 2 and 3). The traditional knowledge about plants, edible fruits, mushrooms and caterpillars is handed over in the form of an oral tradition from generation to generation. Parents already teach their children how to recognize *Gnetum*, wild fruits and yams. The collecting of *Gnetum africanum*, *Ataenidia conferta* as food, and Marantacea leaves for wrapping is practised in all seasons. *Gnetum africanum* occupies a prime place in the traditional food supply with the Central African rain forest. 'Kökö' is often cooked together with game, dry caterpillars, fresh or dry mushrooms or smoked fish. The leaves are eaten raw or boiled in salted water during 10 to 20 minutes. Picking is done by hand or by using a knife while avoiding tearing off the roots. The leaves are rolled up and cut out of fibres before preparation starts. Its food advantage justifies its omnipresence on almost all the markets of the country. Thus, to insist on the stakes of a food safety that the NWFPs represent, one was interested in the

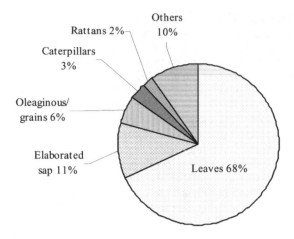

Figure 2. Percentage of different NWFPs aside of agriculture at Ngotto village.

frequency of use and their economic values, as well as their contribution in the income of households.

According to Piri-Dejean (1997), the frequency of use of NWFPs coming from the Ngotto Forest shows a great demand for the leaves (68%), mainly those of *Gnetum*, followed by the sap with 11% (Figure 2). Caterpillars for export represent with only 3,2% a small proportion. The latter low figures for caterpillars could be explained by the perpetual, even annual nature of the collection process of the leaves of *Gnetum* and/or Marantaceae leaves with the detriment of caterpillars that mainly takes place only during three months of the year, particularly in rainy season.

As for the contribution to the income of households (economic value), it appears that NWFPs contribute significantly to the balancing of the level of the income in the studied areas. They induce a degree of added value according of the actors (gatherers, pickers, 'wholeralers', 'retailers', consumers) with consequent benefit. Tchatat (1999) estimated in Kanaré, a village of the ecological 'Reserve of Ngotto', which the NWFPs contribute up to 35% to the income compared to the agriculture which is still the main source of income with 51% (Figure 3). That means that if a villager gains an income of 100.000 FCFA per

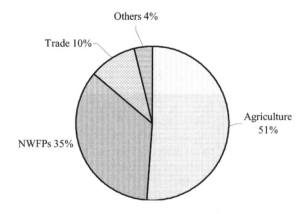

Figure 3. Structure of the income of households at Ngotto village.

annum or Euros 152, the share of the NWFPs would be 35.000 FCFA equivalent with Euros 53 what is already a contribution of second order in the income. From an economic viewpoint the prices of NWFPs vary according to the zone of collection, supply and demand. In periods of abundance or shortage, the individual prices can fluctuate about 100 to 200% as well as in the zones of collection on the urban markets. But it is in the urban centres that the level of speculation does not make too much account for the fluctuation in prices. On each level of exchange, the actors draw a benefit from the product thanks to the degree of transformation and added value (Piri-Dejean, 1997). Moreover, the economic values of NWFPs usually past on the urban markets, are determined by the given quantities collected (Table 4).

Gnetum leaves give in comparison an economic low value, whereas the saleswomen of these leaves still realize a 150 and 200% benefit, even more compared to the purchase price of the wholesalers. Wild yams tubers and honey expose average values in comparison. Nevertheless, if one estimates that with equal volume of the NWFPs considered in table 4, honey would have the highest economic value followed by *Gnetum africanum*.

The Pygmy women are very experienced in hunting and often it is practised by the widows of a community. The space designed for hunting is well organized. It is divided into three suburbs around the dwellings (Figure 4). One notices that there are small hunting spaces surrounding directly the housings. Hunting in this zone normally lasts half a day, and is concentrated to smaller game (rats, cibissi, pangolins). It is practised individually by the installation of traps or collectively by the use of nets. The preys are distributed between the hunters according to a known rule accepted by all of the participants. The gathering during hunting for the nets is a proof of the spirit of solidarity and mutual aid which exists within the communities, in spite of the dispersion of their habitat. It consolidates the family bond because the meat the husbands brought back is quickly prepared by the women and girls. The other great space for hunting is far away from the houses taking one or two days of walking to get there; normally a kind of camp site is set up for each clan or family. The valid men and also some women may stay there for one month. The women go fishing and do some collecting for supplying the camp.

19.6 DEFORESTATION AND ITS EFFECTS ON NWFPs

Deforestation acts directly on the forest cover and indirectly on the Non Woody Forest Products of animal and vegetation origin. Through felling and other forms of extraction

Table 4. Economic values of certain NWFPs collected on urban markets.

NWFPs	Quantity	Weight	Price in FCFA	Price in Euros
Gnetum africanum	3 bunch	0,125 kg	100	0,16
Gnetum africanum	6 bunch	0,250 kg	200	0,31
Gnetum africanum	9 bunch	0,375 kg	300	0,46
Gnetum africanum	12 bunch	0,500 kg	400	0,62
Caterpillars	1 bucket	5 kg	3.000	4,6
Mushrooms	1 bucket	5 kg	3.000	4,6
Wild yam	1 heap	2,5 kg	1.000	1,53
Honey	1 liter	1 kg	1.000	1,53
Elefant meat	1 piece	5 kg	8.000	12,22

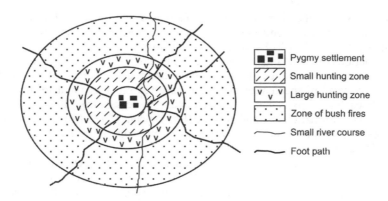

▪▫▪▫	Pygmy settlement
/////	Small hunting zone
v v v	Large hunting zone
⋯⋯	Zone of bush fires
‿	Small river course
▬	Foot path

Figure 4. Organization of the hunting space around a pygmy village (according to Ngana, 2004).

of wood, one witness the gradual disappearance of forest cover and the consequences that are followed: ground being exposed to the erosive risks, threatened nutritive elements, ecological recess and disrupted food chains, density of fauna and the modified flora. On that purpose, the degradation of the forests and its resources represents a real danger to the biocenoses in particular, in terms of specific diversity and regeneration of the species as well as for vegetation. According to natives, the various pressures (bush fires, deforestation and extension of agricultural area) on the forest are the main causes of the distance of game to the villages. They currently estimate that to have game at its own way, it would be necessary to move within a 20 to 40 km radius in the south 'even to penetrate into congolese territory', to claim with a good hunting and considerable spoils, whereas the game was still close to the plantations and villages only two decades ago.

Certain species of caterpillars were much needed by human consumption like Mboyo (*Imbrasia oyomins*), Kanga (*Psendathea discription*), and Guéguéret (*Imbresia obscura*, table 2). They are now endangered owing to the fact that host trees such as Ayous (*Triplochiton scleroxylon*), Sapelli (*Entandrophragma cylindrica*) and Kossipo (*Entandrophragma candollei*) are frequently cut down by logging companies. This phenomenon is also observed for certain NWFPs lianas such as *Gnetum africanum* (Kökö), Ngbîn and Soumba, all dependant on small trees. This shows that deforestation leads implicitly to a fall of supply with NWFPs and causes for this purpose a virtual problem of permanence of these resources. However, the Non Woody Forest Products are an essential element of the strategy of survival and development, and of food safety. They provide a considerable extra food, especially for the joining period where the basic food such as manioc and banana are lacking. Just as the sale of these NWFPs carried out by the villagers, a charge of income gets to them. To see how significant the role played by the wild products of animal and vegetation origin in the human consumption, their decline or shortage would not be a major cause of social imbalances on the food level and income plan for the inhabitants of the concerned zones.

19.7 A RATIONAL MANAGEMENT OF FOREST RESOURCES?

As shown the natural forests and the plantations are rich deposits of non woody forest products. The stakes of rational management of these are manifold: to guarantee food safety by a permanence collection activity for fast growing populations, to ensure a complement of income these populations by the sale of the collected products. The integration between the woody and non woody production could at the same time be advantageous for the local populations and make the extraction of wood less progressive for the environment. The loss of earnings which results from this is compensated by the values of the non woody products (Caldecott, 1988). That is why natural housing constitutes a suitable place for the conservation of biodiversity. A rational exploitation of the non woody forest products would maintain balance between the gatherings and the refresh rate of these resources. This would theoretically express that one should organize oneself to exploit only what nature can replace. This solution appears unrealistic more especially as the non woody forest products are an essential element of the strategy of survival and development necessary to the well being of man, cattle, flora and wild fauna. However, it is important to direct towards an installation of the non woody forest products and to promote their valorisation while maintaining the productivity or by increasing it (Wickens, 1991).

A mode of harvesting tested to apply to the possibilities of valorisation of NWFPs was proposed by the I.C.R.A (Central African Institute of Agronomic Research) at Kenga village with the aim to determine the effects of the extraction on growing back *Gnetum* from August 2003 to September 2004. Five types of different modes were tested: (1) removal of single leave, (2) harvest of two pairs out of three so that there remains a pair of sheets after two nodes without sheet, (3) harvest of all the leaves come ripe, (4) cut of the tops of the lianas above the limit of the leaves become ripe, and (5) cut of all the lianas on the level of the ground. With the exit of these tests, one noted that the lianas on which leaves are left during harvest, produced new leaves and that new lianas pushed on the knots from where the leaves were collected. With the application of this method of prunning, a maximum number of three push back lianas are starting to regrow from the knots located below. This shows that the production of biomass in the case of *Gnetum* can be organized in sustainable way. This experiment gave evidence that the two *Gentum* species can be domesticated which could be an alternative to deforestation in the area.

19.8 CONCLUSIONS

Links between NWFPs and perspectives for food safety in the Lobaye were analyzed. The importance of these products was shown by their variety explaining the frequency of the setting apart on an annual scale and by their contribution in the maintenance of social balance.

From 'immemorial' time, the local populations draw food from the forest, fodder, the products of the traditional pharmacopeia and the income, and can moreover attach a cultural to it. The history of the NWFPs is closely related to the way of life developed by the native populations and the users of the resources. Those are of daily value, in particular in the satisfaction of the vital needs for the local populations especially at certain categories of people not having enough income, nor large agricultural fields. They constitute the base of the food of the stripped populations, rather regularly get income in a relatively short time, compared to the agricultural production and ensure these at last to food safety. As for their contribution in the preservation of forest ecosystems, the valorisation of NWFPs could decrease the rhythm of the extraction of wood; and would limit in return the human pressure on the forests.

ACKNOWLEDGEMENTS

The text proposed was read by Gabriel Danzi, Senior lecturer and dean of the Faculty of Arts and Humanities (French version) and Hermann Jean-Michel Djimapo (English version). Bertin Ouakanga carried out some figures of this article from numerical data base on phytogeography of space research. That they all are thanks, just as the Master student level 1 of Géomatique, François Gonda, who took part in the data gathering on the economic values of NWFPs.

REFERENCES

Bahuchet, S., 1978, Les contraintes écologiques en forêt tropicale humide: l'exemple des Pygmées Aka de la Lobaye (Centrafrique). *Journal d'Agriculture Tropicale et de Botanique Appliquée*, **25 (4)**, pp. 257–285.

Bastin, D., 1996, Forêt de Ngotto, inventaire forestier. *Rapport général Zone 1 et 2 CIRAD-Forêt*, Paris.

Bois, S., 1967, *Les plantes alimentaires chez tous les peuples et à travers les âges*, (Paris: Paul Lechevalier).

Boulvert, Y., 1986, Carte phytogéographique de la République centrafricaine à 1:1.000.000. *ORSTOM, Coll. Notice Explicative*, **104**, Paris, pp. 1–131.

Caldecott, J., 1988, *Proposal for an independent review of forestry policy in Sarwak*, (Londres: Land Associates (Mimeo)).

Carpe, 2001, Les produits forestiers non ligneux. *USAID Central Africa Program for the Environment*, Washington.

Falconer, J., 1990, La forêt, source d'aliments en période de disette. *Unasylva*, **41 (160)**, pp. 14–19.

FAO, 1992, *Ouvrages sur l'aménagement durable des forêts*.

Leclerc, C., 1999, De l'usage social de la forêt tropicale. L'exemple des Pygmées Baka du sud-est Cameroun. In *Nature sauvage-nature sauvée? Peuples autochtones et développement*, Paris, pp. 87–99.

Koko, M. and Runge, J., 2004, La dégradation du milieu naturel en République centrafricaine. *Zeitschrift für Geomorphologie, N.F., Supplement-Band* **133**, pp. 19–47.

Ngana, F., 2004, *Représentations des espaces urbains et processus migratoires des populations marginalisées en Centrafrique*. Thèse doctorat, Géographie, Histoire et Sciences de la société, Université Paris 7-Denis Diderot (France), Paris, p. 438.

Ogden, C., 1990, Intégrer les objectifs nutritionnels dans les projets de développement forestier. *Unasylva*, **41 (160)**, pp. 20–28.

Piri-Dejean, C., 1997, Analyse de la filière des sous produits forestiers de la zone d'intervention du projet ECOFAC-RCA dans une perspective de valorisation de production (cas du Gnetum sp.). Rapport, p. 50. (in edit)

Tchatat, M., 1999, Produits forestiers autres que le bois d'œuvre (PFAB): place dans l'aménagement durable des forêts denses humides d'Afrique centrale. *Projet FORAFRI, Document* **18**, CIRAD, CIFOR, CARPE, IRAD, Yaoundé (Cameroun), p. 103.

Tisserand, R.P. C., 1961, Les brèdes alimentaires de la forêt de la Basse-Lobaye. *Cahiers de la Moboké*, Museum d'Histoire Naturelle, Paris, pp. 360–400.

Wickens, G.E., 1991, Problèmes d'aménagement produits non ligneux, *Unasylva*, **42 (165)**, pp. 3–80.

Yalibanda, Y., 1999, Inventaire de la biodiversité végétale des layons 6 et 1 dans la forêt de Ngotto. *Projet ECOFAC*, Bangui.

Regional/Location Index

Subject Index

.